Michael Klemm, Jim Cownie
High Performance Parallel Runtimes

Also of Interest

Analog and Hybrid Computer Programming
Bernd Ulmann, 2020
ISBN 978-3-11-066207-8, e-ISBN (PDF) 978-3-11-066220-7,
e-ISBN (EPUB) 978-3-11-066224-5

Photonic Reservoir Computing
Optical Recurrent Neural Networks
Ed. by Daniel Brunner, Miguel C. Soriano, Guy Van der Sande, 2019
ISBN 978-3-11-058200-0, e-ISBN (PDF) 978-3-11-058349-6,
e-ISBN (EPUB) 978-3-11-058211-6

Technoscientific Research
Methodological and Ethical Aspects
Roman Z. Morawski, 2019
ISBN 978-3-11-058390-8, e-ISBN (PDF) 978-3-11-058406-6,
e-ISBN (EPUB) 978-3-11-058412-7

Elementary Synchronous Programming
in C++ and Java via algorithms
Ali S. Janfada, 2019
ISBN 978-3-11-061549-4, e-ISBN (PDF) 978-3-11-061648-4,
e-ISBN (EPUB) 978-3-11-061673-6

Dynamic Fuzzy Machine Learning
Fanzhang Li, Li Zhang, Zhao Zhang, 2017
ISBN 978-3-11-051870-2, e-ISBN (PDF) 978-3-11-052065-1,
e-ISBN (EPUB) 978-3-11-051875-7

Michael Klemm, Jim Cownie

High Performance Parallel Runtimes

Design and Implementation

DE GRUYTER

Mathematics Subject Classification 2010
Primary: 34-04, 35-04; Secondary: 92C45, 92D25, 34C28, 37D45

The authors can be reached at authors@parallel-runtimes.info. The source code for this book can be found at https://github.com/parallel-runtimes/lomp.

ISBN 978-3-11-063268-2
e-ISBN (PDF) 978-3-11-063272-9
e-ISBN (EPUB) 978-3-11-063289-7

Library of Congress Control Number: 2020948093

Bibliographic information published by the Deutsche Nationalbibliothek
The Deutsche Nationalbibliothek lists this publication in the Deutsche Nationalbibliografie; detailed bibliographic data are available on the Internet at http://dnb.dnb.de.

© 2021 Walter de Gruyter GmbH, Berlin/Boston
Cover image: MirageC / Moment / gettyimages.de
Typesetting: VTeX UAB, Lithuania
Printing and binding: CPI books GmbH, Leck

www.degruyter.com

Foreword

Parallel programming is the mainstream. From laptops to the largest supercomputers in the world, and from cell phones to high-end medical devices, modern computing systems incorporate architectural parallelism in order to provide high performance for their applications. When multicore computers became the norm, parallel computing also became available to almost everyone with a laptop. Yet in order to benefit from this architectural parallelism, application programs must typically be adapted to express their inherent concurrency. Whether creating a new parallel application or introducing parallelism into an existing one, today's application developer may choose a new, high-level parallel language, exploit a parallel extension of an existing sequential language, or make use of library calls to meet this need. Never before has there been so much active development of parallel languages and experimentation with fresh approaches to obtaining high performance, portability, and performance portability across parallel platforms.

Much has been written about parallel programming language features, the algorithms that exploit them, and the experiences of application developers who deploy them. Less has been written to explain how they are implemented and what goes into the creation of a high-quality compiler and runtime system for a parallel programming interface. A compiler that translates programs that are created using a high-level parallel programming language requires the support of a runtime system that manages the compute resources during execution, assigning work to them, implementing the necessary synchronization, and more. Since the adoption of a high-level programming interface depends to a large extent on the quality of the user experience, each of these must be well designed and their interactions carefully engineered.

Developing a runtime system that provides consistent, high performance to applications that are executed on a multicore platform is about very much more than meeting basic implementation needs. Rather, the components of a runtime system must be highly efficient and carefully crafted. The implementers must select scalable algorithms for key functionality, such as barriers, and implement these and other algorithms in a manner that avoids potential performance problems during execution, such as the false sharing of cached data. They must provide sensible strategies for handling idle threads, help to optimize the performance of parallel loops, and devise methods for using architectural features where possible.

This book focuses on the practice of creating high-performance parallel runtimes. Given the importance of a well-engineered runtime for parallel programming models, it is long overdue. The authors explore in depth many of the design and implementation decisions that go into the creation of a state-of-the-art runtime system for executing parallel application code, provide numerous examples, and discuss several real-world compilers and runtimes. The book describes the salient details of current computer architectures so that the parallel programming requirements and subsequent

https://doi.org/10.1515/9783110632729-201

techniques are well motivated. We learn, for instance, about the details of modern multicore cache management strategies and their performance implications, as well as how atomics and transactional memory may be used to support synchronization.

The authors address general design considerations, but also discuss the implications of contemporary architectures and parallel programming features for the runtime developer. They explain in some detail how the compiler and runtime collaborate to implement the OpenMP[*] standard and Intel[*] Threading Building Blocks (TBB). While these serve admirably to illustrate the interactions between compiler and runtime, the text goes much further than that. It gives a short introduction to the workings of a compiler and describes how it can implement lambda functions that simplify the code for creating TBB tasks. OpenMP has a number of different constructs by means of which an application developer may express parallelism in a code. The authors give an overview of its implementation that is also very much more than an illustration of the role of runtime systems. Their detailed explanation of the compilation steps that precede execution, in addition to details of the runtime components, will allow the book to serve as an excellent introduction to the implementation of these and other high-level parallel programming language features.

There are a variety of different synchronization mechanisms and algorithms for implementing them. Here, we are treated to a comprehensive and practical discussion of hardware support for synchronization, popular synchronization constructs, and strategies for their implementation. There is a treatise on the extraordinary challenges of implementing locks, and a discussion of the perennial problem of what to do with threads that are waiting.

Tasks have grown tremendously in popularity as a means for specifying parallelism. They are inherently dynamic, in the sense that their creation is decoupled from their execution. In contrast to many other language constructs, the runtime oversees all aspects of the execution of tasks, including handling any requirements of their relative ordering and enqueuing them for execution. The discussion of their runtime support here goes far beyond the design of a task pool (or queue) and considers task dependences, scheduling, task stealing, and additional synchronizations.

As our platforms grow in heterogeneity, as well as in scale, the role of the runtime in the implementation is becoming increasingly important. Far from being "merely" a support infrastructure for the compiler, today's parallel runtime must actively manage the workload across multiple NUMA domains, potentially making decisions on where a particular piece of code should be executed or when to attempt to steal a task. Future runtimes are expected to be even more powerful, potentially adapting details of a code's execution to address changes in the execution environment or in the computational needs of the program itself.

I believe that this book will be of great value to systems researchers and practitioners alike. It provides a wealth of information on all the major responsibilities of a modern runtime system, with practical implementation details and potential pitfalls clearly explained. Code for the OpenMP runtime is also available for the reader to try out and experiment with. There is sufficient background on multicore platforms to ensure that newcomers can understand the contents. Indeed, it is also an excellent introduction to the practical workings of such systems. All of this comes in a well-written, entertaining, and up-to-date book, with many examples and illustrations. Despite the depth with which many topics are treated, the narrative remains accessible throughout. The authors are especially well qualified to write on this topic. Both active practitioners, each of them has accumulated extensive experience in the design and implementation of parallel programming libraries, languages, and of course, parallel runtimes. They have a deep appreciation for the challenges and pitfalls that may present themselves in the context of exploiting parallel systems, and for the engineering aspects of any real-world software development effort.

I have been involved in the development of compiler support for OpenMP almost since it was first introduced to the public, and am keenly aware of the importance of a carefully crafted runtime system. A crucial element in any OpenMP implementation is a quality runtime system whose creators have successfully overcome the myriad performance challenges posed by multicore platforms. This requires engineering of the highest order, led by a deep understanding of the nature of the task and its inherent challenges along with some solid solution strategies. This understanding, and the techniques to get the job done, is what this book provides.

I highly recommend it!

Barbara Chapman
Stony Brook University
Brookhaven National Laboratory

Preface

The idea to write this book arose from a lecture that Michael was giving at the Technical University Munich. You know, just for the fun of it. The lecture was called "Parallel Programming Systems" and taught students the basics of how to bring a parallel programming model to a parallel machine. Exactly the topic of this book. What a coincidence!

The topic of parallel runtime systems is a really interesting one. Our day jobs are to work on all sorts of performance questions and low-level machine details on various processors. However, thinking about performance in an application code is quite different from thinking about performance at the lower levels of a software stack and inside a runtime library that supports a parallel programming model. Many more machine details are exposed at this low level and must be taken into account when aiming for the highest possible performance.

After the first couple of series of said lecture, it seemed that the students were really interested in the low-level stuff that Michael was teaching them, so he felt it would be a good idea to move to the next level and write a book! With Jim joining the team of authors, we were ready to write the book. In addition to his many years of experience, Jim brought his extensive knowledge of parallelism and his interest in low-level machine details to the table.

We hope that you'll like the topic as much as we do. It has certainly been fun writing this fine piece of text (and the accompanying code), and we hope that you will enjoy reading it. We use many code examples and figures to illustrate the concepts and to help you to understand the details of how to write critical parts of the runtime system. Most of these code samples will be in C and C++, with some assembly language where we need it to demonstrate what modern compilers do to your source code.

Fortran programmers, or adepts of other languages: don't feel left out; there is still a lot that you can take away from these discussions, and you will certainly be able to translate the key concepts into your language of choice. Our choice of C/C++ is not because we consider other languages uninteresting or irrelevant, but rather because we are considering implementation details that are very close to the machine, where C and C++ are still the dominant languages no matter which language the higher-level user code is written in. So please keep an open mind, even if the trouble with having an open mind is that people will insist on coming along and trying to put things in it.

We would like to thank our publisher, De Gruyter, for accepting the book into their portfolio and their patience while we were busy working on the book. We would like to thank Ute Skambraks at De Gruyter who worked with us patiently to prepare the manuscript. We would also like to express our gratitude to the reviewers who read the early versions of the book and did not run away screaming. They are Mark Bull, Barbara Chapman, Florina M. Ciorba, Chris Dahnken, Alex Duran, Wooyoung Kim, Will Lovett, Larry Meadows, Jennifer Pittman, Carsten Trinitis, and

https://doi.org/10.1515/9783110632729-202

Terry Wilmarth. Others who deserve thanks are our proofreader, Matthew Robertson (https://checkmatteditorial.com), and Randall Munroe, creator of the wonderful https://xkcd.com, for giving us permission to use one of his creations. Of course, all errors that still exist in the book are ours. Finally, this work used the Isambard UK National Tier-2 HPC Service (http://gw4.ac.uk/isambard/) operated by GW4 and the UK Met Office, and funded by EPSRC (EP/P020224/1), and some machines at the University of Bristol, who all deserve special thanks.

Now, without any further ado, please enjoy the book!

Michael & Jim

Contents

List of figures

Listings

List of tables

Glossary

ABI **A**pplication **B**inary **I**nterface. The compiled-code implementation of an API.

Amdahl's law A simple formula to compute the maximum performance of a parallel code depending on the degree of parallelism available in the hardware and the serial fraction of the code. In reality, it is an overly optimistic upper bound.

API **A**pplication **P**rogramming **I**nterface. A stable interface from a program to an underlying layer of software, be that a library or the OS itself.

ASIC **A**pplication-**S**pecific **I**ntegrated **C**ircuit. Die with an integrated circuit that is specifically geared towards a specific function or application, not intended for general-purpose usage.

BasicLockable A named requirement in the C++ standard which describes a class that can be used via `std::lock_guard`.

cache coherence The desirable property that all caches in a machine maintain the same value for any specific cache line.

child task A task that has been created by another task, the parent task.

code synonym for program or application.

coherence fabric The interconnect used to carry the messages used by cores to maintain cache coherence.

compilation The act of converting a human-readable representation of a program into a set of machine instructions (and data) which can be executed by a computer.

core see "physical core".

CPI **C**ycles **p**er **I**nstruction. One way to measure the efficiency of execution of a code on a specific CPU implementation.

CPU **C**entral **P**rocessing **U**nit. See "core". Though in some contexts people use CPU to mean the physical package that holds multiple cores, cache, etc.

descendant task A task that is a child task or a child of a descendant task.

DMA **D**irect **M**emory **A**ccess. A technique that allows devices such as network interfaces or disk controllers to access memory directly without requiring that the CPU moves every byte of data to or from the device.

false sharing The sharing which occurs because a single physical cache line is being used to hold two unrelated variables. False sharing causes cache-to-cache transfers, which can impact performance yet is hard to see in source code because the mapping from variables to locations in store is controlled by the compiler.

futex A Linux[*] system call used to suspend threads or to wake them.

hardware thread See "logical core".

initial thread See "main thread".

IP Internet Protocol. The low-level protocol which allows different machines to communicate over the internet.

IPC Inter-Process Communication. Any method used to communicate between separate processes (which may be on the same machine, or remote). Sometimes also used for Instructions Per Cycle, i. e., $1/CPI$, although we do not use that here.

JIT Just In Time. JIT compilation is an implementation in which compilation of a program occurs as it is executing. This avoids the need for any offline compilation and allows the compiler to know the exact machine for which it is compiling, as well as exploiting profile information to make good optimization choices. The cost is that the compilation has to happen every time the code is run.

jitter Noise injected into the system by the environment or other processes and threads executing on the same system.

LILO time Last In to Last Out time; for a collective operation, the elapsed wallclock time between the last thread arriving and the last thread leaving. For a tree barrier the LILO time will be the sum of the LIRO and RILO times of the join and broadcast operations.

LIRO time Last In to Root Out time; for a join operation where all threads arrive and a single, root thread leaves when it is aware that all have arrived, the elapsed wallclock time between the last thread arriving and the root thread leaving.

logical core A single hardware thread of execution onto which the OS will schedule work. In an SMT machine a logical core is a single SMT thread, whereas a CPU or core would be able to execute multiple SMT threads simultaneously.

logical CPU Same as a "logical core". The "logical CPU" terminology comes from the Linux OS.

main thread The thread which starts execution inside a process when the process is created.

memory fence In an out-of-order CPU, an instruction that enforces the externally visible ordering of memory instructions issued by the local core.

OS Operating System. The combination of an OS kernel and libraries which provide a way for user code to run on a machine while remaining abstracted from underlying hardware complexities.

OS kernel The part of an OS which executes in a privileged mode, controls devices, provides file systems, and allocates resources to user-level processes.

package See "processor package".

parallel fraction The proportion of time in a serial execution of a code that could be executed in parallel.

parent task A task that has created another task.

physical core The hardware entity that executes instructions. In an SMT system a physical core may support more than one logical core. Sometimes also just called a core.

process The protection domain within which code executes. A process contains at least one thread.

processor die A slice of silicon that contains the processor plus supporting infrastructure like caches, memory controllers, etc. Plural "dice".

processor package The physical unit that plugs into the socket of the system. Processor packages may contain multiple dice.

recursion See "recursion".

RILO time **R**oot **I**n to **L**ast **O**ut time; for a broadcast operation, the elapsed wallclock time between the root thread arriving and the last thread leaving.

root thread In a centralizing collective operation (such as a centralizing barrier), the single thread which is the central thread, so that it knows all others have arrived.

sched_yield A Linux system call that enters the kernel to request that it run the scheduler to choose a good thread to run. Do not use this in a polling loop; with modern Linux kernels calling sched_yield() here is futile.

serial fraction The proportion of time in a code that must be executed serially.

sibling task Tasks that have the same parent task.

socket The physical installation on the mainboard of the system to hold and connect the processor package and the mainboard of the computer. Sometimes the term is used as an alternative to "processor package".

task A concurrent unit of work carrying its own data environment and code for execution in a thread.

task pool A container to hold tasks that are waiting for a thread to pick them up for execution.

task queue Misleading term for "task pool".

thread A single entity of execution and its associated state. A thread executes inside a process.

thread pool A set of threads which are kept alive even when they have nothing to do so that they can be rapidly used when there is work for them.

volatile A C keyword that expresses that each load or store to a variable of this type must actually occur, preventing the compiler from optimizing them away if it thought they were unnecessary.

1 Setting the stage

Today's world is a parallel world. Parallelism is ubiquitous. From the smallest devices, like processors that enable the Internet of Things, to the largest supercomputers, almost all devices now provide an execution environment with multiple processing elements. Thus, they require programmers to write parallel code that can exploit the parallelism available in the hardware. This ubiquity also means that it is necessary to implement runtime environments that support such parallel programs. In this book, we discuss the issues involved in building parallel runtime systems so that most (application) programmers don't have to worry about the complicated low-level details of parallel programming, but rather mainly concern themselves with the, unfortunately still complicated, higher-level issues!

We will cover the fundamental building blocks on which a parallel programming language relies, and discuss how they interact with modern machine architectures to help you understand how to provide high-performance implementations of these building blocks. Obviously, this also requires that you understand:
- What the sensible performance measures for each construct are.
- What the theoretical limits of performance are, given the properties of the underlying hardware.
- How to measure the performance of both the hardware and the code.
- How to use measurements of the hardware properties to design software that will perform well.

Throughout the book, we will show some interesting effects of the way modern processors are designed and the (performance) pitfalls that await programmers who have to reason about low-level machine details when they are implementing a high-performance parallel runtime system. You will see that there are some counterintuitive conclusions that will most certainly direct your thoughts about machine performance, but also implementation decisions, in the wrong direction.

1.1 Structure of the book

To better understand the structure of the book, please have a look at Figure 1.1. It shows the typical layers of a parallel runtime system. The application code sits atop the parallel runtime library that implements the key functionality to support the parallelism in the application. The parallel runtime usually relies on a native library that supplies the concept of threading via the operating system (e. g., the POSIX* thread library *pthreads* [34]). The lowest level in the stack is the multi-core processor that executes the code of the parallel runtime system and the application. In many cases, both the threading library and the parallel runtime system will use functionality provided by the multi-core processor for improved efficiency.

https://doi.org/10.1515/9783110632729-001

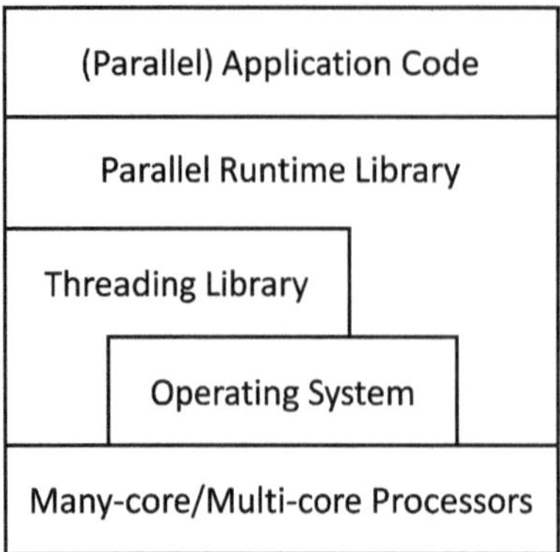

Figure 1.1: Layers of a parallel runtime system.

The remainder of book generally approaches the topic from the top to the bottom. We start with the layer that is accessible to the (application) programmer: the parallel programming model. We discuss some of the design choices and how they affect the general structure of the software stack that implements a parallel programming model. Chapter 2 briefly introduces some of the key concepts of parallel programming models for this book. Don't be disappointed that this is not going to be an in-depth introduction to parallel programming, but rather assumes that you have a basic familiarity with parallel programming already and only scratches the surfaces of this topic.

Chapter 3 describes the basics of multi-core architectures. While this is, for sure, the lowest level (even below the software stack!), covering the machine-level details early on in the book seems useful, as some of the implementation choices and algorithms that we will present are clearly motivated by how modern processors work and how they behave when they are executing a parallel application.

In Chapter 4, we explain how the parallel programming model interacts with the runtime system via the compiler and the runtime entry points. Chapter 5 discusses some cross-cutting aspects that are usually needed in a parallel runtime system, like how to manage parallelism or how to do memory management.

Chapter 6 through Chapter 9 cover the details! These chapters dive deep into the specific aspects and the implementation of key concepts like mutual exclusion, atomic operations, barriers, reductions, and task pools. All of these chapters focus on how the implemented algorithms interact with the machine and what effect they cause in a modern processor. This should provide a clear picture about what you, as a low-level ninja programmer to be, will have to understand to be able to extract parallel performance. Ideally, this will lead to a much better use of the expensive machine that executes your parallel runtime system (and the parallel application on top of it).

1.2 Design space exploration

Before you can think about implementing a parallel runtime system, you have to think about what the parallel programming model should look like. This is, of course, if you start from scratch. If your task is to implement an existing programming model, your options are somewhat more limited, as now the programmer-facing parts of the model are defined, though you may still have some flexibility about what the internals of your implementations will look like.

One of the main questions for the implementer of a parallel programming model is whether your programming model should be implemented as a library that provides an application programming interface (API) or as part of the programming language itself (or an extension of it). To make things even more complicated, you could also think of a hybrid model where parts of the model are expressed in the language while others are covered through API routines. Figure 1.2 shows the three categories and gives a few examples of well-known parallel programming models. Figure 1.3 shows a different categorization by the parallel architecture that is targeted by these programming models.

As you may imagine, each of these designs can bring some benefits, but at the same time these may come at price—that is, the design may have drawbacks with respect to the alternative implementation of the parallel programming model. Here, we review the two main design choices and discuss their benefits and drawbacks.

1.2.1 Parallelism as a library

Injecting parallelism by using a library seems like an obvious choice. Since most programming languages support libraries, you can potentially perform parallel programming from any programming language. One particularly good example of this is the POSIX thread library pthreads, which brings multi-threading to C and other languages on POSIX-compatible systems, e. g., the GNU/Linux[*] operating system. Another example is Intel[*] Threading Building Blocks [151], which adds task-based parallelism to the C++ language.

So, what's wrong with this idea? The main issue with using an API-only approach to parallelism is that the compiler is not normally aware of the special meaning of the calls to the library that are creating parallelism, and therefore has to treat the API routines as black boxes whose content is hidden. Historically, this issue was exposed because the C language did not define a memory model (see Section 3.2.2 for a discussion of memory models). Instead, the compiler assumed that only one thread would be executing the code. If another thread was created through pthreads, then that would also be considered the only thread in the program, even though there would now clearly be more than one. This could be partially overcome by using `volatile` types; however, they were more intended for use in handling memory-mapped devices than for writing

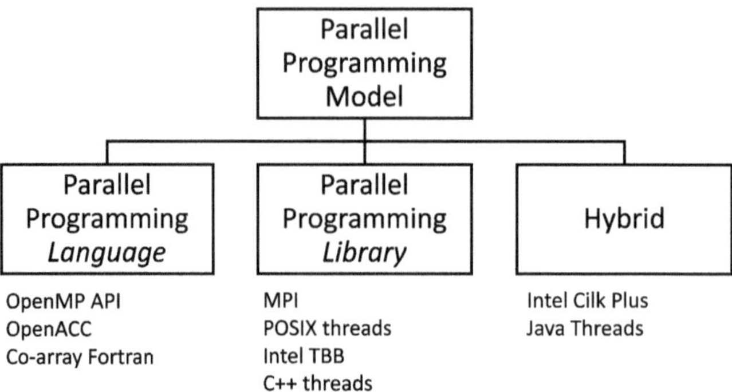

Figure 1.2: Paradigms for parallel programming models.

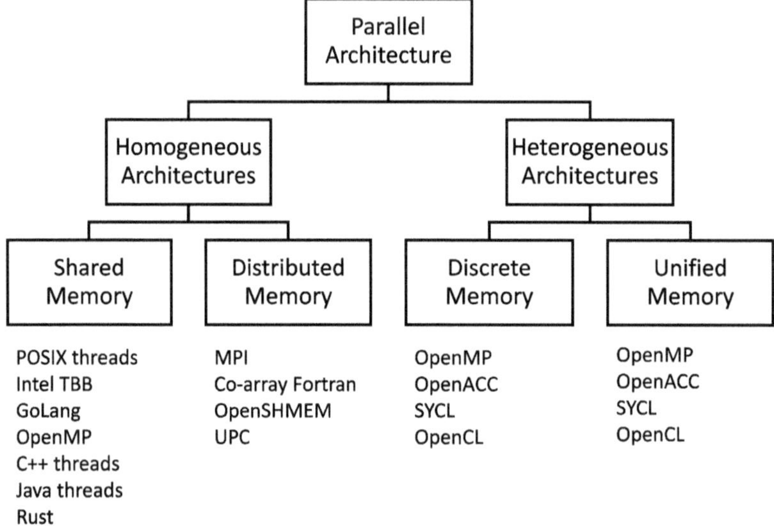

Figure 1.3: Parallel programming models by categorized by memory architecture.

parallel code. Section 3.2.2.2 shows how such compiler assumptions can destroy the user's intent. Although modern versions of the languages provide tools to alleviate the issue, you need to be aware of these issues and use the mitigations in your code.

The example in Listing 1.1 shows a tiny sample of code that uses Intel Threading Building Blocks (Intel TBB). It parallelizes a very simple loop that performs an element-wise addition of two arrays. While the library, Intel TBB, takes care of splitting the loop into smaller chunks to distribute the load across the available threads, the compiler does not know anything about the loop that is being executed in parallel. It therefore has a hard time determining that there are further parallelization opportunities, such as the use of vectorization on machines that support *single instruction*

```
#include <tbb/parallel_for.h>

void array_example(double * a, double * b, double * c,
                   size_t n) {
  tbb::parallel_for(tbb::blocked_range<size_t>(0, n),
          [=](const tbb::blocked_range<size_t> & r) {
             for (size_t i = r.begin(); i != r.end(); ++i) {
               c[i] = a[i] + b[i];
             }
          });
}
```

Listing 1.1: Simple array example using Intel Threading Building Blocks.

multiple data (SIMD) instructions. The simple reason for this is that the for loop iterates from the beginning of a loop chunk to its end and the compiler has to analyze this sequential loop to perform further parallelization.

The main advantage of the library approach is that the parallel programming model is not tightly coupled with the programming language, which makes it more flexible than the compiler approach which we will look at in the next section. The library can be developed independently and add new parallel programming features, while the base language used to call the library can stay the same. Thus, no modification to the compiler is needed, which usually greatly simplifies the roll out of new features. All that is needed to introduce new features is to add additional library calls or extensions to the existing ones, and then to release a new version of the library.

Historically, the advantages of adding features through libraries are clear. "Message Passing Interface" (MPI) [90] libraries are the standard way to build highly parallel codes that use many machines connected by networks in *high-performance computing* (HPC), while "High Performance Fortran" (HPF) [109] is now a marginal historical sidenote, even though it provided all that MPI did and more. As a set of language extensions, HPF faced the large problem of getting compiler vendors to adopt it, while MPI could be implemented as a library by anyone (indeed, there was a prototype MPI library implementation as the standard was being written so that problems in the design could be observed before the standard existed). MPI also demonstrates that it is possible to use a library interface from many different languages, all of which sit on top of an underlying implementation. Thus, MPI is available in compiled languages, such as C, C++, Fortran, and others, but also even in interpreted ones, like Python [148] and R [106], and *just-in-time* (JIT) compiled ones, like Julia [13].

One final remark about the library approach is that it's not always clear where the parallel programming library ends and the internal runtime library begins. In reality, this is a matter of software design that defines which basic features the parallel programming library can rely on and how low-level details of the implementation are abstracted from the higher-level features.

1.2.2 Parallelism as a language

The other end of the spectrum is to implement a parallel programming model inside the programming language—that is, to extend the syntax and semantics of an existing programming language or even design a new parallel programming language from the ground up. As you can probably already guess, since we are now at the other end of the spectrum, parallelism as a language turns the disadvantages of the library approach into advantages and vice versa.

Let's start with the disadvantages to leave the good news for the end. If we want to extend a programming language with new features, we have to modify the compiler to support the new syntax and semantics. While this sounds trivial at first, it quickly turns out to be a showstopper in most cases. Just imagine that you are using a closed-source compiler for your favorite programming language and you have the desire to extend it. Unless you can get access to the compiler's source code to modify it, the only option is to implement your own compiler for that language. This will be heavy lifting!

The picture does not look much better for designing a new parallel language from scratch. While a library can tap into the existing ecosystem of, say, C++, designing a new programming language means that you first have to establish an ecosystem for the new programming language to make it useful. What often happens is that people stick to the n existing languages and you just created the $(n + 1)$-th language. The xkcd.com comic shown in Figure 1.4 (see [93]) nicely summarizes this effect.

At the same time, there's an important advantage of using a compiler-enabled parallel programming model: the compiler knows about parallelism and can thus use this additional information in its analysis of the code and, later, to transform and optimize the code using this additional context information from the programming model. It does not have to infer it from the (sequential) code by analyzing the code as it had to in the library case.

Let's look at the Fortran example in Listing 1.2. Fortran allows us to do element-wise operations on arrays. In the example, we use this to express that each element of array a should be added to the corresponding element in array b and stored in the right place in array c. As you can see, there is no need to write explicit loops; the compiler automatically generates the code to process the two-dimensional arrays. Because the operation on each array element can be executed independently from the others, the compiler can see that these can be executed concurrently, and it can use this knowledge to perform parallelization, e. g., SIMD code generation or multi-threading.

Figure 1.4: XKCD 927: "Standards" (© xkcd.com; used with permission).

```fortran
subroutine array_example(a, b, c)
  implicit none
  double precision, dimension(:,:) :: a, b, c
  c = a + b
end subroutine
```

Listing 1.2: Simple array example using Fortran array syntax.

1.3 Code samples

Unless there are specific language-dependent requirements (for instance, to show how vectorization can be more easily achieved in Fortran than in C/C++), the code examples we use here will be in C or C++. Of course, if you are implementing your own parallel runtime, then you may be implementing it for some other language entirely. You might want to pick a language that allows for accessing low-level machine details without too much hassle and at sufficient speed without crossing too many boundaries (e. g., interpreted-to-native execution).

We certainly do not want to sell C++ (C++17 precisely) to anyone, but it provides a nice balance between modern programming paradigms, while providing easy access to low-level machine details. Since our choice will lead you to learn enough C++ to understand those examples, it is natural to also use C++ for other examples that are not taken from the runtime code. The OpenMP[*] Application Programming Interface (API) together with Intel TBB will be our main examples for a parallel language and a parallel library, respectively.

Most code examples that we use to show the implementation aspects of a parallel runtime system in this book are available as part of a small OpenMP runtime system that you can download at https://github.com/parallel-runtimes/lomp ("Little OpenMP Runtime"). The implementation is compatible with the internal APIs that the Intel[*] Compiler for C/C++ and Fortran as well as Clang use to implement the OpenMP language. Even though our implementation is certainly not complete and in fact only covers a tiny aspect of the OpenMP Application Programming Interface, it is sufficiently capable to cover all the OpenMP language features that we show in this book, including a subset of OpenMP tasking.

1.4 Machine configurations

One important topic is the hardware that we have used to benchmark the key implementation concepts that we are going to present throughout the book. We have tried to pick contemporary processors that resemble the state of the art of multi-core and many-core processors, yet exhibit the different architectural choices made by their respective vendors. The selection is arbitrary and is certainly not extensive, given the number of parallel processors on the market; however, by choosing processors from three different hardware vendors (AMD[*], Intel, and Marvell[*])—and, therefore, design teams—which implement two different *instruction set architectures* (ISAs) (x86_64, Arm[*] V8.1a), we believe that we are providing a reasonable sample of machines.

To show how the different primitive operations behave, we will measure them using identical benchmarks on each of the machines. The goal is *not* to conduct a bake off between vendors or architectures to show that one processor is better than another (that is for the market to decide). The benchmarks we present are meant to show that the issues we discuss are common to these different architectures and implementations, and therefore need to be considered when designing a parallel runtime system that will have to run on these machines or similar ones. Obviously, one can measure a specific machine and then design code that works well on it; however, if the code is to succeed, then it will have a longer life than any specific processor implementation, and therefore we want it to work on a variety of current and future hardware. Since we don't have a time machine, we can't be sure what the performance quirks of such future machines will be; however, by looking at a variety of current machines, we can see what performance characteristics are common and, therefore, likely to persist.

We hope that Table 1.1 through Table 1.3 is the most boring part of this book. Those tables list the key architectural features of the three processor implementations that we will use as examples; some of the details will make more sense when you have worked through Chapter 3. We describe the Arm machine as a "Marvell ThunderX2"; however, it was initially designed by Cavium[*], before they were acquired by Marvell, and therefore, if you are searching for more details about it, it's also worth searching for "Cavium ThunderX2" (or "Cavium TX2").

We provide this configuration data here both to set the stage, to have the gory details off the table for the remainder of the book, and for reproducibility purposes. If

Table 1.1: Machine configuration: Intel Xeon[*] Scalable Processor.

Component	Specification
Processor	Intel[*] Xeon[*] Platinum 8260L Processor
Instruction Set Architecture	x86_64
Cores per processor	24 cores each with 2 hardware threads
Clock frequency	2.40 GHz (turbo: 2.90 GHz)
Cache	L1: 768 KiB, L2: 24 MiB, L3: 35.75 MiB
Processor dice	1
Number of sockets	2
Main memory	12× 16 GiB DDR4 at 2666 MT/sec
Operating system	CentOS Linux release 7.7.1908 (Core)
Kernel version	3.10.0-1062.4.3.el7.x86_64

Table 1.2: Machine configuration: Marvell ThunderX2[*] Processor.

Component	Specification
Processor	Marvell ThunderX2 CN9980
Instruction set architecture	Arm V8.1a
Cores per processor	32 cores each with 4 hardware threads
Clock frequency	2.00 GHz (turbo: 2.50 GHz)
Cache	L1: 1 MiB, L2: 8 MiB, L3: 32 MiB
Processor dice	1
Number of sockets	2
Main memory	16×16 GiB DDR4 at 2666 MT/sec
Operating system	SUSE Linux Enterprise Server 15
Kernel version	4.12.14-150.17_5.0.90-cray_ari_s

Table 1.3: Machine configuration: AMD EPYC[*] Processor.

Component	Specification
Processor	AMD EPYC 7742
Instruction set architecture	x86_64
Cores per processor	64 cores each with 2 hardware threads
Clock frequency	2.25 GHz (turbo: 3.40 GHz)
Cache	L1: 2 MiB, L2: 32 MiB, L3: 256 MiB
Processor dice	4
Number of sockets	2
Main memory	16×64 GiB DDR4 at 3200 MT/sec
Operating system	CentOS Linux release 7.8.2003 (Core)
Kernel version	3.10.0-1127.el7.x86_64

you know what we measured, then you can potentially measure the same thing to convince yourself that we are not making all of this up as some kind of weird joke. (Note that the micro-benchmarks that we used to make these measurements are available as part of the source release of the small, incomplete, OpenMP runtime that accompanies the book, so you can also run them on your own machines should you wish to.)

2 Parallel programming models and concepts

As explained in Chapter 1, a good compiler and runtime system for parallel hardware must support the efficient execution of a parallel program on that hardware. This book assumes that you're already familiar with parallel programming in general. Still, it seems like a good idea to review some of the key concepts and features of parallel programming models that we will need for the rest of the book. This will greatly help us to understand what the lower levels of the parallel software stack need to do.

Although we discuss parallel programming models and their implementation in this book, our scope is limited to a shared-memory environment. We do not discuss the implementation of message-passing systems such as the Message Passing Interface (MPI), or languages and libraries that provide the programmer with a partitioned global address space (PGAS), such as Unified Parallel C (UPC) [145] or the Coarray features of modern Fortran [98]. We do not omit these models because they are poor or unnecessary (indeed, they are essential for exploiting large machines, and the combination of "MPI+X" is the canonical way to program the largest machines in the world), but rather because, like Andrew Marvell [86], our time is limited.

Even with this limitation, we will not be able to cover all aspects of parallel programming in a single chapter. Instead, we will focus on the most important concepts that we will need for the rest of this book. We have picked two parallel programming models to show these concepts: the OpenMP[*] API [100] and Intel[*] Threading Building Blocks [151]. There are many books and papers on both these and other parallel programming models that you should certainly read if you want to know more about them. The bibliography at the end of this book provides some references, but can only scratch the surface of what is available in any good online or offline library.

The journey will start with a very brief introduction to multi-processing. We will not spend much time on this topic, as the focus of this book is on efficient runtime systems for multi-threading, so we will cover that in more depth. Finally, we will investigate the fundamental Amdahl's law and discuss its implications for performance.

2.1 Multi-processing and multi-threading

Multi-processing (sometimes also called multi-tasking) was invented in the 1950s and implemented in the early 1960s to allow more than one user to access the same, expensive, computing system to improve resource usage. On a uni-processor system with a single central processing unit (CPU), each process runs on the hardware for a short period of time (a *time slice*) before another process is allowed to run. If one process is blocked (for instance, in the case of slow I/O), another can execute so that the CPU is kept busy. This gives each user of a *time sharing* system the illusion that they have a machine to themselves, while in reality, there is only one physical machine that is switching between the different available jobs at high frequency. It is the job of the

https://doi.org/10.1515/9783110632729-002

operating system (OS) to interrupt a process when its time slice has expired, save its execution context (machine registers and other state), choose which process to run next (which could, of course, be the one that was just interrupted), and finally restore the protection state of that process and its machine registers to allow it to run.

On a multi-core system, several processes can genuinely run simultaneously, since now there are multiple physical cores available, and the OS aims to keep them all busy.

Historically, since the machines only had a single core, there was no incentive to support more than a single point of execution in each process. Therefore, a process was thought of as being both a protection domain and a set of registers (including the program counter or PC). However, in a modern OS, we separate the concepts of a thread and a process [126]:

- **Process:** A process represents a protection domain. It owns an address space, open file descriptors, and all other privileges derived from the user account to which it belongs. However, it does *not* have any execution context and cannot execute any instructions.
- **Thread:** A *thread* represents an execution context, i. e., a set of machine registers including a program counter. A thread can only execute in the context of a process, which provides the address space within which the thread resides. As well as accessing the process' resources, each thread will also own execution-specific properties, such as the set of cores on which it may execute, and accounting information, such as the amount of CPU time it has consumed.

Since a process with no threads in it cannot do anything, when creating a new process, the OS will create it with a single thread (the *main thread* or *initial thread*). Since it can execute, that thread can then make OS calls to create additional threads should they be required. It is important, however, to separate the concepts of *process* and *thread*, since some properties belong to a thread (e. g., execution priority, the set of cores on which it can run), while others (e. g., file access permissions, memory, open file descriptors) belong to the process.

The idea of multi-processing is to use more than one process (probably each with a single thread) to solve a problem in parallel. Obviously, to do this we will need some form of *interprocess communication* (IPC), since by default, processes share no resources with other processes.

In the wider world, using multiple processes to solve a problem is common in cloud computing and web services. A process on the user's phone runs a web browser that communicates over the Internet sending requests to multiple processes running in the cloud, each of which may then communicate with others running databases or other services. When dealing with *software as a service* (SAAS), even what seem to be simple function calls may be directed to other machines somewhere on this planet. All of this happens using interprocess communication that does not require shared-memory segments.

There are many ways to implement IPC. In UNIX[*]-like operating systems, IPC can happen with pipes between processes in the same machine (i. e., the same operating system instance) or via communication using networking protocols between processes on different systems (or even on the same system). Protocols like the *user datagram protocol* (UDP) [103] and *transmission control protocol* (TCP) [77] using the *internet protocol* (IP) are typically used here. IP in turn uses an underlying physical transport layer such as Ethernet or InfiniBand[*]. If you are OK with a "slightly" increased network latency, then a network based on IP datagrams on avian carriers [152] could be your choice.

The most relevant form of IPC in the context of this book is through shared memory. Modern OSes provide ways to create memory segments that are common between two or more processes on the same system. The virtual memory page tables of participating processes are modified so that some of the physical memory pages are shared between the processes. On Linux,[*] you can use the `mmap()` system call [74, 131] to create memory segments that can be accessed by other processes by using the `MAP_SHARED` [122] flag. Another option is System V shared-memory segments (system calls `shmget()`, `shmat()`, and `shmdt()`) [74]. Both types of shared-memory segments can then span multiple processes and each process can equally read or write data in the shared-memory segment. Of course, if these accesses are to the same memory location, appropriate synchronization will be required to avoid incorrect results. Chapter 6 explains the issues and solutions to this problem.

2.1.1 Threading basics

With the introduction of commodity multi-processor systems in the 1970s and 1980s, it became clear that it would be useful to have many threads executing inside a single address space (*multi-threading*) so that a single process could exploit more of the available hardware and code could be written more easily, as by sharing everything the effort of specifying what should be shared is removed. This led to the introduction of the POSIX[*] threads API, or *pthreads* [59], in the 1980s to provide a standard way to access this new functionality, and, more recently, language standards have evolved to bring threading concepts into programming languages.

Of course, the simplicity of sharing everything comes with associated costs, since now care must be taken to ensure that each thread has a consistent state, and threads do not accidentally corrupt each other's state (see Chapter 6). It is an interesting philosophical language-design question, whether it is better to enforce separation of data and use message passing (as in MPI or Communicating Sequential Processes, CSP [57]), or to allow arbitrary sharing. While the message-passing models force one to think about data movement and sharing, which seems like extra work, one may feel that the investment of time was worthwhile after one has spent a week looking for a rare race condition in a shared-memory code.

```c
#include <stdio.h>
#include <stdlib.h>
#include <pthread.h>

#define NUM_THREADS 8

static pthread_mutex_t mtx = PTHREAD_MUTEX_INITIALIZER;

static void * run(void *) {
  pthread_mutex_lock(&mtx);
  printf("Hello World!\n");
  pthread_mutex_unlock(&mtx);
  return NULL;
}

int main(int argc, char * argv[]) {
  pthread_t threads[NUM_THREADS];

  for (size_t i = 0; i < (NUM_THREADS - 1); ++i)
    pthread_create(&threads[i], NULL, run, NULL);
  run(NULL); // Print from the main thread, too!
  for (size_t i = 0; i < (NUM_THREADS - 1); ++i)
    pthread_join(threads[i], NULL);

  return EXIT_SUCCESS;
}
```

Listing 2.1: Classic "Hello World" example using the POSIX thread API.

Listing 2.1 shows a version of the classical "Hello World" program that uses pthreads to launch a few threads to print the message. The call to pthread_create() stores a pthread_t object (a handle to allow the new thread to be identified in later calls) into the storage passed in via the first argument (see also [122]). The second argument defines attributes for the thread (e. g., the size of the stack to be allocated). As the third argument, pthread_create() receives a function pointer to the code that the thread should be executing (called the *start function*). The last argument is a void * pointer, which can be recast to provide access to arguments to be passed into that function. We ignore this argument, as we don't need any arguments.

The run() function in this example has a single parameter (the void * pointer to the packed arguments) and can pass data back by returning a void * pointer to a set of packed results. This data can be retrieved when the pthread_join() function is

```cpp
#include <iostream>
#include <vector>
#include <thread>
#include <mutex>

int main(int argc, char * argv[]) {
  const auto nthreads = std::thread::hardware_concurrency();
  std::mutex mtx;
  std::vector<std::thread> threads;

  for (auto i = 1; i < (nthreads - 1); i++) {
    threads.push_back(std::thread([&] {
      const std::lock_guard<std::mutex> guard(mtx);
      std::cout << "Hello World" << std::endl;
    }));
  }
  { // Print from the main thread, too!
    const std::lock_guard<std::mutex> guard(mtx);
    std::cout << "Hello World" << std::endl;
  }
  for (auto & t : threads) {
    t.join();
  }

  return EXIT_SUCCESS;
}
```

Listing 2.2: Classic "Hello World" example with C++ threads.

called by the application. The second `pthread_join()` argument is a pointer pointing to a space into which the result of the function that is called in the thread being joined can be placed. Since we're not using that result, we can pass a null pointer. Note that if the `pthread_join()` function is not called to clean up a thread's state at the end of its lifetime that is an error and may lead to undefined behavior.

You can see that this code hardwires the number of threads to be created, since pthreads provides no easy way to discover the properties of the machine on which it is running and then execute one thread for each logical core that is available, and that we have introduced explicit locking around the `printf()` calls to ensure that the output to the shared `stdout` stream is correctly interleaved.

C++11 introduced basic parallel concepts and now offers language and library support for threading via the `std::thread` class. You can see from Listing 2.2 that the level

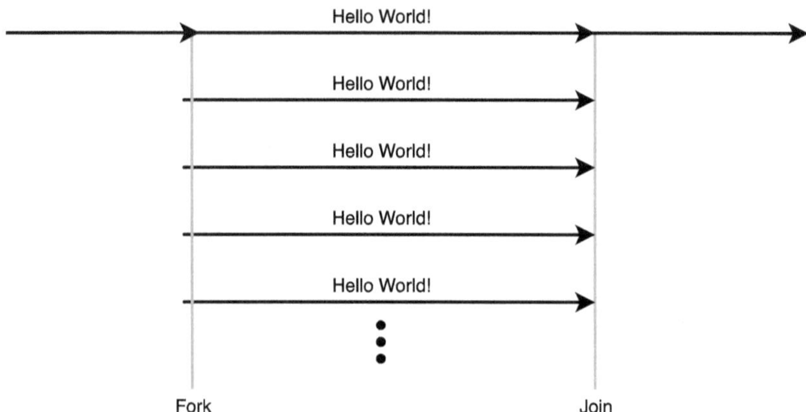

Figure 2.1: Fork/join model to spawn and terminate parallel execution.

of abstraction is a bit higher than it is when using the C interface to directly access the pthreads library. Since this is C++, we can use a lambda expression [123] rather than a function pointer to pass the code we want to be executed in the thread into the `std::thread` constructor. We must record the thread in a `std::vector` so that the main thread can call the `join()` method for each of the threads in the same way that we did for the pthreads API. With C++20, there will be a new thread class (`std::jthread`) that does not require an explicit join operation, because its destructor implicitly performs the join operation and all needed cleanup. You can also see that C++11 includes an inquiry function `std::thread::hardware_concurrency()` that tells us how many concurrent threads make sense for us to use, and that the locking around the output has become easier since we can use a scoped lock to ensure that the lock is released when the scope is left. We will revisit this in Section 2.3.1.

This model of spawning threads and continuing with serial execution when all threads have completed is called the *fork/join* model. Figure 2.1 visualizes what is going on. When the main thread reaches the *fork* point, it launches a set of worker threads to execute in parallel. In Listing 2.1, this is the first loop with the calls to `pthread_create`. At the end of the parallel region, the loop with `pthread_join` calls constitutes the *join* point, at which the main thread awaits the completion of all of the spawned threads.

Although this code now looks relatively simple, and it seems as if we could replicate it in each place where we have parallelism that we want to exploit, we need to tread carefully, since creating and destroying threads are both relatively expensive operations. On our Arm[*] machine, creating and then joining a pthread that does no work takes ~28 µs. The solution to this is to use a *thread pool* so that the OS threads are only created and destroyed once. However, this is all more code to write, and this is not trivial code, since the threads must now be sensibly put to sleep during serial periods of execution, and then woken up when there is parallel work available (see Section 5.1.2).

While it is certainly possible to use a low-level API to implement parallelism in your application code, higher-level programming models are normally much simpler to use, provide higher productivity, and can enable optimizations in ways that are otherwise cumbersome to achieve. You will see various examples of such optimizations throughout this book. By writing directly to such a low-level API, you are forcing yourself to re-implement a lot of functionality, which you will gradually discover that you need, and whose implementation is usually non-trivial (e. g., a thread pool). In effect, most of the code we describe in this book exists to allow you to deal with shared-memory parallelism at a higher level than that presented by low-level interfaces. Always remember that "The best code is the code I do not have to write."

2.1.2 Thread affinity

We saw above that a typical piece of modern hardware will almost certainly support multiple hardware threads of execution, either because it has a core that supports multiple hardware threads or because it has many cores (or both), and that our underlying software model is to create multiple threads so that we can exploit the parallelism in the hardware. However, we also need to consider where the software threads execute. From a high level, there seems to be no particular reason why a specific software thread should not run on any available hardware thread, and—in a multi-processing, time-sharing machine—we would expect there to be many more software than hardware threads, with the OS choosing where to run each software thread and how long to allow it to run.

In an HPC environment, though, we are trying to achieve the highest performance, and the high-level view that any hardware thread can execute any software thread at the same rate may not be true, since if we can avoid re-fetching variables from main memory because they are still in a nearer cache, then this will allow the code to run faster. Similarly, if memory has been allocated near to the core executing the thread, then if the OS were to execute the thread on a core further from that memory, the code would run more slowly. While operating systems factor these considerations into their scheduling decisions, most also explicitly allow user code to control which hardware threads on which a specific software thread will execute. This is called *thread affinity*.

Since many of the measurements we are making here are designed to help us understand the impact of communication between threads, we explicitly use a single thread per core, and tightly bind threads to their cores when performing our performance and hardware measurements.

```
#include <stdio.h>
#include <stdlib.h>
#include <omp.h>

int main(int argc, char * argv[]) {
#pragma omp parallel
  {
#pragma omp critical
    printf("Hello World!\n");
  }
  return EXIT_SUCCESS;
}
```

Listing 2.3: Classic "Hello World" example using the OpenMP API.

2.1.3 The OpenMP API for thread-based programming

One high-level programming model for parallel programming with threads is the OpenMP Application Programming Interface [100], or OpenMP API for short. The OpenMP API (despite having API in its name) is a parallel language that uses compiler pragmas for C/C++ and compiler directives in Fortran to decorate serial code with parallelization information that describes to the compiler how to turn it into a parallel code. The OpenMP API also specifies a set of runtime routines and environment variables to control execution (e. g., number of threads). Good introductions to the OpenMP API can be found in [21] and [147].

Listing 2.3 shows the same "Hello World" example implemented using the OpenMP API. You can see that it is effectively the serial "Hello World" code with the addition of three lines (two pragmas, and one #include, which is not actually required in this code). Thread creation has moved away from the programmer into the underlying parallel runtime library and the compiler translates the code to make use of the runtime library (see Chapter 4). Similarly, there is no need to declare a lock or manipulate it; it is enough to inform the compiler that a region of code is a critical section, while no work at all is required to choose an appropriate number of threads to fork, since, by default, OpenMP runtimes will do that for you and choose a number appropriate to the hardware on which the code is running. Note, too, that any reasonable OpenMP runtime will implement a thread pool for you; indeed, the OpenMP semantics related to thread-private variables almost require the use of a thread pool, as shown in Listing 5.3 and the associated discussion.

The fork happens when the main thread encounters the parallel construct. It then forks (creates or wakes) a *team* of threads to execute the code in the parallel region (similar to what the pthreads version did). However, instead of writing a new function (as was required in the pthreads code), the parallel region's code is defined as the structured block that directly follows the `parallel` directive. For C and C++, a *structured block* is defined as a single statement or a sequence of statements that is enclosed in curly braces. In Fortran, there are `end` directives to indicate the end of the structured block. The code may not branch into this sequence (e. g., with `goto`) or out of the sequence (e. g., by using `goto` or `longjmp`, or by throwing an exception in C++). This is also sometimes called "Single Entry, Single Exit" (SESE).

In the OpenMP model, the join operation happens automatically at the end of the `parallel` construct. The main thread will wait for the other threads in the team to finish executing the structured block. In Chapter 5, we will see how using a parallel runtime system can be smarter than terminating all worker threads at the end of the parallel execution, as we did in the pthreads example in Listing 2.1.

In a multi-threaded programming model, threads are typically assigned a *thread ID*, so that each thread can be identified in the system. The pthreads API uses an opaque handle of type `pthread_t`. The OpenMP API abstracts away from such low-level thread IDs and uses a logical numbering scheme that assigns numbers from zero to $N_{\text{Threads}} - 1$. This makes things a lot easier for a programmer, especially when the thread ID is needed as part of a computation to assign work to a particular thread.

2.1.4 Worksharing

So far, each of the threads has executed the same work, which is not, generally, useful. Let's do something more meaningful and distribute work across the available threads. This is called *worksharing* and can have very different forms, depending on the parallel algorithm to be distributed and the capabilities of the parallel programming model. In the OpenMP API, worksharing refers to distributing a parallel loop across a team of threads, and to constructs that assign work to only a subset of the available threads.

Since you have probably seen the "Hello World" program enough, let's do something more slightly interesting and implement a parallel matrix-matrix multiply (MxM) [48]. Listing 2.4 shows simple OpenMP code to do this for two square matrices A and B each with $n \times n$ elements. It uses a very naive implementation, as we want to show parallel-programming concepts and not go into the complexities of how one should implement such an algorithm for high performance. This example also has neither vectorization nor cache-blocking, both of which are required to achieve good performance, even for this apparently simple operation. Hence, you should *not* use this code other than for educational purposes, but rather find a library that provides a tuned version of the mathematical algorithms you're looking for. Examples of such libraries are ATLAS [9], Intel[*] Math Kernel Library [63], and BLIS [149].

The loop in Listing 2.4 uses static scheduling with a chunk size of eight iterations. For example, the loop iterations assigned to chunk 3 are $[24, 32)$, which are the iterations from 24 to 31 (32 is excluded from the open interval). If the loop is executed with

```
void matmul_par(float * C, float * A, float * B, size_t n) {
#pragma omp parallel for schedule(static, 8) firstprivate(n)
  for (size_t i = 0; i < n; ++i)
    for (size_t k = 0; k < n; ++k)
      for (size_t j = 0; j < n; ++j)
        C[i * n + j] += A[i * n + k] * B[k * n + j];
}
```

Listing 2.4: Worksharing loop in the OpenMP API.

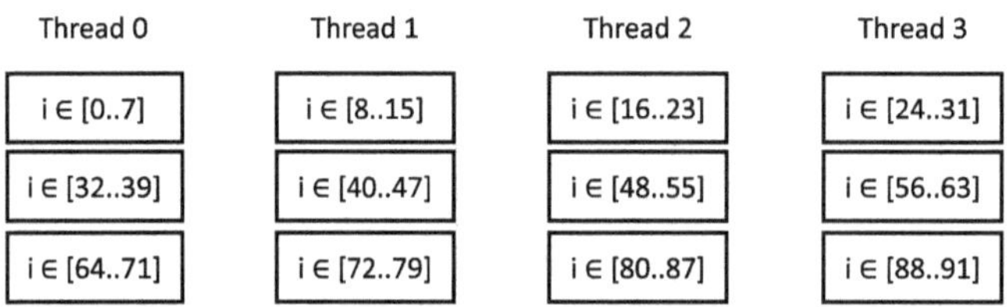

Figure 2.2: Distributing 92 iterations with static schedule with chunk size eight.

eight threads, each of them will receive an equal amount of work to perform. Figure 2.2 shows this for the case of $n = 92$ and four threads. The last loop chunk may be an exception, since it will receive fewer iterations if the chunk size does not evenly divide the number of loop iterations (as you can see in Figure 2.2, where the last chunk only has four iterations).

Parallel programming models may offer different scheduling options or may even offer to pick from a set of loop-scheduling kinds themselves. The OpenMP interface for worksharing loops offers the basic forms of static scheduling, dynamic scheduling, and guided scheduling, all of which can optionally have a chunk size. With dynamic scheduling, a thread that has completed a loop chunk will request the next chunk from a conceptual central pile of loop chunks. Guided scheduling is similar, except that when there is a lot of remaining work, the thread will take larger pieces of it. We will plow through the implementation details of loop scheduling in Chapter 8.

Other forms of worksharing involve data decomposition, producer/consumer patterns, pipelining, parallel tree traversals, and, more generically, parallel graph traversals [88]. We cannot cover all of the aspects of these patterns, but we will go through producer/consumer patterns as part of the discussion of task-based parallelism in Section 2.2.

Another type of worksharing is to allow only a subset of the threads to execute a specific code fragment. Listing 2.5 shows four examples of such constructs. In future versions of the OpenMP API, the master construct will be deprecated and replaced with the more generic masked construct whose filter clause allows the programmer

```c
void main_thread_only() {
#pragma omp master
  printf("This code runs only in the main thread!\n");
}

void masked_construct() {
#pragma omp masked filter(2)
  printf("This code runs only in the thread with ID 2!\n");
}

void single_construct() {
#pragma omp single
  printf("This code runs only in one thread!\n");
}

void parallel_sections() {
#pragma omp sections
  {
#pragma omp section
    { // section 1
      code_for_section_1();
    }
#pragma omp section
    { // section 2
      code_for_section_2();
    }
#pragma omp section
    { // section 3
      code_for_section_3();
    }
  }
}
```

Listing 2.5: The OpenMP master, masked, single, and sections constructs.

to specify which subset of threads (which could be a single thread, of course) should
execute the guarded code region (see Listing 2.5) [101]. The master construct can then
be expressed as masked filter(0) or masked without any filter clause. However,
at the time of writing, version 5.1 of the OpenMP API Specification is still in development, so we will focus on the master construct. The implementation of the master and
single constructs is described in Section 6.9.

The `sections` construct in Listing 2.5 is a worksharing construct that performs *static tasking*. The three structured blocks of the three `sections` directives can be executed concurrently. If the number of threads in the team is larger than the number of `section` blocks, some threads may idle. If fewer threads than blocks are available, then some of the structured blocks will be executed by the same thread. We call this static tasking (in contrast to the task-based model of Section 2.2), as the number of tasks is statically defined at compile time and cannot change.

One aspect of the OpenMP syntax is the use of optional *clauses* that refine the behavior of the directive to which they are applied. You have already seen the `schedule` clause being used to specify the type of loop scheduling to be used for a parallel loop. The OpenMP API also provides clauses to control which data should be visible to the threads of a team. You may have noticed that in Listing 2.4, the `parallel for` directive has a `firstprivate` clause for the variable n.

The OpenMP API provides the following clauses to control the visibility of data between threads that can be specified on a parallel directive:

- **shared:** A variable that is declared as shared is visible to all the threads in the team. There's a single instance of the variable that each thread can access.
- **private:** A private variable is only visible to a single thread. Each thread creates its own uninitialized instance of the variable and only the creating thread has that instance in scope.
- **firstprivate:** This is the same as a private variable, but the private instance is initialized by assigning each private instance the value of the original variable from the thread that created the parallel region.
- **lastprivate:** This is a private instance of a variable for each of the threads. At the end of the parallel region, the variable in the thread that created the parallel region is updated to the value from the private instance of the variable in the thread that executed the last iteration of the loop or lexically last section.

The `lastprivate` directive can only be applied to parallel-loop directives or static-tasking directives (since there is no concept of the "last execution" in the general case). Where valid it returns a value to the outer-scope variable.

It is usually the programmer's responsibility to ensure that the data sharing is correctly specified; the `default(none)` clause can be used to force the programmer to specify the sharing explicitly. If two threads are sharing a variable or memory location (e. g., via an array pointer) concurrent accesses may lead to inconsistent results if at least one of the accesses involves an update of the shared memory location. This is called a *race condition* and needs to be avoided by either privatizing the threads' memory or by adding proper synchronization (see Section 2.3) to avoid the race condition.

Since race conditions can be very hard to find (as their appearance may be timing dependent), there are a number of tools, such as Intel Inspector [62] or Clang's ThreadSanitizer [28], which can and should be used to track them down.

2.1.5 OpenMP thread affinity

As we have seen, the OpenMP API explicitly names threads, so that in any parallel region with N_{Threads} threads, the threads are numbered $[0, N_{\text{Threads}})$. At first glance, you might assume that this also implies a tight binding to the underlying hardware, since operating systems also tend to enumerate hardware threads in a contiguous range from zero. However, it is not that simple. In fact there is no relationship at all between the OpenMP thread enumeration and the underlying operating system's hardware enumeration. While you can provide an OpenMP code with an explicit set of resources to use by giving an integer enumeration, how that maps to the underlying hardware remains implementation defined. While this may seem perverse, it is a reasonable choice, since outer-level resource-control systems may already have partitioned the available hardware and the executing process may not have access to hardware thread zero, for instance.

The OpenMP API therefore provides control of the granularity of the resources being accessed through the use of the `OMP_PLACES` environment variable, which allows a specification of `threads`, `cores`, or `sockets`, though without defining exactly what those terms mean. There is also the choice of how to map the available hardware resources to the thread-number enumeration. To control this, one can use the `OMP_PROC_BIND` environment variable (or the `proc_bind` clause on the `parallel` directive). This allows you to control whether the threads are all placed on the same resource as the main thread that is creating the parallel region (by using `primary`), nearby (by using `close`), or far away (by using `spread`).

The correct choice of affinity can affect the performance of an OpenMP code significantly. For instance, if one were running a code using SMT threads (which share all levels of the cache) and had a loop accessing arrays with a `static(n)` (i. e., block cyclic) loop schedule so that neighboring threads will be executing nearby iterations, then a `close` thread binding may perform well because the closely enumerated threads are likely to be accessing similar array elements, so they can usefully share a cache.

When comparing the performance of different affinity policies, be very careful to compare them fairly if you are also using SMT threads. As you increase the number of threads being used, a `spread` affinity will move on to use another core before it has filled all of the hardware threads in the first one, whereas the `close` affinity policy will use all hardware threads before moving into the next core. Therefore, the `spread` affinity policy may be using twice as many physical cores (in a machine with two hardware threads per core) as the `close` one. That will obviously give the `spread` affinity a big advantage, since it will have twice the physical resources for the software threads! For a fair comparison you should use "cores" as the x-axis (rather than "threads") and show separate lines for the different affinity policies.

```c
#include <stdio.h>
#include <stdlib.h>
#include <omp.h>

int main(int argc, char * argv[]) {
#pragma omp parallel
  {
#pragma omp task
    {
#pragma omp critical
        printf("Hello World!\n");
    }
  }

  return EXIT_SUCCESS;
}
```

Listing 2.6: Using OpenMP tasks to print "Hello World".

2.2 Task-based parallel programming

Task-based parallel programming generalizes the idea of static tasking by providing a more dynamic mechanism than the restrictive sections construct in parallel languages like the OpenMP API. Instead of the narrow syntactic scope of the sections construct with its nested section constructs, a task-based programming model typically offers a task-creation feature that can be used (almost) anywhere in the code to create concurrent work.

An OpenMP task version of the "Hello World" example is shown in Listing 2.6. Once a team of threads has been created, a task-generating construct (in the OpenMP API one of them is the task directive) can decorate a structured block and mark it for concurrent execution. This may involve clauses to control the specific behavior of the created task (e. g., what data is privatized for the task). Later on, Listing 2.11 will use the task construct for a recursive algorithm.

What happens at a conceptual level is shown in Figure 2.3. Whenever a thread encounters a task-generating construct, it may create a new task and store it in a *task pool* for execution (it is also allowed to execute the task immediately). When a thread runs out of work, it looks in the task pool, and, if there is available work, it grabs one of the tasks from the pool and starts to execute it. The executing tasks may in turn generate new tasks that may be deferred in the task pool and will eventually be picked up by one of the threads for execution. This will continue until the parallel algorithm stops generating new tasks, all existing tasks complete their execution, and the parallel region comes to an end.

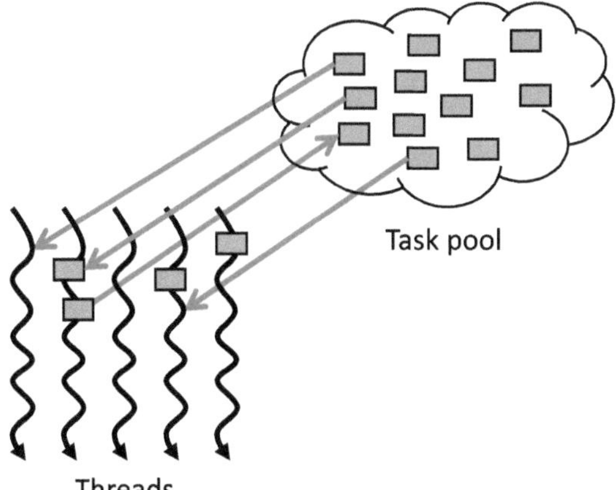

Figure 2.3: Conceptual execution scheme for tasks using a task pool.

```cpp
#include <iostream>
#include <tbb/task_group.h>
#include <tbb/spin_mutex.h>

const size_t num_tasks = 8;

int main(int argc, char * argv[]) {
  tbb::task_group grp;
  tbb::spin_mutex mtx;

  for (auto i = 0; i < num_tasks; ++i) {
    grp.run([&] {
      mtx.lock();
      std::cout << "Hello World!" << std::endl;
      mtx.unlock();
    });
  }
  grp.wait();

  return EXIT_SUCCESS;
}
```

Listing 2.7: Classic "Hello World" example using Intel Threading Building Blocks.

Listing 2.7 shows the same example, using the Intel Threading Building Blocks (TBB) library [66, 151]. TBB is a C++ template library that provides a powerful task-based programming model on top of C++. The run() method of the tbb::task_group class

```
void matmul_taskloop(float * C, float * A, float * B,
                     size_t n) {
#pragma omp parallel firstprivate(n)
#pragma omp master
#pragma omp taskloop firstprivate(n) grainsize(8)
  for (size_t i = 0; i < n; ++i)
    for (size_t k = 0; k < n; ++k)
      for (size_t j = 0; j < n; ++j)
        C[i * n + j] += A[i * n + k] * B[k * n + j];
}
```

Listing 2.8: OpenMP taskloop Matrix Multiplication.

is similar to the OpenMP task directive in that it creates a task object that holds the associated lambda expression so that it can later be executed.

Another key difference between the OpenMP API and TBB is how parallelism is managed. In the case of an OpenMP program, the programmer is responsible for creating the parallel region so that there are threads available to execute the concurrent tasks. In a TBB program, threads are created by the runtime library, and thus, programmers can immediately start creating tasks without worrying about having to create the team of threads first. When writing code that will live in a library but that wants to exploit parallelism, this is an advantage, since that code need not worry about whether it was called from a serial or parallel region.

This points to a philosophical difference between the two models. When using the OpenMP API, programmers can discover the current number of threads (via omp_get_num_threads()) and which thread is executing a piece of code (via omp_get_thread_num()), and the operation of loop scheduling is described in terms of which iterations are mapped to which specific thread.

In TBB, there is no specific concept of a thread's identity and the programmers are not expected to concern themselves with the number of threads, or the detail of how tasks are allocated to threads—rather, that is handled by the TBB runtime library, leaving the programmer to worry only about expressing the appropriate amount of parallelism. In this worldview, threads are an implementation detail that should not normally be of concern to the application programmer.

Listing 2.8 shows a task-parallel version of the MxM algorithm. It uses the OpenMP taskloop construct, which splits the iteration space of a loop into OpenMP tasks. While this looks similar to the worksharing construct that was shown in Listing 2.4, the behavior is fundamentally different. When using the OpenMP for construct, all threads in the parallel region have to encounter the construct so that they can split up the work, whereas the taskloop construct need only be executed by a single thread.

"Why is this needed?", you may ask. The answer is that the `taskloop` construct is defined in a way that is similar to the definition of a regular task (by the way, this is the same for TBB's `tbb::parallel_for` template). This means that if N threads encounter the construct, each of the threads will start executing the same loop, so the loop will be executed N times rather than having a single incarnation of the loop split between them. Since here we only want the whole loop to be executed once, we must ensure that only one thread encounters the `taskloop` construct. To ensure this, we use the `master` construct (though we could equally have used a `single` construct; there is no requirement that the thread that enters the taskloop is any particular thread).

The `grainsize` clause defines how many iterations should be executed per task. Funnily enough, the OpenMP standard allows this to be an interval, which in this example will be eight to sixteen iterations, so an implementation has some flexibility in choosing the exact number of iterations (and, therefore, tasks). The construct also supports the `num_tasks` clause, which specifies exactly how many tasks should be created, and then adjusts the chunk size accordingly.

It's worth noting that although it looks as if this design is serializing task creation by having only a single thread execute the `taskloop` construct, the OpenMP semantics do not enforce this. An implementation that performs recursive iteration-space splitting is allowed and may well be used by the OpenMP runtime library.

One specialty of OpenMP tasks is that they can either be *tied* or *untied* tasks (through the `untied` clause). The designers of the OpenMP API had to solve a problem that may arise from how OpenMP tasks were integrated with the explicit threading model. Since the OpenMP language contains the `threadprivate` directive to make a global variable a thread-local one (see Section 5.3.5). When a thread starts to execute a task, the task may read such a `threadprivate` variable. If the task is put on hold and later resumed, the task could re-read that same `threadprivate` variable. If the task has been picked up by a different thread, the task may see two different versions of the `threadprivate` variable. Note that similar issues could occur if code executing in a task claimed a lock and was then suspended and resumed in a different thread. Since the lock is owned by the thread, not the task, unpredictable behavior would follow.

That's why OpenMP tasks are tied by default, so that this case does not happen. For a tied task, the OpenMP implementation will make sure that, should the task ever be suspended and resumed, it will always resume on the same thread from which it was suspended. This ensures that the entire execution of the task is performed by a single, specific thread. By marking a thread as `untied`, programmers can assure the OpenMP implementation that the `threadprivate`-variable problem does not happen or that they don't care about it happening. The OpenMP implementation is thus free to choose a thread to resume a suspended, `untied` task.

Before we finish this basic introduction to tasks, let's recap a few important terms that we will need for the remainder of this book. When a task creates another task, the creating task becomes the *parent task* of the new task. The new task then is called a *child task* of its parent task. The term *sibling tasks* refers to all tasks that have the same parent. A *descendant task* is a task in the ancestor chain of a parent, so either a child task or a task created by a descendant task (e. g., a child task of a child task). If a new task is put into the task pool, it is said to be *deferred* while if it is executed straight away, it is *undeferred*. A task is described as *completed* when it has been scheduled for execution and that execution has finished.

2.3 Synchronization constructs

Synchronization is very important in parallel programs, as it ensures that the individual threads or tasks do not execute concurrently without proper coordination. One example is that the main thread of a parallel region waits for all of the other threads in the parallel team to reach the end of the parallel region before the main thread continues with the sequential part of the program.

But there are more synchronization constructs that come in different flavors and provide different scopes of how many threads and tasks participate in the synchronization operation. Regardless of this, programmers have to be careful when applying synchronization in their code. Too little synchronization and the code will suffer from race conditions, non-determinism, and ultimately, incorrect execution. Conversely, too much synchronization will overly constrain execution, limit the amount of parallelism exposed, and reduce performance.

Let's walk through the most important synchronization primitives that we need for the remainder of this book.

2.3.1 Mutual exclusion with locks

One of the most fundamental synchronization constructs is a *lock* that implements mutual exclusion between a set of threads and that protects code from being executed concurrently. The basic functionality is that once a thread has entered a protected region of code, no other thread can enter a region of code that is protected with the same lock until the current thread (the current *owner*) has released the lock. This can be used to secure shared data structures against concurrent modification and thus avoid race conditions that would otherwise occur due to the conflicting accesses to data.

Locks typically offer two operations: `lock()` to acquire the lock if it is free and `unlock()` to release the lock when the critical section of code being protected by the lock should end. The `lock()` operation typically blocks the calling thread until the lock becomes available. Whether or not these operations are exposed as regular functions (with a lock as a formal parameter) or as part of a lock object, does not really affect the implementation. Some interfaces for locks also offer a `try_lock()` function that checks the lock state, and if it is owned by another thread, immediately returns and does not wait for the lock to become free.

```cpp
void lock_example() {
  omp_lock_t lock;
  omp_init_lock(&lock);

#pragma omp parallel
  {
    omp_set_lock(&lock);
    mutually_exclusive_code();
    omp_unset_lock(&lock);
  }

  omp_destroy_lock(&lock);
}

void critical_construct() {
#pragma omp parallel
  {
#pragma omp critical(example)
    {
      mutually_exclusive_code();
    }
  }
}
```

Listing 2.9: Mutual exclusion features in the OpenMP API.

All of our "Hello World" examples above already include lock operations. Listing 2.7 already showed the `lock()` and `unlock()` operations. In this example, mutual exclusion was needed as the output through `std::cout` using the << operator is not thread safe. If the lock was removed, then the output might be garbled. You may want to try this yourself to see what could happen.

Earlier in this chapter, Listing 2.2 used a `std::mutex` object to achieve the same. That code uses the `std::lock_guard` template that was introduced with C++11. It nicely wraps the passed `std::mutex` objects and implements a RAII-style pattern [123]. When the `lock_guard` is constructed it calls the `lock()` method of the lock object. When the scope ends and the destructor of the `lock_guard` object is called, the lock object is released. This helps to release the lock in the light of complex control flows and exceptions that may be thrown by called functions.

Listing 2.9 at the top shows one option for mutual exclusion in the OpenMP API. The interface for locks offers the `omp_set_lock()` and `omp_unset_lock()` routines to acquire and release a lock, respectively. In the OpenMP design, a lock is an opaque object that the programmer has to create as a variable of type `omp_lock_t` and has to initialize it using `omp_init_lock()`. Consequently, at the end of its lifetime, the lock must be destroyed by calling `omp_destroy_lock()` to tell the OpenMP implementation that the lock is no longer needed.

The OpenMP API specification also provides pragma syntax in C/C++ and directives for Fortran to work with mutual exclusion (see Listing 2.9, bottom). The `critical` construct acquires a lock when a thread encounters the construct and automatically releases the lock at the end of the construct. The management of the lock initialization is done by the compiler and the underlying runtime system. The `critical` construct can take a name in parentheses, which can be used to use different locks associated with different names. Only `critical` constructs with the same name will exhibit mutual exclusion.

One advanced class of locks is *speculative locks*. When acquiring the lock, these locks do not await the release of the lock by the current lock owner. Instead, the acquiring thread assumes that the lock is free and starts to execute the critical section speculatively, while guarding against conflicting memory accesses. This kind of speculative execution can be implemented in software, but high performance requires explicit hardware support.

This is helpful if the mutually exclusive code region needs to avoid race conditions to be correct, but the chance of conflicting memory accesses is low. One particular example is concurrent accesses to hash tables. The odds that two threads will access the same hash bucket are very low (for good hash functions), but such conflicts cannot be ruled out completely. Thus, the hash table needs a lock for protection, but a speculative lock can dramatically speed up execution, as unless two threads happen to use the same hash bucket, execution can just proceed without "real" mutual exclusion taking place.

Of course, you need a mechanism that allows the threads, the compiler, or the runtime system to detect such access conflicts. They typically occur if multiple threads are writing to the same memory locations or if there's at least one thread that writes to a memory location, while other threads are reading from it. Typical implementations rely on the concept of *transactional memory* [54], which provides the capabilities to perform the detection of such conflicts. Transactional memory can also be implemented in hardware, e. g., with the Intel[*] Transactional Synchronization Extensions [68] and the Arm[*] Transactional Memory Extension [8].

We cover this in much more detail in Chapter 6 and specifically Section 6.8, where we show the performance of a speculative lock when applied to a `std::unordered_map`.

2.3.2 Barriers, reductions, and latches

A *barrier* is a more coarse-grained feature for the synchronization of threads. While a lock can be viewed as a fine-grained mechanism that mostly involves synchronization between the current lock owner and the waiting threads, a barrier typically synchronizes all threads in a parallel region.

When a thread executing in a parallel region encounters a barrier, it has to wait until all other threads in the parallel region have also reached the same barrier. Only then does the barrier become permeable and the threads are allowed to leave it and continue executing the code that follows the barrier.

Barriers can be *implicit*, that is, they occur as part of another surrounding construct, or they can be *explicit*, written by the programmer. Listing 2.10 has some OpenMP examples.

The first example contains code that takes an array and computes a second array which represents the fraction of the total of the first array that each element represents. For instance, if the input was (0, 4, 3, 3), then the result would be (0.0, 0.4, 0.3, 0.3). It uses an initial `parallel for` loop with a reduction to compute the total, and then a second one to compute the fractions. Clearly, the second loop cannot start until the first has finished, since it is only then that the total of the initial vector will be known. The implicit barrier at the end of the first loop ensures this for us. The implicit barrier at the end of the second loop is effectively unnecessary, since it occurs immediately before the end of the parallel region; we could remove it (by adding a `nowait` clause to the loop pragma) or hope that this is so obvious that the compiler will remove it anyway.

The second function shows explicit `barrier` constructs, which programmers can add to make sure that all threads have completed the code above the barrier before they start executing the code below the barrier.

When thinking about barriers, it's also worth considering *reductions*, in which all threads combine results into a single answer. Since that answer cannot be finally computed until all threads have finished executing the relevant code, reductions complete at barriers, and their implementation can be related to that of the barrier.

The function `scalar_product()` in Listing 2.10 shows the use of a reduction operation. All threads can compute a partial result from the assigned work, usually stored in a thread-local variable. When the construct ends, the partial result of each thread is gathered and aggregated into a single, global result. In the example of Listing 2.10, each thread sums up the products of the array elements in its assigned loop chunk. When the parallel region ends, all of these intermediate results are added together and the final result is stored in the `sum` variable before it is returned.

A *latch* is a special, strictly decreasing counter with waiting functionality. At creation time, it is initialized with a starting value. A thread can interact with the latch in three basic ways. First, it can ask the latch to decrement. Second, a thread may enter the latch and wait for it to reach zero (while other threads may decrement the latch). Third, it may decrement and wait in a single operation. When the latch reaches zero it will release the waiting threads. The main difference between a barrier and latch is that the barrier's goal is to wait for all threads (of the parallel region), while the latch can handle cases in which only a subset of threads needs to wait since the count can

```c
void compute_fraction(double * frac, const double * src,
                      int n) {
  double total = 0.0;
#pragma omp parallel shared(frac,src,total) firstprivate(n)
  {
#pragma omp for reduction(+:total)
    for (int i = 0; i < n; ++i) {
      total += src[i];
    }
    // implicit barrier
#pragma omp for
    for (int i = 0; i < n; ++i) {
      frac[i] = src[i]/total;
    }
    // implicit barrier
  }
}


void two_phases() {
#pragma omp parallel
  {
    printf("Before the barrier\n");
    stuff_before_barrier();
#pragma omp barrier
    printf("After the barrier\n");
    stuff_after_barrier();
  }
}


double scalar_prod(double * a, double * b, size_t n) {
  double sum = 0.0;
#pragma omp parallel for reduction(+ : sum)
  for (size_t i = 0; i < n; ++i)
    sum += a[i] * b[i];
  return sum;
}
```

Listing 2.10: Implicit and explicit barriers in the OpenMP API.

be set to an arbitrary value. While the OpenMP API does not have a corresponding concept, C++20 includes the `std::latch` class to implement latches.

Chapter 7 will cover the implementation aspects of barriers and reductions while we skip the implementation details of latches.

2.3.3 Task barriers

In task-centric programming models, programmers also need synchronization features that help tame task execution and keep the tasks from going wild. We will call these synchronization mechanisms *task barriers* in this section, as they constitute barrier-like semantics in that they await the completion of (a subset of) the created tasks. However, one key difference though is that for a regular (thread) barrier, the synchronizing entities are still executing, while for a task barrier, some of the created tasks may have completed execution already.

The two scenarios that are typical in task-parallel programming models are:

– **Wait for immediate child tasks:** The parent task waits for the child tasks it created and continues execution once the last child has finished executing. Descendant tasks created by the child are not part of the synchronization set, so they may still be executing.
– **Wait for a subset of tasks:** Execution will wait for a subset of the tasks that are described by a grouping construct. Tasks that are not part of the group are not part of the synchronization set.

The first waiting pattern can be seen in Listing 2.11. It shows a terribly inefficient, yet illustrative version of the calculation of the nth Fibonacci number. To compute the number, the code performs a recursion that breaks down the problem and launches a new task for each of the recursion branches to compute fib($n-1$) and fib($n-2$). The task-parallel recursion stops when n reaches ten and switches to a sequential version of the algorithm. This is called a *cutoff*, and helps to avoid creating too many fine-grained tasks. Overall, the recursion finally stops at $n = 2$.

At each recursion level, the `fib_task()` function has to wait for the two child tasks to complete. Each computes a value that `fib_task()` must sum up, so the function cannot continue until the two tasks are done and have updated x and y, respectively. This is accomplished by the `taskwait` construct. It stops the encountering task until all of its child tasks have completed, but it does not wait for other descendant tasks that have been created by any of the child tasks (although in this case, we know from the recursive structure of the code that no child can return until its children have finished executing, and that therefore there can be no descendants at any level that are still executing).

The second, more general, waiting pattern needs some syntax to describe that some tasks should belong to the synchronization set, while others do not. Typical task-

```
size_t fib_seq(size_t n) {
  if (n < 2)
    return n;
  return fib_seq(n - 1) + fib_seq(n - 2);
}

size_t fib_task(size_t n) {
  size_t x, y;

  if (n < 2)
    return n;

  if (n < 10) {
    x = fib_seq(n - 1);
    y = fib_seq(n - 2);
  }
  else {
#pragma omp task shared(x)
    x = fib_task(n - 1);
#pragma omp task shared(y)
    y = fib_task(n - 2);
#pragma omp taskwait
  }
  return x + y;
}
```

Listing 2.11: Calculation of Fibonacci numbers to illustrate waiting for child tasks.

parallel programming models offer the concept of a *task group*. All (descendant) tasks that are created from within in the construct then participate in the synchronization, while all other tasks do not.

Listing 2.12 shows a short piece of code that traverses a linked list and creates a child task for each element to process it in parallel using a `taskloop` construct. When this code executes, the parent task that is running the `while` loop will create a new child task for each list element. Once it has finished creating tasks, it will wait at the closing curly brace of the `taskgroup` construct until all child tasks have been completed. As each of the child tasks creates additional tasks when encountering the `taskloop` construct, the waiting task will also await the completion of those tasks.

Intel TBB offers a similar concept as part of the `tbb::task_group` class. Listing 2.7 used this already, as the tasks to print "Hello World" were created as a part of such a task group, and therefore the parent task must already wait for the children to complete printing the message on the console.

```
void process_element(data_t * data) {
#pragma omp taskloop simd nogroup
  for (size_t i = 0; i < data->length; ++i)
    data->c[i] = data->a[i] + data->b[i];
}


void linked_list(linked_list_t * list) {
#pragma omp taskgroup
  {
    while (list) {
      data_t * data = list->data;
#pragma omp task firstprivate(data)
        process_element(data);
      list = list->next;
    }
  }
}
```

Listing 2.12: Traversal of a linked list with task barrier.

Task groups can also be an implicit synchronization mechanism for other tasking features. For instance, the OpenMP `taskloop` construct has an implicit synchronization point at the end of the parallel loop that is caused by the creation of an implicit `taskgroup`. Programmers can remove this implicit task group by using the `nogroup` clause, as the code in Listing 2.12 does.

Task-parallel programming models can add additional features to task groups beyond using them for synchronization and waiting. The OpenMP API provides *task reductions* (via the `task_reduction` clause, in see [100]) that generalize the reduction of worksharing constructs to arbitrary task graphs. Programmers can specify it for a task group to which tasks can contribute a partial result by using the `in_reduction` clause. When the waiting thread leaves the `taskgroup` construct, the reduction variable will have been updated with the aggregated values.

Another feature that the OpenMP language offers is the cancellation of tasks. A `cancel` construct can trigger the termination of running tasks. The `taskgroup` construct then defines the set of tasks that are subject to cancellation.

One final point to make is that programming models that offer both thread synchronization and task synchronization may tie task synchronization to thread synchronization. A particular example is the OpenMP `barrier` construct, which not only requires all threads to arrive before any can leave, but also awaits the completion of all tasks that have been created by the threads before they reached the barrier.

```
void task_deps () {
  int a, b, c, d;

#pragma omp task depend{out : a, b } // T1
  a = b = 0;

#pragma omp task depend(in : a) depend(out : c) // T2
  c = computation_1(a);

#pragma omp task depend(in : b) depend(out : d) // T3
  d = computation_2(b);

#pragma omp task depend(in : c, d) // T4
  computation_3(c, d);
}
```

Listing 2.13: OpenMP code with task dependences.

2.3.4 Task dependences

The last, but not least, synchronization concept in this chapter is *task dependences*. Instead of making one task wait for (a subset) of other tasks, a task dependence introduces a partial order in the execution of the tasks that the worker threads use to pick tasks that are ready for execution, because their dependences are fulfilled.

The OpenMP syntax for task dependences uses the depend clause that receives a list of entities that model the dependence between two tasks:

```
#pragma omp task depend(type: list-items)
```

The argument to the depend clause is a comma-separated list of either variables (like a or b in Listing 2.13) or array sections (e. g., a[0:99]) with a start index, optional length, and optional stride. The type of dependence for each entry in the depend clause must be in, out, or inout.

If an in dependence is present at the task directive for task A, then task A cannot execute until all sibling tasks B that have been created before and that have a non-empty intersection with the *out* items of the dependent task have completed. The *inout* dependence type combines *in* and *out*.

What this means is shown in Figure 2.4 for the code in Listing 2.13. The code creates four tasks in sequence. Task T_1 has an *out* dependence for the variables a and b because it initializes both to zero. Tasks T_2 and T_3 need these variables as inputs, so they have the corresponding *in* dependence for a and b. At the same time, they pro-

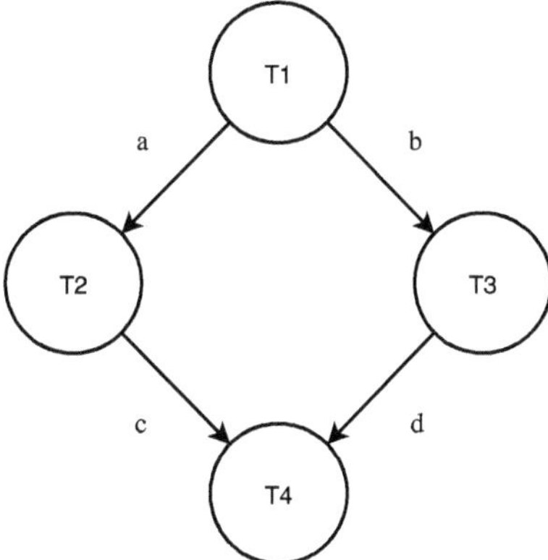

Figure 2.4: Visualization of the task dependences in Listing 2.13.

duce values in c and d, which means that T_2 and T_3 need to have a *out* dependence for these variables. Finally, T_4 uses c and d, so it specifies an *in* dependence.

The resulting, simple *task dependence graph* (TDG) is shown in Figure 2.4. A task dependence can only be a forward-looking dependence—that is, a later task can only depend on a previous task, but not vice versa. Thus, the task graphs in this dependence model form a *directed acyclic graph* (DAG), where the nodes are the created tasks with an edge between two nodes if and only if there is a task dependence between the two tasks. In Figure 2.4, we have shown the list item that generates an edge as a label at the arrow that represents it. As the DAG does not have cycles, the task graph cannot cause a deadlock.

The graph for the example code shows that T_2 and T_3 can only execute after T_1 has been completed. As T_2 and T_3 do not have a dependence on each other, their relative order of execution is not defined. The only constraint is that they must both have executed before T_4 can be scheduled for execution. This is in line with our expected execution order, considering how the tasks read and update their data.

The task dependences and the resulting TDG induce a partial ordering relationship without the need for coarser-grained synchronization, like `taskgroup` or `taskwait`. When a worker thread wants to execute a task from the task pool, it can only pick tasks that have been marked as executable, as all of their task dependences have been satisfied, and it will not consider tasks with unsatisfied dependences. Thus, the threads will respect the dynamic ordering of the tasks enforced by their dependences.

Listing 2.14 shows another version of the matrix-matrix multiplication example that we used earlier in the chapter. What is different now is that we have applied a blocking factor *bf* to the code. For each block of size *bf* × *bf*, the code now generates an OpenMP task to multiply two blocks of A and B and update the corresponding block in the C matrix.

```
void matmul_task(float * C, float * A, float * B,
                 size_t n) {
  const size_t bf = 512;
  assert(n % bf == 0);
#pragma omp parallel master firstprivate(n, bf)
    for (size_t ib = 0; ib < n; ib += bf)
      for (size_t kb = 0; kb < n; kb += bf)
        for (size_t jb = 0; jb < n; jb += bf) {
#pragma omp task firstprivate(ib, kb, jb)                  \
                 firstprivate(n, bf)                       \
                 depend(inout : C [ib * n + jb:bf]) \
                 depend(in    : A [ib * n + kb:bf]) \
                 depend(in    : B [kb * n + jb:bf])
          for (size_t i = ib; i < (ib + bf); ++i)
            for (size_t k = kb; k < (kb + bf); ++k)
              for (size_t j = jb; j < (jb + bf); ++j)
                C[i * n + j] += A[i * n + k] * B[k * n + j];
        }
}
```

Listing 2.14: Matrix-matrix multiplication using task dependences.

As the MxM algorithm corresponds to a repeated (n^2 for a $n \times n$ result matrix) scalar product of row and column vectors [48], one can parallelize along one dimension, as in Listing 2.8. If we want to create more parallelism, however, we also have to parallelize the scalar product. Thus, a reduction will be required to aggregate the partial sums of the scalar products of the partial vectors assigned to threads or tasks. Similarly, in a block-wise distribution, like that in Listing 2.14, this means that multiple tasks need to update a block in the C matrix. So, two tasks that multiply blocks that update the same block in C must serialize. That is what is done in the code by adding the proper task dependences.

Figure 2.5 visualizes the resulting TDG for matrices of $3,072 \times 3,072$ elements and a block size of $1,024 \times 1,024$ elements. This means a total of nine blocks per matrix, or 27 tasks that will be created at runtime. As you can see from the TDG, nine tasks can run in parallel, one for each of the blocks in the C matrix. For each block, three tasks have to execute sequentially. These are the ones that process the three blocks of a row in A and a column in B.

While implementations with task dependences may seem more complicated than simple worksharing or the taskloop construct, they can expose additional parallelism (allowing the code to run well on larger machines) and allow the code to perform better when dealing with jitter or noise in the system.

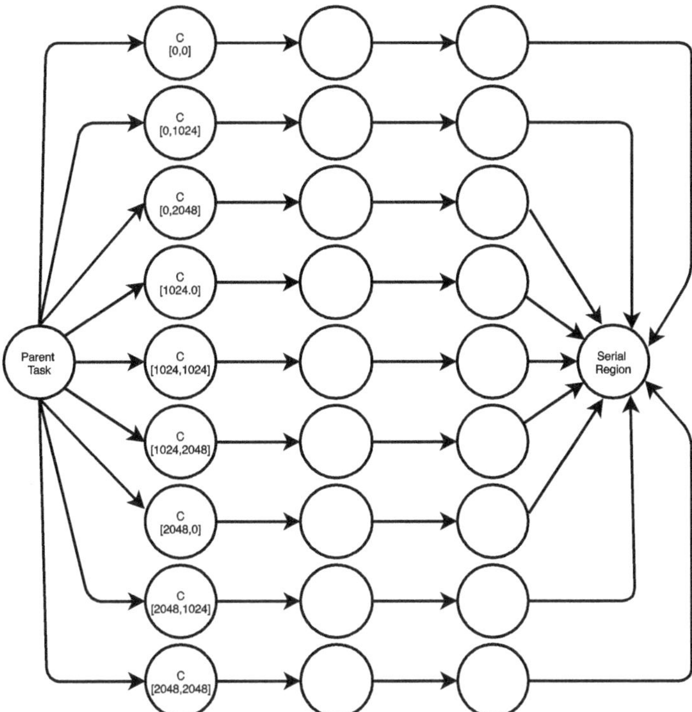

Figure 2.5: Visualization of task dependences for the MxM code in Listing 2.14.

2.4 Amdahl's law

We have been discussing parallel programming models and their properties, It's also worth stepping up a level and considering the fundamental limits that no model can overcome. We will therefore now discuss Amdahl's law [6], and also use that discussion to illustrate how we choose to display performance results in this book, and why we make the choice we do.

Amdahl's law is a simple, yet consequential, model of a parallel program. Let's assume that there are some aspects of the code that necessarily have to be serial (the *serial fraction*), while other parts can be parallelized to some degree (the *parallel fraction*), which is bounded either by the available hardware or the grain size of the chunks of parallel work. Since here we're dealing with shared-memory parallelism, we will express the available degree of parallelism with N as the number of threads. Of course, the same rule applies equally well to any parallel execution model.

We can then easily see that if we have a serial fraction F_{Serial}, and the serial execution time is T_{Serial}, when running the parallel fraction with N threads the parallel execution time, T_{Parallel}, can be computed like this:

$$T_{\text{Parallel}}(N) = T_{\text{Serial}} \cdot \left(F_{\text{Serial}} + \frac{1 - F_{\text{Serial}}}{N} \right).$$

If we consider the case where $N \to \infty$ (i. e., we have infinite hardware parallelism and can split the parallel work into infinitesimally small pieces), we can see that the best performance we can possibly achieve is

$$T_{\text{Parallel}} = F_{\text{Serial}} \cdot T_{\text{Serial}}.$$

Amdahl argued that since the available parallelism is always limited, we can never achieve the performance we desire solely by using parallelism, but rather, we also need to improve the performance of the serial code. Gustafson [53] argued that Amdahl's law is not useful for HPC problems because Amdahl's law applies only to fixed-sized problems (*strong scaling*), whereas in HPC, we are often interested in solving larger problems, and increasing the problem size often leads to a smaller serial fraction of execution (*weak scaling*). While Gustafson is undoubtedly correct, from our point of view as runtime implementers, we have to deal with the parallelism that is exposed by the user application, and do not control the problem size, so what interests us is what performance we can expect at different scales.

2.4.1 Presenting performance results

When dealing with parallel programs, we must be careful about how we perform our experiments, and how we present our results. It is much too easy to run experiments that are not comparable, or to present results in a way that leads to incorrect conclusions. David Bailey gives examples in his classic paper "Twelve Ways to Fool the Masses When Giving Performance Results on Parallel Computers" [11]. Some of the details there are no longer appropriate for current machines; however it is possible even now to find people who compare the performance of code that has been heavily optimized and parallelized to run on an expensive accelerator with that which runs serially on a single CPU core (and which was probably compiled with -O0). It is always important to remember that there are two ways to increase the value of a ratio $\frac{A}{B}$; one can either reduce B or increase A. If A is "performance after my optimizations" while B is "performance before them", then making initial performance poor can be significantly easier than making performance after optimization fast!

One classic way to display parallel performance is to use the *speedup*, or $S(N)$, for N threads. If we measure the elapsed time $T(N)$, and have a normalizing T_{Ref} (which is often the time measured when running with a single thread, i. e., $T(1)$) then the speedup is defined as

$$S(N) = \frac{T_{\text{Ref}}}{T(N)} = \frac{T(1)}{T(N)}.$$

However, since it is a ratio, speedup is vulnerable to the "choose a slow initial implementation" trick; therefore, when we present normalized results, we try to be clear in what we are normalizing against. Usually, when dealing with thread scaling, that will be the best performance of any of the cases tested when running on a single thread. One can, reasonably, argue that we should use the performance of the best serial implementation; however, since here we are normally comparing implementations of parallel aspects of a runtime, and not considering whether code should be run in parallel or not, our decision seems reasonable for the comparisons we want to make. When looking at quoted speedups, it is important always to consider what T_{Ref} has been used.

Expressed in terms of speedup, Amdahl's law looks like this:

$$S(N) = \frac{1}{F_{\mathrm{Serial}} + \frac{1-F_{\mathrm{Serial}}}{N}}.$$

Figure 2.6 shows the speedup (including the perfect speedup when the serial fraction is zero). There are a number of problems in this display, though:

1. **Wasted space:** The whole upper-left side of the graph contains no data, and is unlikely to, since super-linear speedups are rare. Wasting half of the plot area goes against one of Edward Tufte's cardinal rules (in his classic "The Visual Display of Quantitative Information" [139]) that one should "Maximize the data ink" and not waste ink or space on decorative fripperies.
2. **Legibility at low thread count:** Since the y-axis bounds are set by the highest values, the data for low thread counts is all compressed into a small amount of space, and it is effectively impossible to see what the differences are between the different low thread count cases.

We can overcome these limitations by plotting the *parallel efficiency*, $E(N)$, rather than the speedup. Parallel efficiency is defined as

$$E(N) = \frac{T_{\mathrm{Ref}}}{N \cdot T(N)} = \frac{T(1)}{N \cdot T(N)}.$$

It shows us how much of the available resources we are exploiting productively, and how much we are wasting. You can also see that it is the speedup divided by the number of threads. Since it represents an efficiency, it can be expressed as a percentage that is normally <100 %. Graphically, this is useful because our y-axis will now span the full range from 0 % to 100 %, no matter how large a machine we have, and we can therefore still look in detail at one or two threads, and also see the large-scale trend at high thread counts.

Figure 2.7 shows the same Amdahl data shown in Figure 2.6, omitting the, now redundant, perfect speedup line, which would sit at 100 %. You can see that most of the graph area is occupied with useful data, and that we can make useful comparisons

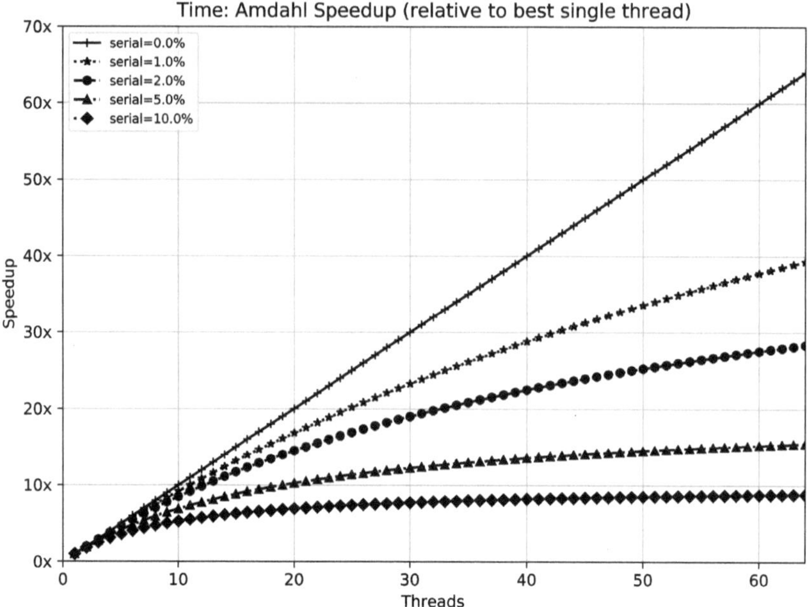

Figure 2.6: Amdahl's law speedup curves.

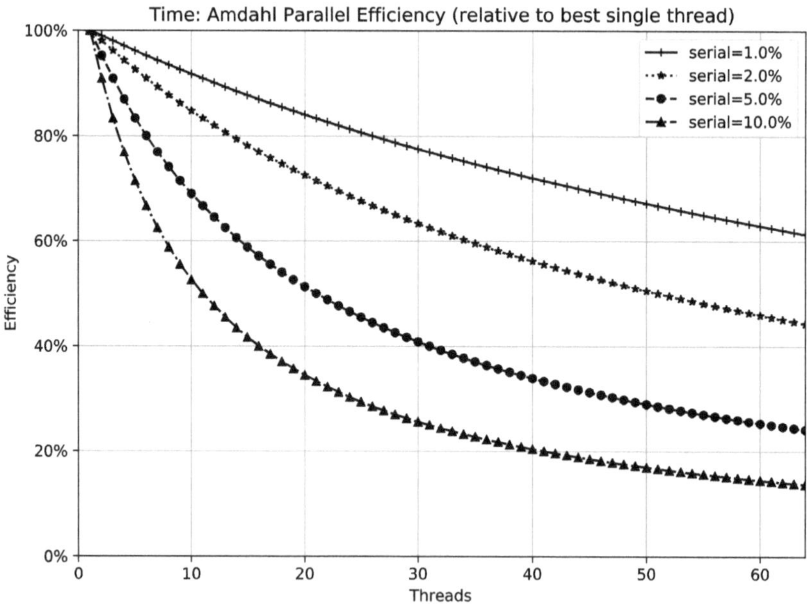

Figure 2.7: Amdahl's law efficiency curves.

between the different cases when there are few threads being used, as well as for the cases with many threads.

Since the parallel efficiency makes us focus on how poorly (or well if you want to be more optimistic) we are doing ("We only got 60 % efficiency, so we've lost the equivalent of almost 26 CPUs in our 64-core machine!") rather than speedup ("Yee-ha, we got a 39× speedup over one core when we use the 64-core machine!"), it is useful to us engineers, whereas marketing folk (and some managers) would much rather see a graph that goes up toward the right, so they tend to prefer speedup. (Any speedup curve that doesn't manage to go up to the right really is bad, since it means that adding resources is not improving performance!) Note that the two cases we commented on above represent the same result, since a parallel efficiency of 60 % at 64 cores is effectively the same as a speedup of 39×, since $64 \times 0.6 = 38.4$, which would round up to 39×, or even 40×, faster when being described by a salesperson.

2.4.2 Effect on performance

We can, of course, turn Amdahl's law around and compute the required serial fraction if we are to reach a given efficiency (E_{Target}) at a particular level of parallelism N:

$$F_{\text{Serial}}(N) \leq \frac{1 - E_{\text{Target}}}{E_{\text{Target}} \cdot (N - 1)}.$$

Figure 2.8 shows those results for various required efficiencies in the range of [10,64] threads. You can see that even if you are satisfied with only 70 % efficiency, the serial fraction of the code must be below 1 % above 44 threads. If you want to achieve 95 % efficiency, then you must be <0.5 % serial at 12 threads and <0.1 % at 54 threads.

Amdahl cuts hard!

2.4.3 Mapping overheads to Amdahl

So far, we have been considering Amdahl's law as though we can easily see where serial code and parallel code are. However, in reality, things are not quite that simple, since even inside parallel code we can have serializing constructs (such as the OpenMP `master`, `single`, or `critical` directives) or regions protected by locks, which can also force serialization. The complexity here is that the amount of serialization is not easy to determine simply by looking at the source code, since the degree of contention of a given lock may very well depend on the specific data that the program is processing. The only real solution here is to use performance profiling tools, which show the reality of specific code execution.

In addition, any overheads in the runtime implementation of operations such as fork and join could be considered as serial fractions, since they are periods during

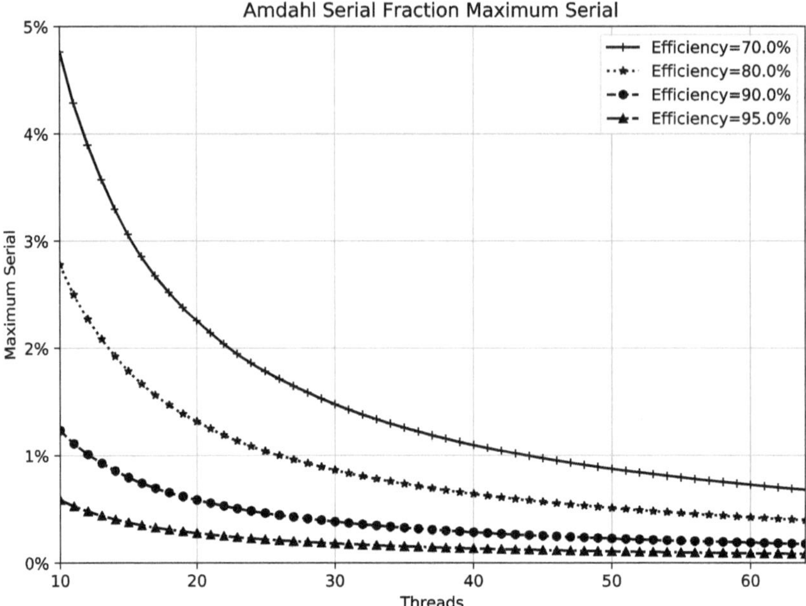

Figure 2.8: Amdahl's law serial fraction limits.

which parallelism is not being exploited. This kind of serial fraction usually depends on the number of threads, and thus increases when we add more and more threads. In reality, such overheads are even worse than being serial fractions, since as well as preventing parallelism, they are not allowing productive execution in the serial code; however, adding them to the serial fraction gives us at least some useful information.

2.4.4 Other variants of Amdahl's law

As well as the straightforward version of Amdahl's law that we have been discussing, it is possible to enhance the law to investigate other system configurations that may be interesting. For instance, we have seen that many of the hardware platforms that we have looked at support simultaneous multi-threading (SMT) (see Section 3.1.6). This raises the question of whether it is better to enable SMT to provide us with 2× or 4× more threads that all run slower or stick with using only a single thread/core that we hope will run at the highest available speed.

Clearly, there is no possible gain if enabling SMT2 gave you two logical cores, each of which executed at $\frac{1}{2}$ the performance; however, the hardware architects are not that stupid, and do achieve higher overall throughput. At HotChips 2020, Marvell[*] published performance data for the ThunderX3[*] Arm core [124], which is shown in Table 2.1. Their presentation showed "Instructions Per Cycle"; we have taken the re-

Table 2.1: Marvell TX3 SMT overall throughput.

Test Case	SMT Threads		
	1	**2**	**4**
High CPI (~2)	1.00	1.79	2.21
Medium CPI(~0.8)	1.00	1.38	1.73
Low CPI(<0.5)	1.00	1.18	1.28

Table 2.2: Marvell TX3 SMT per-thread performance.

Test Case	SMT Threads		
	1	**2**	**4**
High CPI (~2)	1.00	0.90	0.55
Medium CPI(~0.8)	1.00	0.69	0.43
Low CPI(<0.5)	1.00	0.59	0.32

ciprocal to show Cycles Per Instruction (CPI), since that is the notation we are using in this book.

Table 2.1 shows the per-core throughput (i. e., performance), however the SMT per-thread performance which we can deduce from that (shown in Table 2.2) is more useful as the input to our Amdahl calculations.

If we assume that the level of parallelism in our parallel execution phases is un-bounded (the normal Amdahl assumption) and we believe that our hardware design-ers are clever, and so have designed hardware that automagically chooses the appro-priate SMT level for the number of hardware threads that have work to do, then things are easy and using more SMT threads should always win. However, if the hardware cannot easily switch back into the SMT1 state for our serial code, then we have an "Am-dahl with Slowdown" problem, since the serial code will now take more time than it did before, and this can negate the gains made from executing the parallel code faster.

Figure 2.9 shows the expected performance with 1% serial time for the SMT1, SMT2, and SMT4 options with low, medium, and high-CPI codes. Since we want to compare the performance using the same hardware resources, the x-axis here is "Cores", not "Threads". We saw the same logic when discussing how to plot OpenMP performance with different affinity policies in Section 2.1.5. As a result, the SMT1 case appears to perform badly in the 1-core case, because it can only run serially, whereas the SMT2 or SMT4 cases can run 2 or 4 threads.

However, we can see that enabling SMT is not a guaranteed win, and that by the time we get to 64 cores, only a high-CPI code would show any gain over the SMT1 case unless the hardware architects (or possibly the runtime implementers) can rapidly switch from SMTn back to SMT1 to execute serial code.

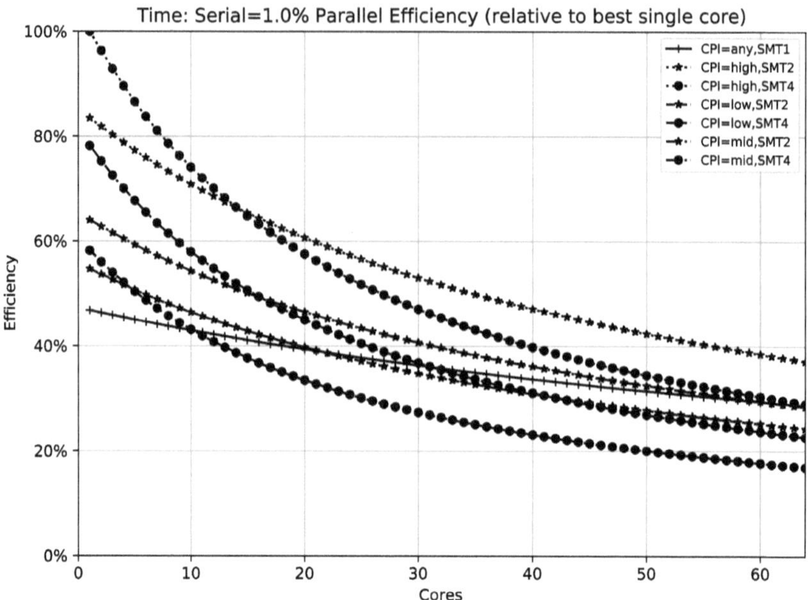

Figure 2.9: Amdahl's law with slowdown.

Note that a few weeks after publishing the technical details of the ThunderX3, Marvell announced that they won't be selling it as a commodity part, but only as a component of ASICs. However, although this means that this specific calculation does not matter to anyone, the real point here is that simple variants of Amdahl's law can be useful to investigate decisions like these from a high-level point of view, and that it can provide insight into the best performance we can expect to achieve, even if we have a zero-cost parallel runtime implementation. This can be helpful when dealing with hardware designers, or application programmers, who are convinced that poor performance must be the result of a poor runtime implementation.

2.5 Conclusions

This chapter is by no means an extensive introduction to parallel programming and did not cover all aspects in sufficient depth. As this book is about parallel runtimes, all we could do was to introduce the most important concepts to set the stage for what we need throughout this book. There's certainly more to parallel programming than meets the eye. In addition to what we referred to during the chapter, some books that we can recommend for reading (some of them being a bit dated, but still useful) are

- Tim Mattson's et al. book about parallel patterns [88],
- Michael Quinn's books about "Parallel Computing: Theory and Practice" [104] and "Parallel Programming in C with MPI and OpenMP" [105],

–   Rauber's and Rünger's "Parallel Programming: for Multicore and Cluster Systems Hardcover" [107], and
–   the book "Introduction to Parallel Computing' [138].

Ultimately, there are many ways to express parallelism in a program. Which is right for any given problem will depend on more than what you may have experience with. Though if you are working on a serious pre-existing code the choice will likely already have been taken by someone else.

What you've also seen in this chapter are some fundamental limits in what a parallel code can achieve in terms of maximum performance. There are simple mathematical models (such as Amdahl's law) which are useful when designing code and which should not be ignored, because they cannot be defeated. Unfortunately, Amdahl's law is overly optimistic (despite setting some tough limits for your code), as it ignores additional, even stricter limits that are introduced by the hardware of the system such as memory latency, memory bandwidth, compute capabilities, and many more.

We will take a closer look at hardware in the next chapter and discuss some of these limits, as they will greatly affect how we have to think about the machine, when we attempt to write code for a parallel runtime system.

3 Many-core and multi-core computer architectures

Whether you have a multi-core processor or a many-core processor is a bit fuzzy. In fact, there's no agreed-upon and well-established threshold that separates the multi-core from the many-core world. Some use the term many-core to describe processors with tens to thousands of processing cores [146], which, at the time of writing, already seems to indicate that we have entered many-core territory. So this issue is clearly a moving target, and we will leave it up to you to make your own decision as to whether your favorite processor is multi-core or many-core.

Regardless of how you categorize your favorite processors, we are focusing on parallel architectures, and will use the term multi-core from now on. In the following, we will construct a processor from the very basic elements so that we have a common ground on how to build a parallel, multi-core processor out of individual cores plus extra hardware to connect all the cores to form the parallel processor efficiently.

However, we cannot cover all aspects of the individual topics of computer architecture. For a deeper understanding of this subject, refer to books such as Hennessy and Patterson's [56] or Tanenbaum's [125].

3.1 Execution mechanisms

The discussion of how to build a parallel machine starts with examining how early (single-core) processors worked. We will then incrementally refine the design and make it more complex to extract more parallelism within the individual cores to achieve higher execution speeds. Please note that this kind of parallelism is usually transparent to the programmer, as, since the commercial failure of *very long instruction word* (VLIW) machines, the processor does not expose these kind of features as part of the ISA. However, generally, you will notice a measurable speed up when executing instructions on more modern machines that exploit *instruction-level parallelism* (ILP).

As we will see in Section 3.2.2, some of the architectural concepts embodied in processors do have an effect when we consider multi-core machines and how they interact with memory. They might cause the reordering of memory operations, which can be observed from other cores, and that must be considered when programming a parallel system. However, let's put this topic on the back burner for a bit and revisit it after we have understood the basics of processor architectures.

3.1.1 Von Neumann architecture and in-order execution

Let's start easily and build a simple execution engine for machine instructions. The prevalent designs of modern processors are all based on the ideas of the von Neumann

https://doi.org/10.1515/9783110632729-003

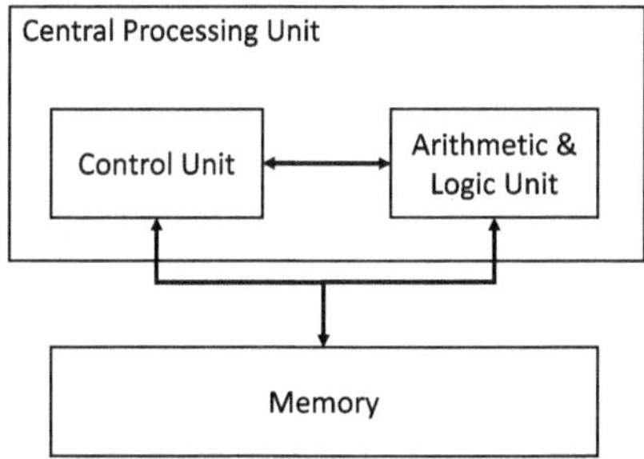

Figure 3.1: Von Neumann processor architecture.

architecture [19, 46, 150]. Before the von Neumann architecture, processors were usually fixed-function devices that could only execute a single hardwired task. Instead, a processor with the von Neumann architecture can execute streams of instructions stored in memory that can easily be changed to execute a different program.

Figure 3.1 shows the main components of the von Neumann architecture. The central processing unit (CPU) is the engine that executes the program code and contains two conceptual components, the *control unit* and the *arithmetic logic unit* (ALU). The task of the control unit is to retrieve the instructions of the program code from the *memory* and decode them. It then issues operations to the ALU to perform the actual work. If the operations refer to data in memory, the data is fetched by accessing the memory. What is called the *CPU* in the von Neumann architecture, we will call the *processor core*, or simply *core*, from now on. One thing to note is that the von Neumann architecture also includes a component for input and output, which we have left out for the sake of simplicity.

The key improvement of the von Neumann architecture is that to-be-executed code is just some special form of data that is coming from the same memory as any other data. Thus, the computation performed by the core can easily be changed by loading a different program into memory (or even by modifying the code in memory like viruses and self-modifying code do). Earlier electronic computers (such as the "Colossus", which was used for code breaking at Bletchley Park during World War Two) were programmed using switches and plugs to rewire the machine to solve a particular problem [115].

The format of the instructions stored in memory is documented as part of the ISA specification. Each instruction consists of an *opcode* (e. g., load, addition, or branch) and a set of *operands* (e. g., registers or memory addresses) encoded as data in memory. The control unit loads that data from memory and then pulls it apart as a set of bit fields. In a complex instruction set computer (CISC) machine, this may be a byte stream that has to be parsed carefully, as different instructions may have different lengths. In a reduced instruction set computer (RISC) machine, the decoding is normally simpler because instructions are packed into a few 32-bit or 64-bit words.

We can see the different instruction encoding styles if we look at a trivial code fragment, like this:

```
int zero() {
  return 0;
}
```

In the CISC x86_64 architecture that fragment compiles into three bytes of code, shown below. The xor instruction takes two bytes in memory and the ret instruction only one:

```
zero:
  31 c0     xor eax, eax
  c3        ret
```

In contrast, on the RISC Arm[*] V8 architecture, the instruction sequence consist of two 32-bit words of code, one for each instruction:

```
zero:
  e0 03 1f 2a    mov w0, wzr
  c0 03 5f d6    ret
```

Armed with the relevant architecture manuals, one can see the details of the encodings, but that is a level of detail and complexity that we don't need to go into here, and that you are unlikely to need anyway (unless you are writing an assembler or disassembler, or are creating instruction-decoding logic in a processor implementation).

The main task of the control unit is to interpret the code stream that is loaded from the memory subsystem and decipher it so that it can feed the different operational units (e. g., load, store, integer arithmetic, and floating-point arithmetic) with the right kind of work. Thus, every instruction goes through a series of operations:

1. **Instruction Fetch (IF):** Before it can be decoded, the data that represents an instruction must be brought to the decode unit from memory.
2. **Instruction Decode (ID):** Once the control unit has the data that represents an instruction, it must decode it to decide which functional units can execute it, and extract the specific operation that the functional unit must perform. (For instance, all integer arithmetic operations feed into an integer ALU, but which operation is performed will be determined by giving the ALU additional information.)
3. **Load Operands (LO):** Before a functional unit like the ALU can operate on the values, they must be fetched from wherever they are stored. In a RISC machine performing arithmetic operations, the operands are normally limited to being stored in registers; in a CISC machine, this step may require fetching data from memory (possibly from more than one memory operand). By the way, in von Neumann's initial design, there was only a single accumulator, which was always used as one operand and the target of the operation.

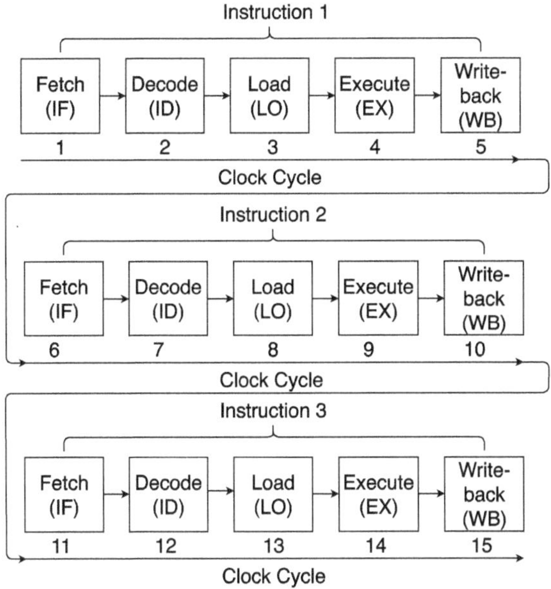

Figure 3.2: In-order execution of instructions in a simple core.

4. **Execute (EX):** Once all of the operands are available, the functional unit can perform the operation on the operands.
5. **Writeback (WB):** After the operation is complete, the result must be stored somewhere; this is done in the writeback phase, which puts the result wherever it is required. In a RISC architecture, that must normally be a register; in CISC machines, the result can often be written directly to memory. If the architecture has them, then condition flags may also be set by the operation (for instance, to show that the result is zero).

This "Fetch, Decode, Load, Execute, Writeback" sequence is similar to the classic description of the stages of instruction execution discussed in textbooks such as [56]. Processors like MIPS [55] and the educational DLX [114] processor demonstrate similar ideas.

This cycle repeats for the next instruction to be executed. This may simply be the next instruction in memory or, if a branch instruction has just executed, an instruction at a completely different address. So, the *IF* maintains a counter, usually called the *program counter* (PC), or *instruction pointer* (IP), that keeps track of the location from which the next instruction must be fetched.

The execution for a stream of instructions is shown in Figure 3.2. As you can see, the sequence of *IF-ID-LO-EX-WB* repeats for each of the executed instructions. If we assume that each of the steps takes precisely one cycle, we can see that it takes five cycles to execute a single instruction or a total of fifteen cycles for all three instructions.

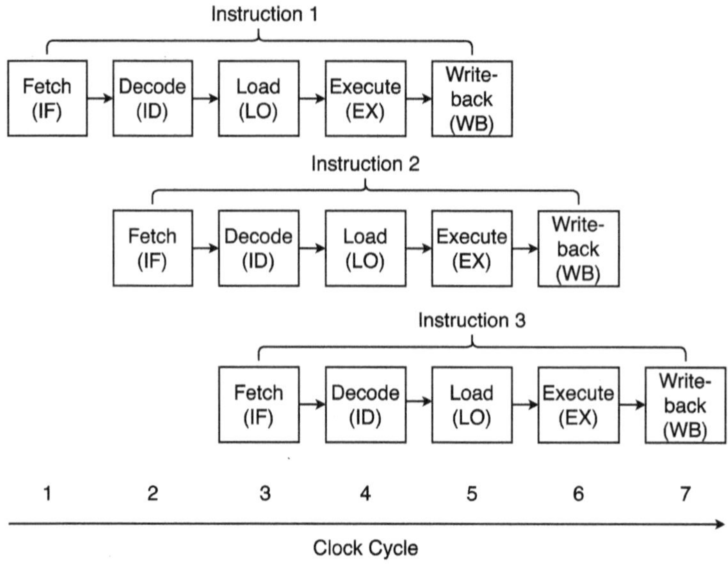

Figure 3.3: Pipelined execution processing of instructions.

3.1.2 In-order pipelined execution

One observation about this sequential execution engine is that we can overlap some of the execution steps. If the *IF* step has loaded an instruction from memory and passed it on to the *ID* stage to decipher it, then the *IF* logic in the core will idle until the processing of the instruction completes at the very end of the execution cycle. The same logic applies to the *ID* step: once an instruction has been decoded and pushed to the *LO* stage, the *ID* logic will be idle for the remainder of the instruction processing. This carries on for all of the processing steps.

This means that one can think about overlapping the steps in a pipelined fashion. The steps will then be stages in this pipeline, handing off work (the instruction) to the next stage and grabbing the next piece of work. Once the *IF* stage starts to idle and the *ID* decodes an instruction, the *IF* stage can already load the next instruction from memory. As Figure 3.3 shows, at this point, we can have multiple instructions in flight. Whenever a stage becomes free, the previous stage pushes its result into that stage.

If we still assume that each stage in our simple DLX-like pipeline example takes exactly one clock cycle to complete its work, we can expect a speed up of five in steady state—that is, with an indefinitely long instruction stream. There's a ramp-up phase for the first five instructions, when the later pipeline stages do not yet have work. And for a finite instruction stream, there's also a ramp-down phase, when the early pipeline stages run out of work when the stream ends. In Figure 3.3, you can see that the number of cycles needed to execute the three instructions is already a lot shorter despite the ramp up and ramp down of the execution pipeline.

However, there are two fundamental problems with this pipeline idea. Both lead to what people call a *pipeline hazard*, *pipeline stalls*, or *execution bubbles*. We will use the term pipeline stalls, or simply stalls, throughout this chapter.

First, the pipeline may stall if an operation takes longer than one clock cycle. It's rather unrealistic to assume that all operations take exactly one clock cycle in each of the stages. One example of this is a memory access that may have a latency that is longer than the time for a single clock cycle, which is rather typical for almost all modern processors. This can affect the *LO* or the *WB* stages. The *EX* stage may be affected if an operation takes longer than a single clock cycle (e. g., instructions for integer division or a floating-point square root). In this case, the *WB* stage has to idle until the instruction completes execution and is handed over to *WB* from *EX*.

Second, branch instructions changing the control flow cause real pipeline trouble. The *ID* stage decodes the branch instruction, but it is not until the *EX* stage that the branch instruction is executed and the program counter for the *IF* stage points to the next instruction to be loaded. This applies to both indirect branches that load the branch target from a register or memory and conditional branches that have to evaluate whether or not a branch has to be taken. Until this happens, the pipeline has to stall and stop executing. As a consequence, the *IF*, *ID*, and *LO* stages may have to discard their work. This is called a (partial) *pipeline flush*, and it occurs so that the instructions that are already in the pipeline, but which should not be executed as they are after a taken branch, are not executed and do not have an effect.

Many early RISC machines, such as the Hewlett-Packard[*] PA-RISC[*] [144] design, did not clear the pipeline when they executed a branch instruction. Instead, the regular execution of instructions simply continued despite the branch instruction remaining in flight. Compilers (or the programmer) either had to add no-operation instructions to fill these pipeline slots or issue useful work that had to be done regardless of the branch instruction. Similarly, the SUN[*] SPARC[*] architecture always had a "delay slot" after each branch and the instruction there would always be executed whether the branch was taken or not. More modern designs (such as RISC-V[*] [110]) do not have architectural delay slots. In effect, a design with delay slots moves the details of a particular implementation into the architectural specification. While this is helpful for the initial implementation, it means that later implementations (which may have deeper pipelines or out-of-order execution) have to emulate the delay-slot behavior, while gaining no benefit from it.

The Intel[*] i860 processor [67] went even further in exposing its internal pipelines, since not only did it have delay slots after branches, but it also provided "pipelined load" instructions for floating-point data, in which a `pfld` instruction would return the result of the third previous `pfld`. And finally, some versions of the Intel Pentium 4 processor design had super-deep pipelines of up to 31 stages that helped to make the individual pipeline stages a lot simpler, but also meant that flushing the pipeline was rather expensive and had to be avoided if at all possible.

3.1.3 Out-of-order execution

To tackle the first cause of pipeline stalls, we need to modify the pipeline stages so that we can pull in work that can be performed while an instruction is waiting for operands to arrive or the next pipeline stage to clear and become available. In other words, we would like to change the execution order of instructions, so that fewer (or ideally no) pipeline stalls occur. This concept is called *out-of-order execution*, or OOO.

One key observation is that a processor core cannot arbitrarily rearrange the instruction sequence as there are dependences between instructions that need to be maintained. These create an execution order that the processor core cannot ignore. Consider the following small assembly snippet in Intel assembly syntax, where the left-most operand is the target of the operation and the operand on the right is the source operand; square brackets indicate a memory access that can involve an address calculation:

```
1    lea  eax, [rsi + 4]
2    mov  dword ptr [rdi], eax
3    add  esi, edx
4    imul eax, esi, 42
5    mov  dword ptr [rdi + 4], eax
```

In this small code fragment, the store instruction on line 2 clearly cannot precede the lea instruction on line 1 (which added 4 to the value in rsi and stored the result in eax), since the first instruction writes eax and the second reads it.

Therefore, if the instructions were executed in reverse order, eax would not contain the correct value when it was used by the mov instruction. However, the add and imul instructions on lines 3 and 4 have no data dependence on the first two instructions, so the code would still work correctly if we swapped the first two and last three instructions.

To reorder the instructions, the *ID* feeds the decoded instructions into a new pipeline stage called *SC*, which is short for *scheduler* (see Figure 3.4). Between *ID* and *SC*, there's now a queue of instructions that the scheduler analyzes for dependences between the incoming instructions. It relies on a data structure called a *reorder buffer* (ROB) to keep track of instructions in flight, their original order, and their dependences with other instructions. A typical approach to implement this dynamic reordering of instructions is the Tomasulo algorithm [56].

After the execution stage *EX*, we introduce a new pipeline stage called *RT* (*Retire*). It takes care of the instructions that are marked as completed in the ROB and materializes their effects so that the out-of-order execution appears to have executed

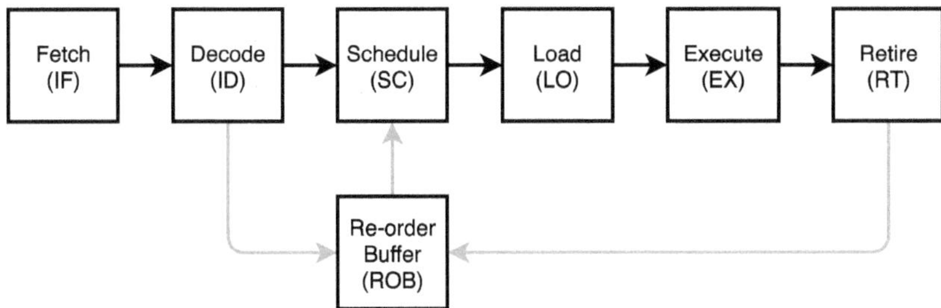

Figure 3.4: Out-of-order execution of instructions.

everything in the original program order. This ensures that the programmer is not surprised by any reordering effects that may change program behavior beyond what the ISA specifies. Section 3.2.2.2 will come back to this.

As soon as a core has OOO capabilities, it must determine a safe ordering of the instructions. In effect, the constraints here are similar to those that are required to avoid race conditions when a parallel program is accessing a shared variable or the dependences that a compiler has to track.

The three possible dependences that the OOO engine must ensure remain satisfied are:

– **Read After Write (RAW):** This is the most obvious data dependence since it represents a clear data flow. The first instruction places a value somewhere (e. g., a register) and the dependent instruction uses that value.
– **Write After Read (WAR):** This is the reverse of a RAW hazard. Here, the problem is that a write operation cannot update a location with a new value if there is still an instruction that needs to read the value in that location that has not yet completed.
– **Write After Write (WAW):** The problem here is that a later write operation to a location will overtake the result of an earlier write operation. If the two writes are reordered, then the final result will be wrong.

The location can be a register in the register file of the processor core, an overlapping partial register (e. g., full register eax and lower 16-bit part ax in eax for x86 processors), or (overlapping) memory addresses.

Table 3.1 shows the small code snippet we saw previously in Listing 3.1.3, but annotated with the dependences between instructions. We can see that there are some spurious dependences (such as the one on eax between instruction 4 and instructions 1, 2, and 3) that are the result of the compiler using the eax register to evaluate the two expressions. To overcome such dependences, the hardware usually performs register renaming, and converts the register name used by the machine code into an internal register name. This effectively allows it to rewrite the code so that it uses different registers for the first three and last two instructions. This also allows it to use a much larger

Table 3.1: Dependences between instructions.

Seq	Instruction	Reads	Writes	Dependences
1	`lea eax, [rsi + 4]`	rsi	eax	None
2	`mov dword ptr [rdi],eax`	eax,rdi	Memory	RAW[1]
3	`add esi,edx`	edx	esi	WAR[1]
4	`imul eax,esi,42`	esi	eax	RAW[3], WAR[2], WAW[1]
5	`mov dword ptr [rdi+4], eax`	eax,rdi	Memory	RAW[4]

register file than can be encoded in the available space for the register operand in the instruction encoding, though, of course, this comes at the cost of needing additional hardware to detect the dependences and perform the register renaming.

For those unfamiliar with the x86_64 architecture, instruction 3 has a WAR dependence on instruction 1 because the esi register is the low 32 bits of the rsi register. This points out the complexity of such analyses; the analysis requires knowledge about the semantics of the instruction set, and must also track states (such as condition codes) that are often not explicitly exposed as part of the instruction encoding, but are part of the semantics as specified by the ISA.

One additional complexity that is introduced with OOO execution is the handling of load and store operations. Clearly, it is a minimum requirement of any CPU implementation that it can execute existing single-threaded code correctly. However, as described so far, our OOO core may not achieve this. Consider code like this for the x86_64 architecture:

```
1  func:
2      mov   rax, qword ptr [rdx]
3      mov   qword ptr [rdi], rax
4      mov   rax, qword ptr [rsi]
5      mov   qword ptr [rdx], rax
6      ret
```

It is generated by the clang compiler from this code:

```
1  void func(int64_t *a, int64_t *b, int64_t *c) {
2      *a = *c;
3      *c = *b;
4  }
```

It certainly appears that the CPU could apply register renaming to avoid the WAR dependence on rax and then reorder the code to move the two apparently independent loads up in the executed instruction sequence while still respecting the need to load from qword ptr [rdx] (*c) before storing there. However, this is C or C++; nothing prevents b from pointing to the same memory location as a, (in which case rdi and `rsi` will hold the same value). Therefore, there is a hidden dependence, which can only be detected at runtime, between the store to $*a$ and the load from $*b$, which has to be resolved.

As usual, there are a number of solutions to overcome this problem:

1. **Do not reorder any loads or stores:** While this is a simple approach, it removes a lot of potential parallelism, because loads and stores are quite common operations and so not reordering them may have a drastic effect on execution performance.
2. **Detect loads and stores that overlap in memory and order only those:** This is more complicated, since now the core needs an additional structure to store some additional state (to hold the address/size information for loads/stores) and logic to check whether any given memory operation conflicts with another one.

Either of these solutions can fix the problem of ensuring that serial code executes correctly, but as soon as we allow any load or store to execute (and complete) out of order, we need more rules and control of how the CPU enforces ordering on memory operations. The processor then either has to handle parallel execution (where another thread can observe the memory state) or has to deal with device drivers that control a device by writing to memory-mapped device registers.

For instance, if there is a device that can perform direct memory access (DMA) transfers, we might have to write a pointer to the base of the DMA buffer, a size, and then a control word to initiate a transfer; each of these will be to a separate device register, but if the write that triggers the transfer completes before either of those that set the parameters, bad things will happen. Equally, in a threaded code, we must ensure that all store operations that take place inside a critical section are globally visible before the store operation that releases the lock, and, similarly, that no loads that read values written inside the critical section float above the load that observed that the lock had been released.

Different architectures have different rules about which memory access operations can overtake each other and which instructions force ordering by draining the memory-operation pipelines and preventing new instructions from entering them. Some also introduce different classes of load/store operations that have different memory-ordering properties. For instance, the normally rather strictly ordered x86_64 architecture has *non-temporal store* instructions with more relaxed semantics [68].

Architectures with a relaxed memory model also have to introduce specific instructions (*memory fences*) to enforce the ordering of loads and stores, and/or have additional load and store instructions that also have memory fence semantics. Section 3.2.2.2 discusses how we can handle these issues in more detail.

Describing and rationalizing memory-consistency models is a large field, justifying, for instance, the need for a 294-page book ("A Primer on Memory Consistency and Cache Coherence, Second Edition" [95]). We do not intend to delve too deeply into this subject here, but in order to be efficient, we do need to be aware that we must step carefully when writing important parts of a parallel runtime system that deal with these kind of low-level issues.

3.1.4 Branch prediction

While out-of-order execution is one way to deal with pipeline stalls that arise from long-latency instructions, branches are a different problem. The problem here is to "know" which instruction will be targeted by the branch before the execution of the branch instruction happens in the *EX* stage. There are several kinds of branch instruction that are relevant here.

First, unconditional branch instructions change the control flow so that it is at the target instruction. They seem easy to handle, but also require parts of the pipeline to be stopped. The branch is in the *ID* stage when the core realizes that it is dealing with a branch instruction. If the branch is a relative branch to the current program counter, then the execution unit first needs to compute the final address to inform the *IF* stage as to where it should fetch new instructions from.

Second, conditional branches and indirect branches have to go all the way through into execution before the branch target is known. When a branch is conditional, the execution unit first has to determine the branch condition to decide whether or not the branch has to be taken. This may involve either reading from a condition register or performing a computation. An indirect branch involves fetching the branch target either from memory or a register that stores the (relative) target address.

All modern processors thus contain a branch-target prediction (BTP) unit that tries to predict the branch target and provide the *IF* stage with an address from which to load the next instruction. As you can imagine, the accuracy here is key to performance. If the predictor mispredicts a branch, the core will have to flush the pipeline and restart execution at the correct target program counter after the branch instruction has been executed. This also requires that the processor introduces speculative execution; although it is fetching and executing from the predicted target of the branch instruction, it cannot commit any of the computations it is doing there until after it knows for sure that this code path will execute. This therefore requires the introduction of an additional new state to hold results that cannot yet be committed, as well as additional buffering so that store operations that are executing speculatively do not have any visible effects.

Figure 3.5 shows how the BTP unit integrates with our simple execution pipeline. The BTP stores the prediction database and informs the *IF* stage about the next value of the program counter. The retire stage at the end of the pipeline updates the BTP unit's database when a branch instruction completes execution and passes along information about the outcome of the branch (taken or not taken).

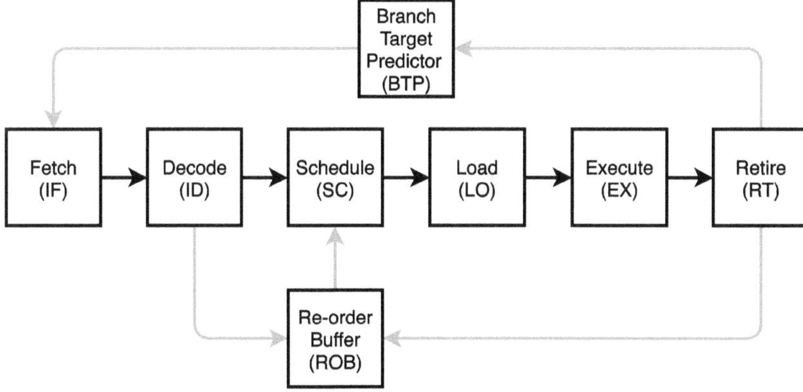

Figure 3.5: Branch prediction in an out-of-order execution engine.

There are many algorithms that can be used to perform branch target prediction. They range from very simple 1-bit predictors (e. g., if branch was taken, then predict taken again) to more sophisticated 2-bit predictors that use a saturating counter to tolerate one misprediction before changing future predictions. There are even branch-target predictors that include neural networks and machine-learning techniques to improve prediction quality.

3.1.5 Superscalar execution

We can add more parallelism to the execution pipeline. The introduction of the pipeline concept already provided more overlap between the execution stages. If we take a look at typical instruction sequences, we will find that the instruction sequence usually contains a mix of instructions: load/store instructions, branch instructions, arithmetic instructions, and so on. At the same time, we will find that an execution unit that can execute all these instructions could be split into smaller, more specialized execution units.

Such a *superscalar* execution pipeline with multiple *execution ports* is shown in Figure 3.6. With this design, we can adjust the number of execution ports so that their ratio corresponds to the expected instruction mix. In our example, we have broken up the single *EX* stage into two load ports, one store port, and an ALU for integer arithmetics, branching, and other stuff. We also added one unit for adding and one unit for multiplying floating-point numbers, respectively. Note that the ratio that we used here is not an arbitrary choice, but coincides with the instruction ratio of the dgemm() function [29] that is one important component of the HPL benchmark [31] used to rank the Top 500 systems [30]. Of course, the precise choice of what execution ports to implement, and how many of each to have, is a micro-architectural decision, so will very likely differ even between CPUs that have the same instruction set and are designed by the same vendor.

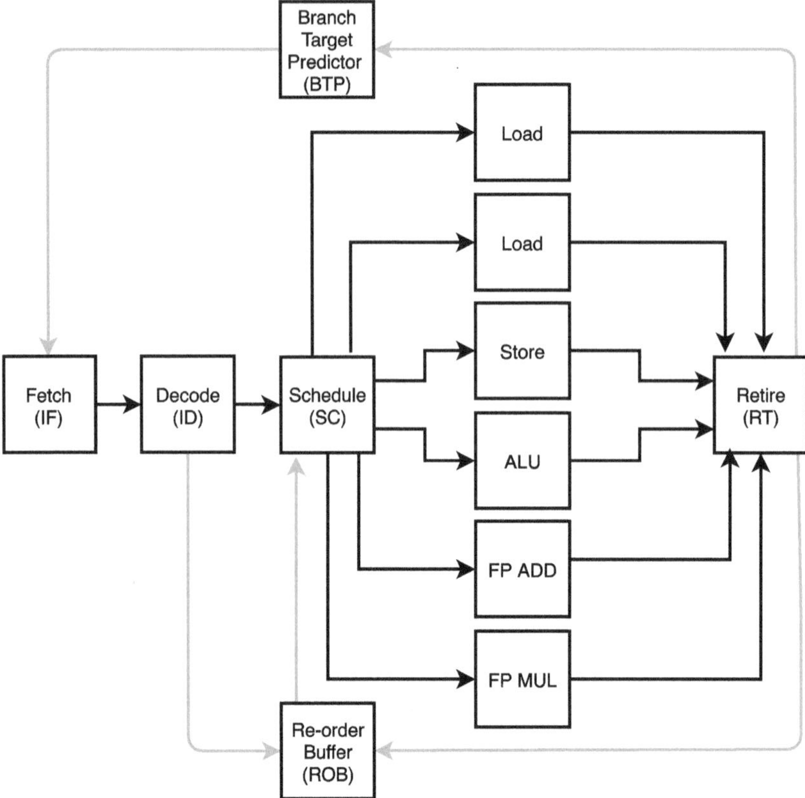

Figure 3.6: Superscalar, parallel pipeline with multiple execution ports.

When an instruction arrives at the instruction scheduler *SC*, the scheduling algorithm will check whether the instruction is ready for execution by determining whether all dependences to other instructions are fulfilled. If this is the case, then the scheduler finds a free execution port that can execute the instruction and assigns the instruction to that port. If all matching ports are busy, then the instruction has to wait until a port becomes available. If the program code can keep all execution units busy with work, then the core can have six instructions in flight in the execution unit in addition to the instructions that are being worked on in other pipeline stages.

3.1.6 Simultaneous multi-threading

In the previous sections, you have seen that latency is one of the key problems that we encounter when executing code in a modern processor. For instance, memory latency and the pipeline stalls that it causes are a major source of execution issues. Long-latency instructions are another source of pipeline stalls that need to be addressed to increase execution performance.

In general, there are two basic ways to deal with latency issues and the pipeline stalls they are causing:

1. **Latency Reduction:** The core can employ additional logic and structures that implement a smart solution to shorten the time for the operation that causes the pipeline stall.
2. **Overlapping Operations:** The core introduces additional execution capabilities so that it can execute other work while it is waiting for long-latency operations to complete.

For the example of memory accesses, one typical solution is to introduce caches (see Section 3.2 for an in-depth discussion) that have a much lower access latency than regular memory. However, if cache misses occur, then the core still has to wait until the requested memory operation has completed and the item has arrived in the internal buffers. Another solution is to add prefetching logic to the core so that the core can try to predict future memory accesses and bring the data into the cache, effectively reducing the latency for these future accesses.

Out-of-order and superscalar execution are one way to tolerate latency through overlapping operations. If, for instance, a memory access happens, the core can reorder the instruction stream and keep on executing other operations. This can also be used to compensate for long-latency arithmetic operations, such as floating-point divisions, by feeding the other execution units with other instructions. Some cores also have (sub-)pipelines within the individual execution units, so that they can start a new instruction while another one is in a later execution stage in the same unit.

However, in many cases, the processor cannot extract enough parallelism, and thus overlap, from the instruction sequence to compensate for large latencies. It is also difficult to compensate for the stalls that happen because of high load/store pressure, as the scheduler may not be able to find enough other instructions that it can schedule in the meantime. So another concept will be needed to feed the core with more (independent) work.

This is when simultaneous multi-threading (SMT) [140] comes into play. The key idea here is that if one instruction stream does not provide enough opportunities to reorder instructions to tolerate (memory) latency, why not feed another instruction stream from a different process or software thread into the pipeline? Now, when a pipeline stall happens and the scheduler cannot feed another instruction into execution from the executing thread, the *IF* stage switches to another thread and starts to execute instructions from there. Most implementations of SMT either offer two-way or four-way SMT.

This idea goes back to at least the I/O processor designed by Seymour Cray for the CDC 6600 processor in 1964 [25]. Then, the implementation was simplified to avoid

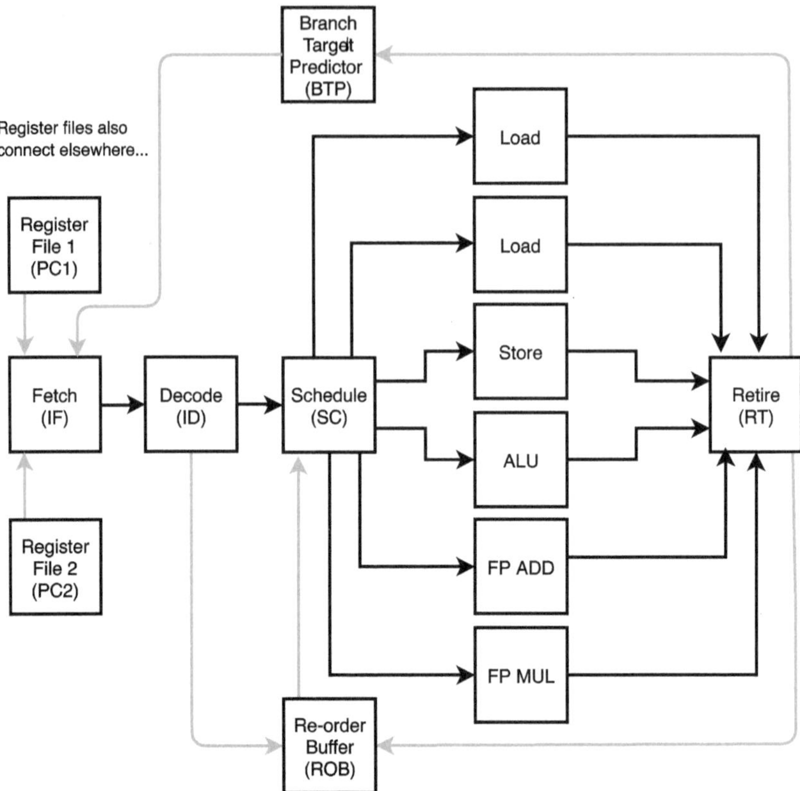

Figure 3.7: Two-way SMT pipeline (extended register file not shown).

much of the complex hardware we have been describing to handle reordering and data dependences by having a ten-deep pipeline with ten different sets of user registers. Each notional I/O processor could only ever have one instruction in the pipeline, and therefore there were never data-dependence issues. This extreme SMT style is often known as a *barrel processor*, and is an idea that has re-surfaced in some of the modern graphics processing units (GPUs).

As ever, the processor core needs to be extended with additional logic and data structures. It now needs to keep track of the execution state of two (or more) instruction streams at the same time. Most modern processors already have a much larger internal register file that they use to implement advanced execution concepts like register renaming and speculation (Section 3.1.3); the processor can use these additional resources to provide (virtual) registers to several instruction streams. While instructions are in the pipeline, the processor must tag them, so that all execution units and pipeline stages know which instruction stream a particular instruction belongs to, since memory accesses must use the appropriate set of page tables. In addition, each SMT thread clearly needs its own program counter (see Figure 3.7).

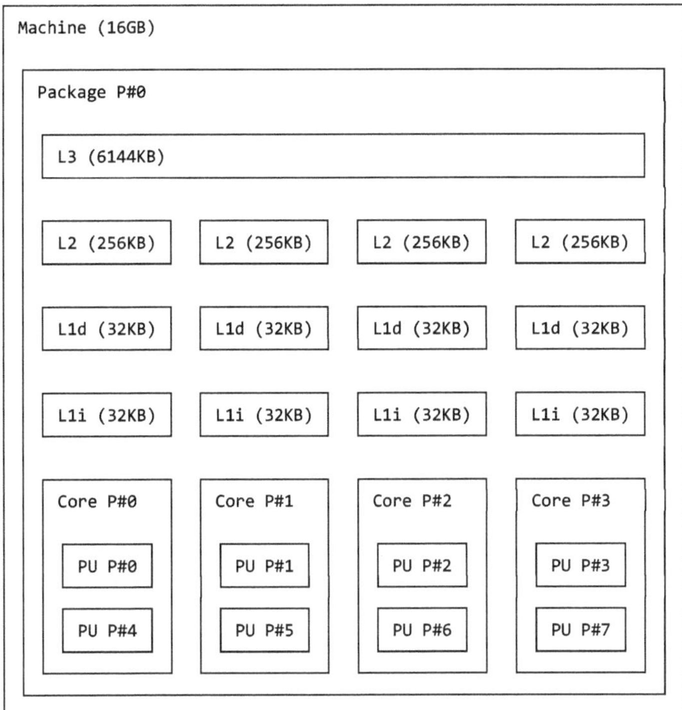

Figure 3.8: Output of `hwloc-ls` for an Intel Core i7-6670HQ processor.

A typical nomenclature is to call the processor core a *physical core* and call each instruction stream that the physical core can execute a *logical core*, *logical CPU*, or *hardware thread*.

Figure 3.8 shows the graphical output of the `hwloc-ls` command [99]. The topology consists of four physical cores (P#0 through P#3), each with two logical cores that are denoted as PU P#0 through PU P#8. Linux[*] and Microsoft[*] Windows[*] will then expose these as cores 0 through 7.

In an operating system, these logical cores then show up as regular cores that the OS can use to schedule processes and threads. Of course, their presence adds to the complexity of the job that the OS and parallel runtime system have to perform. Multiple hardware threads that are multiplexed in a single physical core also share caches at all levels of the hierarchy. Therefore, the performance of code running on one hardware thread can have a significant effect on code running on another on the same core. Thus, the OS needs to take the SMT structure of the system into account when scheduling runnable processes and threads.

3.1.7 Single-instruction multiple-data

The final task that will add even more parallelism to the processor core is to add support for *single instruction multiple data* (SIMD) instructions to the execution units, introducing *data parallelism* according to Flynn's taxonomy [39].

Instead of making the pipeline wider or deeper, the instruction set architecture is changed to offer instructions that process multiple data elements in one go. Common choices for the SIMD width of modern processors are 128-, 256-, and 512-bit SIMD registers. The Scalable Vector Extension (SVE) [121] for Arm processors goes even further and supports up to 2,048-bit SIMD.

Besides providing a register file that can hold the wide SIMD registers, this requires us to replicate the arithmetic logic units until the desired SIMD length is achieved. For, say, a 512-bit wide SIMD execution unit, the logic to add double-precision floating-point numbers will be present eight times, once for each double-precision floating-point number that can be stored in the SIMD register. This number will vary if the instruction set supports different data types for SIMD execution, e. g., `int8_t`, `int16_t`, `float`, and so on.

Figure 3.9 shows a few examples of SIMD instructions that the Intel Advanced Vector Extensions 512 (AVX-512) instruction set offers. We show 512-bit instructions for 64-bit values in the figure, but the instruction set also supports 128-bit and 256-bit instructions and word sizes from 8 to 64 bits. Other architectures typically have similar categories of SIMD instructions.

The first instruction at the top is a simple load instruction that retrieves 512 bits of data from memory and loads them into an SIMD register. The example is for 64-bit values (e. g., double-precision floating point), so eight values will fit into the SIMD register. The second instruction in the figure shows the class of element-wise operations that take a number of operands (here: two source operands) and produce an output operand by performing an operation for each element of the SIMD operands.

In the middle, there is an element-wise instruction, which adds SIMD registers a and b and stores the results in the destination register `dest`. Typically, instruction sets offer a wide range of these instructions for all sorts of arithmetic operations, ranging from simple (e. g., addition) to more complex ones (e. g., fused-multiply addition or reciprocal square root).

Some instruction sets also offer to perform mask computations in the SIMD register. Conditional instructions may fill a so-called mask register as the outcome of a (element-wise) comparison on a SIMD register. This mask register (`k1` in Figure 3.9) can then be used to restrict the effect of an SIMD instruction to only those elements in the SIMD registers that have a `true` value in the corresponding bit of the mask register. Depending on the instruction set, the elements with a `false` bit in the mask may be set to 0, left unchanged, or may even be taken from another source.

Finally, there's often a large set of instructions that deal with rearranging the contents of an SIMD register. Figure 3.9 shows an example of the `vpermilpd` instruction

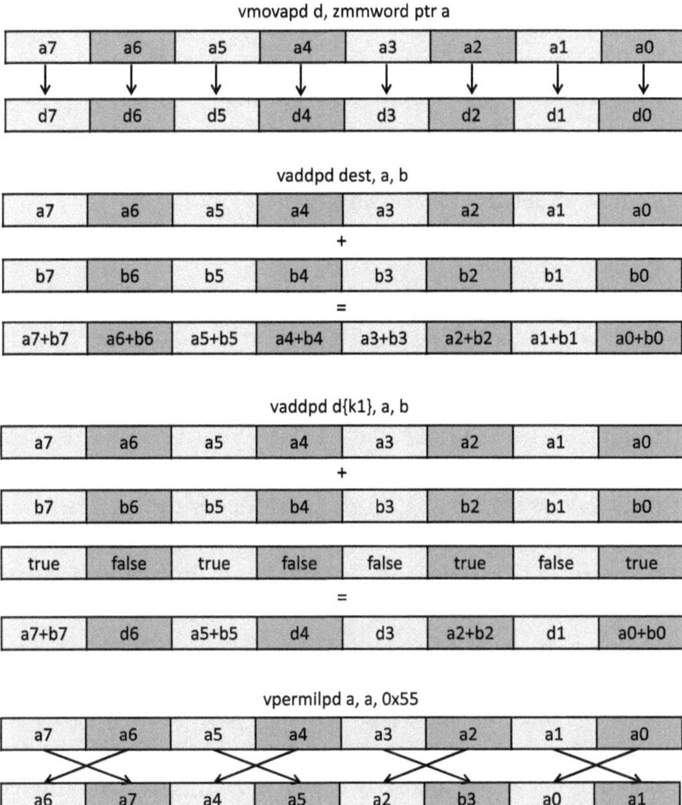

Figure 3.9: Examples of SIMD instructions of the Intel AVX-512 instruction set.

(for "intra-lane vector permute packed double") to flip two neighboring elements. The immediate value 0x55 that is encoded in the instruction describes the action taken on the elements and how they should be moved around. These kinds of instructions are important to format the contents of the SIMD register so that the data-parallel execution on the elements can take place. One example would be to separate the real and imaginary parts of a complex number for computation.

One of the benefits of SIMD instructions is that they explicitly tell the CPU that a set of operations are independent of each other, so the CPU does not need to perform dependence analysis to determine that it can execute, for instance, eight floating-point additions simultaneously. This information also allows an implementation in which a wide SIMD instruction is split into multiple narrower operations. Thus, a four-wide SIMD operation could be executed as four single-width operations one after the other, one for each SIMD lane. Such an implementation will not have the peak performance that could be achieved by providing the full four-wide operation, but may still reduce energy and improve performance over decoding and scheduling four separate instructions. Implementations like this are known as "double pumped" implementations (when splitting the SIMD operation into two). This approach also allows a vendor to maintain binary compatibility between their CPUs, even when a given CPU implementation is hardware-constrained to meet a cost point and cannot afford to have four or eight floating-point units. By n-way pumping the ALU, it can still implement the wide SIMD instructions, even if that provides no large performance gain.

3.2 The modern memory subsystem

We are now leaving the guts of the processor core and will focus on how the memory subsystem works, so that we can bring in data from the main memory to the core. When the *LO* stage in the pipeline has to retrieve data or the *WB* stage has to write data back, a request leaves the processor core and is handled by a memory controller that attaches whatever physical memory technology (e. g., synchronous dynamic random-access memory with double data rate, or DDR-SDRAM) to the processor package. The memory controller can be viewed as a translator that interprets the request for data and issues the electrical signals on the wires that connect the memory modules to the processor package.

However, memory systems are slower than processors. This is ultimately inevitable because the speed of light is finite, and matter is not infinitely divisible. Of course, we are still a long way from reaching the ultimate physical limits of what the hardware can do. But in recent years, there has been a growing gap of how much faster processing data has become in comparison to the speed of memory delivering data. While in the late 1970s, SRAM ran at the same speed as a micro-processor, and could satisfy a request in one cycle, now, accessing memory may take hundreds of CPU cycles. This DRAM latency is on the order of 100 ns, which can be measured with tools like the Intel Memory Latency Checker (MLC) [64]. Even though and because our OOO cores can now execute more than one instruction per cycle, the impact of such a 100-cycle pause to fetch data from memory on performance is huge. Therefore, this has led to the introduction of a hierarchy of different (faster) memory components that we will describe next.

3.2.1 Memory hierarchy

Figure 3.8 has already provided a glimpse into the memory hierarchy of a typical multi-core processor. The data paths to and from the individual physical cores (P#0 through P#3) are not directly connected to the 16 GiB of main memory that is shown at the very top. Instead, the processor has multiple caches between the memory and the cores. Let's first take a look at this hierarchy of caches. We will then focus on how to build even larger multi-core machines that consist of multiple processor packages.

3.2.1.1 Cache hierarchy

Since fast memory is very expensive to build and integrate with the processor package, the trend over the past decades has been to move the memory away from the processor, making it easier to upgrade by not soldering it to the computer's mainboard. Also, the average access latency of typical and affordable memory technologies has not decreased a lot, and in some cases it has even increased.

A high-level summary of the topic of memory could be that with today's memory technology, one can either build small, yet expensive, fast memory or less expensive, but larger, slow memory. This led to the classic and widespread solution to be to introduce smaller amounts of fast memory between the slow main memory of the computer and the processor. These units of memory are called *caches*; they transparently provide intermediate storage for data that the processor requests. The different levels of the hierarchy are usually denoted as *Lx*, with *x* denoting the number of the level (typically 1, 2, and 3).

When we look back at Figure 3.8, we see three levels of caches for the processor. The first cache level *L1* is the smallest and fastest, and is actually built into the processor cores for fastest access speed. Usually, processor cores have one cache for data (*L1d*) and one for instructions (*L1i*). As instructions in the von Neumann design also come from the main memory, they are subject to the same latency problem; using a cache greatly reduces the potential of having to wait for the next instruction that comes from memory.

In our example processor, the next level of the cache hierarchy is *L2*, which is again a core-private cache that is associated with a particular physical core. It may be tightly integrated with the core or it may be a separate hardware component on the die. At this level, the cache is larger, but has a longer access latency.

Lastly, there's an *L3* cache in the system that is shared across all the physical cores of the processor package. This is the largest cache in the system, but also has the highest access latency. Yet, the latency for accessing this cache is still an order of magnitude better than for accesses to the main memory.

Caches are typically organized in *cache lines* that contain several data words. Typical sizes are 32 bytes or 64 bytes. While this does reduce the complexity of the memory subsystem and the logic needed to implement the cache, it means that if a core requests a single byte, the memory subsystem will always move around a full cache line and move it through the caches. As you may recall, some of the caches in the hierarchy are private to a core, which has implications if two cores want to modify neighboring data words that happen to be in the same cache line. Section 3.2.4 will revisit this topic and discuss it in more depth.

Caches can be organized either as *inclusive* or *exclusive* caches. If a cache hierarchy is inclusive, then a cache line that is loaded from the main memory is replicated throughout the hierarchy. So, for instance, if the cache line has been loaded to the L1 cache, a copy of the cache line will remain in the L2 and L3 caches. An exclusive cache hierarchy may remove these copies from the outer-level caches and only maintain a single copy. Of course, there can be hybrid approaches. For instance, the Intel Xeon[*] Scalable Processors [78] use inclusive caches for L1 and L2, while L3 is an exclusive cache.

A natural consequence of the individual caches being of a smaller size than the main memory capacity is that they cannot hold all the data that the memory may contain. At some point, a cache in the hierarchy might be full, and, to buffer additional memory, it will have to evict cache lines that it deems are no longer being used. There are several different strategies to deal with this. See [56, 97] for more information.

There are a variety of possible reasons why an access may cause a cache miss— that is, data not being present in the cache. The first scenario is a *compulsory miss*, which always happens when a cache line is accessed for the first time. If we ignore prefetching, it's clear that a cache cannot store a cache line that has never been requested before. A *capacity miss* happens when the cache is full and a cache line needs to be evicted to make room for incoming cache lines. Finally, a *conflict miss* happens when a new cache line occupies the same location in the cache and thus the currently present cache line must be expelled.

Organizing the cache in cache lines has another interesting consequence. As we saw above, when a core loads a single byte from memory, the memory system will return a full cache line to the core. This means that the other content in the cache line is already close to the core. If the next load operation is to a nearby word in the same cache line (e. g., when iterating through an array in sequence), the access latency for that word drops dramatically. This is called *spatial locality*, and should be exploited as much as possible. Its counterpart, *temporal locality*, describes the fact that an access to one memory location is likely to happen again in the near future, and unless an eviction takes place, the data will still reside in the cache.

3.2.1.2 Non-uniform memory access architectures

The processor architecture that we have seen in Figure 3.8 consisted of a single processor package that contained a number of cores. We can extend this concept to a multi-package system that contains several processor packages that are connected to each other via a network, or fabric, so that each package can exchange data with all other packages in the system. Usually, this also means that all processors can access the entire memory within the system, so that it appears to software that it runs on a machine with a lot of cores and a lot of memory.

The design that would connect a single memory sub-system to the system's internal network has proven not to scale well for larger numbers of processor packages. When more and more processor packages are added to the system, the single path to the memory becomes a bottleneck and the central switches required to route the traffic from the packages to the memory become prohibitively complex [56]. Thus, most modern machines follow a different design idea that is exemplified in Figure 3.10.

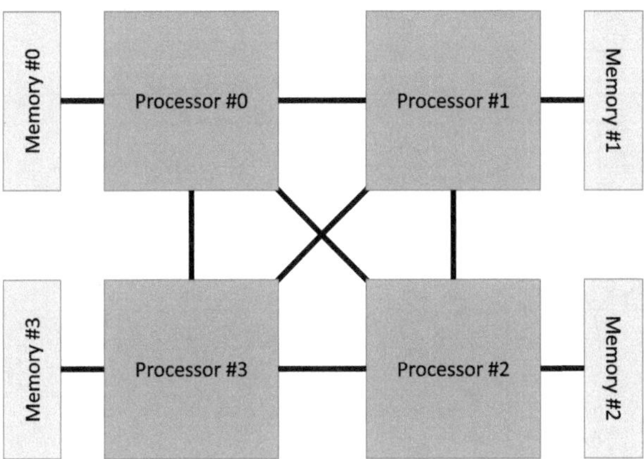

Figure 3.10: Four fully connected processor packages and their local memory.

As mentioned, all processors are connected to an on-board fabric so that they can exchange messages. Instead of a single memory block, each processor attaches locally to a part of the memory through the memory controller that is located in the same processor package. The clear benefit of this design is that the path to the memory now scales with the number of processor packages in the system. If we double their number, we also automatically double the number of memory components, including the data paths to the memory.

The downside of this design is that access to the memory is no longer uniform; that is, different memory locations now have different access latencies. If, for instance, Processor 2 wants to access data (which corresponds to the amount of data in a cache line) in Memory 0, then Processor 2 has to request that memory from Processor 0, which will forward the request to Memory 0. The requested data then travels back to the requesting processor that will bring it into its local caches. This kind of architecture is called *non-uniform memory access* (NUMA). The memory now exposes a non-uniform latency for accesses depending on the processor from which the request is made and on the *NUMA domain* from which it is being served.

Figure 3.11 contains the output of the hwloc-ls command for a dual-socket system that is equipped with Intel Xeon Platinum 8260L Processors. The individual physical cores P#0 through P#23 (per-package numbering) for each package provide the logical cores that are numbered PU#0 through PU#95 for the entire system. Each physical core has two hardware threads. The numbering scheme for this system is PU#0 through PU#47 for the first hardware threads, then the numbering wraps around and all second hardware threads for each core are then numbered PU#48 through PU#95.

With respect to the NUMA structure of the system, the two processors constitute one NUMA domain each. However, the system in Figure 3.11 has been configured with sub-NUMA clustering (SNC) [116]. Thus, each physical NUMA domain is further split in half by changing the mappings in the processor's internal communication mesh.

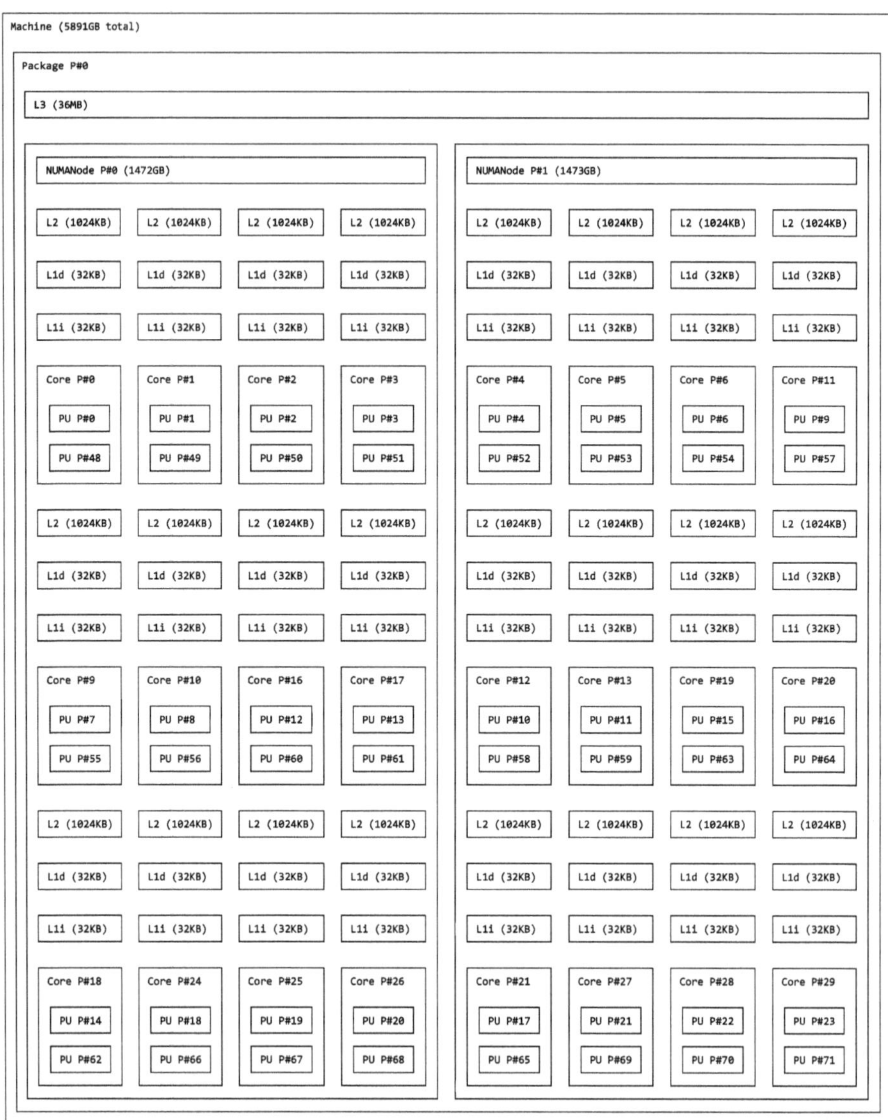

Figure 3.11: Dual-socket Intel Xeon Platinum 8260L Processor system.

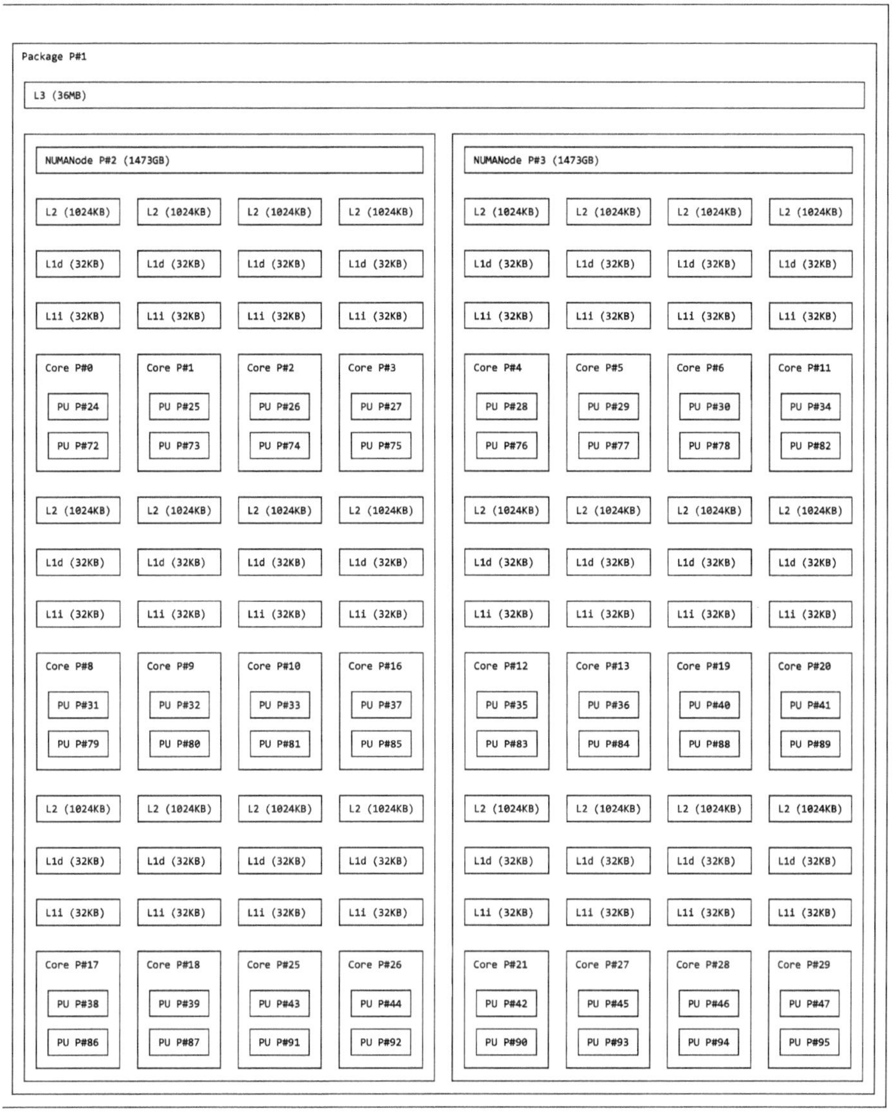

```
Package P#1

L3 (36MB)

NUMANode P#2 (1473GB)                          NUMANode P#3 (1473GB)

L2 (1024KB) L2 (1024KB) L2 (1024KB) L2 (1024KB)   L2 (1024KB) L2 (1024KB) L2 (1024KB) L2 (1024KB)

L1d (32KB)  L1d (32KB)  L1d (32KB)  L1d (32KB)     L1d (32KB)  L1d (32KB)  L1d (32KB)  L1d (32KB)

L1i (32KB)  L1i (32KB)  L1i (32KB)  L1i (32KB)     L1i (32KB)  L1i (32KB)  L1i (32KB)  L1i (32KB)

Core P#0    Core P#1    Core P#2    Core P#3       Core P#4    Core P#5    Core P#6    Core P#11

PU P#24     PU P#25     PU P#26     PU P#27         PU P#28     PU P#29     PU P#30     PU P#34

PU P#72     PU P#73     PU P#74     PU P#75         PU P#76     PU P#77     PU P#78     PU P#82


L2 (1024KB) L2 (1024KB) L2 (1024KB) L2 (1024KB)   L2 (1024KB) L2 (1024KB) L2 (1024KB) L2 (1024KB)

L1d (32KB)  L1d (32KB)  L1d (32KB)  L1d (32KB)     L1d (32KB)  L1d (32KB)  L1d (32KB)  L1d (32KB)

L1i (32KB)  L1i (32KB)  L1i (32KB)  L1i (32KB)     L1i (32KB)  L1i (32KB)  L1i (32KB)  L1i (32KB)

Core P#8    Core P#9    Core P#10   Core P#16      Core P#12   Core P#13   Core P#19   Core P#20

PU P#31     PU P#32     PU P#33     PU P#37         PU P#35     PU P#36     PU P#40     PU P#41

PU P#79     PU P#80     PU P#81     PU P#85         PU P#83     PU P#84     PU P#88     PU P#89


L2 (1024KB) L2 (1024KB) L2 (1024KB) L2 (1024KB)   L2 (1024KB) L2 (1024KB) L2 (1024KB) L2 (1024KB)

L1d (32KB)  L1d (32KB)  L1d (32KB)  L1d (32KB)     L1d (32KB)  L1d (32KB)  L1d (32KB)  L1d (32KB)

L1i (32KB)  L1i (32KB)  L1i (32KB)  L1i (32KB)     L1i (32KB)  L1i (32KB)  L1i (32KB)  L1i (32KB)

Core P#17   Core P#18   Core P#25   Core P#26      Core P#21   Core P#27   Core P#28   Core P#29

PU P#38     PU P#39     PU P#43     PU P#44         PU P#42     PU P#45     PU P#46     PU P#47

PU P#86     PU P#87     PU P#91     PU P#92         PU P#90     PU P#93     PU P#94     PU P#95
```

Figure 3.11 shows this as additional NUMA domains within each processor package. Other processor architectures can also expose multiple NUMA domains, if the processor package is composed of multiple dice that each have local memory controllers. Examples of such a design are the Intel Xeon Platinum 9282 Processor or the AMD* EPYC* 7742 64-core Processor, where the hwloc lstopo tool [99] shows four NUMA domains per socket.

3.2.2 Memory models and memory consistency

When we have multiple cores with multiple (private) caches that may be spread across multiple processor packages, the threads they are executing can access shared memory either because they are inside a single process where the whole address space is shared or, if they are in multiple processes, they may have explicitly shared part of the address space. We therefore have to consider how the threads will interact through memory and how the caches interact with the memory traffic.

There are two different aspects to this:
1. **Atomicity**
2. **Visibility/Ordering**

3.2.2.1 Atomicity

The issues here are concerned with ensuring that what appears to one thread to be a single operation is also seen as a single operation by all other threads. For instance, suppose that two threads write different values to a single 32-bit storage location "at the same time." Is there any guarantee that if one thread writes `0xffffffff` and the other writes `0x00000000`, then the result in memory is either `0xffffffff` or `0x00000000`, but not `0xff00ff00ff`, or any other value that is mixed from the two values that were written?

While it initially seems obvious that a single write will happen atomically, this is not guaranteed by some architectures under some circumstances; for instance, the x86_64 architecture allows unaligned memory-access operations, and makes no guarantees about such atomicity if an operation crosses a cache-line boundary (the formal guarantees are even less stringent, but cache-line crossing is something that may occur in reality).

Similarly, there are update operations that are used to implement locks and other synchronization operations. In some architectures, there are explicit atomic instructions for this (e. g., `lock` prefixed instructions for the x86_64 architecture); in others, the atomicity of these operations is achieved using speculation and the detection of failure (e. g., using load-locked/store-conditional instructions, such as Arm V8 `ldxr`/`stxr`) or using a single atomic compare-exchange instruction.

The important aspect of both of these types of operation is that they require that a thread has exclusive access to a single cache line for a period of time; however, they do not have to enforce any specific ordering of the operation with respect to other loads or stores generated by the same, or other threads, so with an OOO CPU with a relaxed memory model this atomicity does not require that these operations be ordered with respect to other local memory reads or writes, so they do not also have to be memory fences.

3.2.2.2 Visibility/ordering

We have seen previously (Section 3.1.3) that modern high-performance cores execute out of order, and can therefore potentially reorder memory accesses. While we also saw that the OOO engine goes to some lengths to ensure that a single thread cannot detect such reordering, the order in which store operations become globally visible may not be what one would expect from looking at the assembly code.

For instance, code like that shown in Listing 3.1 will not work as written because it makes incorrect assumptions about the language and hardware memory models that allow memory operations to become visible to other cores in an order that is not what a naive programmer expects. This code is trying to implement a point-to-point communication channel, but it relies on the store of `payload` in the `send()` function being completed and globally visible before the store of `go` is visible. However, with an out-of-order core, there is no guarantee that that's true. As mentioned earlier, C++ semantics even allow the compiler to treat each polling loop as a test before an infinite loop, since, in a single-threaded code, once the value has been read, the compiler can validly assume it cannot change. We can see this happening in the assembly code in Listing 3.2.

In both the `send()` and `recv()` function implementations the compiled code checks the state of the go flag and then branches to an infinite loop (at labels `.LBB0_1` and `.LBB1_1`).

To make this code work correctly, we have to tell the compiler that:

1. The go field is being used for communication, so it must be updated atomically.
2. When the assignment `go = true` is executed, we also require that the assignment `payload = data` is visible to the other thread.

These two requirements are exactly the atomicity and visibility constraints discussed above. Once the compiler understands our requirements, it can ensure that the appropriate instructions to enforce memory ordering are generated so that the CPU cores do what is required. We can tell the compiler this by including the `<atomic>` header file and by ensuring that the go flag has an atomic type. Let's take a short look at the C++ memory model to understand how this works under the hood.

```cpp
// This example is broken!
class channel
{
  bool go;
  void * payload;
public:
  channel() : go(false) {}
  void send (void * data) {
    while (go)
      ;                 /* Wait for data to be consumed */
    payload = data;
    go = true;
  }
  void * recv() {
    while (!go)
      ;                 /* Wait for data to be written */
    void * result = payload;
    go = false;
    return result;
  }
};
```

Listing 3.1: Broken point-to-point channel.

```asm
# Compiled by clang 9.0.0 -O3

send(channel*, void*): # @send(channel*, void*)
  cmp byte ptr [rdi], 0
  je .LBB0_2
.LBB0_1: # =>This Inner Loop Header: Depth=1
  jmp .LBB0_1
.LBB0_2:
  mov qword ptr [rdi + 8], rsi
  mov byte ptr [rdi], 1
  ret
recv(channel*): # @recv(channel*)
  cmp byte ptr [rdi], 0
  je .LBB1_1
  mov rax, qword ptr [rdi + 8]
  mov byte ptr [rdi], 0
  ret
.LBB1_1: # =>This Inner Loop Header: Depth=1
  jmp .LBB1_1
```

Listing 3.2: Broken point to point channel, x86_64 assembly code.

```
#include <atomic>

class channel
{
  std::atomic<bool> go;
  void * payload;
public:
  channel() : go(false) {}
  void send(void * data) {
    while (go.load(std::memory_order_acquire))
      ;                /* Wait for data to be consumed */
    payload = data;
    go.store(true, std::memory_order_release);
  }
  void * recv() {
    while (!go.load(std::memory_order_acquire))
      ;                 /* Wait for data to be written */
    void * result = payload;
    go.store(false, std::memory_order_release);
    return result;
  }
};
```

Listing 3.3: Correct point-to-point channel.

3.2.2.3 C/C++ memory model

Since the C and C++ programming languages allow programming close to the hardware and they are used for tasks such as implementing OS kernels, they have a memory model that reflects the properties of out-of-order hardware while also allowing the programmer to enforce ordering where required.

By giving a variable (or class member) a type that is an instance of the std::atomic template, we immediately guarantee that accesses to the variable will happen atomically. As the default for accesses to an atomic is std::memory_order_seq_cst, by default all accesses will also enforce strict ordering constraints, i. e., sequential consistency that guarantees that all preceding loads and stores have completed and no subsequent ones have started.

In cases where this is more restrictive (and expensive) than what the code requires, it is possible to use the load() and store() methods, which allow you to explicitly specify the desired memory ordering.

In our example code above, simply declaring the variable go with type std::atomic<bool> is sufficient to give us correct code. However, with x86_64 pro-

```
# Compiled by clang 9.0.0 -O3
_ZN7channel4sendEPv: # channel::send(void*)
.LBB0_1:
        movzx     eax, byte ptr [rdi]
        test      al, 1
        jne       .LBB0_1
        mov       qword ptr [rdi + 8], rsi
        mov       byte ptr [rdi], 1
        ret


_ZN7channel4recvEv: # channel::recv():
.LBB1_1:
        movzx     eax, byte ptr [rdi]
        test      al, 1
        je        .LBB1_1
        mov       rax, qword ptr [rdi + 8]
        mov       byte ptr [rdi], 0
        ret
```

Listing 3.4: Fixed point-to-point channel x86_64 assembly code.

cessors, the compiler then introduces a full memory fence at each of the stores to go. The g++ compiler adds an `mfence` instruction and `clang++` uses an `xchg` instruction for the operation that is always locked and, like all locked instructions in the x86 architecture, is also a full memory fence.

If, however, we express the semantics that we really require, which is that each load of go is an acquire operation and each store is a release, then code is correctly generated without the additional memory-fencing operation for the x86_64 architecture.

We can see that the fixes have worked if we look at the assembly code generated for that code in Listing 3.4. We can now see that the polling loops are testing the loaded value and leaving the loop when appropriate, and that the loads and stores are correctly ordered.

Full details of the C and C++ memory models, including the formal logic, can be found online, for instance, in the `std::memory_order` section at the *CPP Reference* website [1].

The *acquire* and *release* descriptions of memory ordering come from thinking about what memory ordering is required to enable critical sections implemented with locks to operate correctly.

3.2.2.4 Acquire and release

The *acquire* memory ordering applies to a load, which checks whether a thread has acquired a lock. Since the lock acquisition does not expose any state to other threads, this is a restriction to ensure that operations that should happen inside the critical section do actually happen there. In other words, no loads or stores that are lexically after the load with acquire semantics can happen before it. It does not enforce any restrictions on operations that happen lexically before the acquire operation.

The *release* ordering applies to a store that signals that a thread is leaving a critical section. Here, we must ensure that all of the store operations that occurred inside the critical section are visible to some other thread that claims the lock. It therefore prevents the release operation from completing and becoming visible until all preceding stores (from inside the critical section) have completed. In effect, it prevents stores from dropping below the releasing store, while the acquire operation prevents loads or stores from floating above the acquiring load.

In our channel example, we can see that the ordering semantics we require are *release* on the sending side and *acquire* on the receiving side.

3.2.3 Caches

In a memory system with no caches, there is only one place from which the value of a given memory location can be read: the memory. However, when we introduce a cache, the whole intent is to allow some memory locations (that have recently been accessed or that a prefetcher expects to be accessed soon) to be accessible closer to the core, so they are now in both the cache and memory. This immediately introduces a problem: how do we ensure that all cores see a consistent view of that replicated data?

To ensure that all cores see the same value when required by the architecture's memory model, a cache coherence protocol is required that guarantees that operations are not lost and that the hardware supports the memory model of the architecture. There are various different coherence protocols that can be used, and that have slightly different performance implications. Here, we focus mostly on the simplest MESI protocol [56]. MESI is short for the four states that a cache line can be in: Modified, Exclusive, Shared, and Invalid. The fundamental issues are not affected by this choice, and you can find details of more complicated models (such as MOESI [56] for the states that are Modified, Owned, Exclusive, Shared, and Invalid).

3.2.4 Cache coherence: overview

The aim of any cache coherence protocol is to ensure that the presence of the caches does not cause an architecture's memory model to be violated.

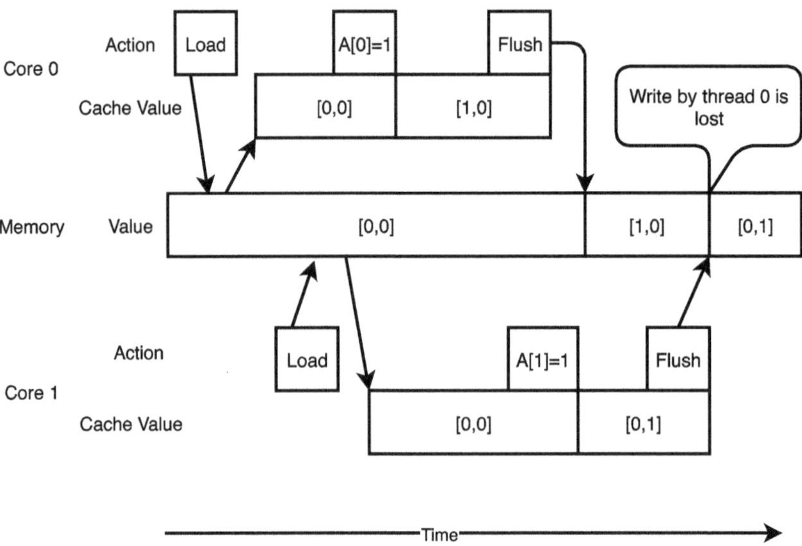

Figure 3.12: Example of a lost write.

To see how that could happen, consider what might occur in a system where memory transfers are always in the granularity of cache lines, and exclusivity is not properly tracked. Under these circumstances, two caches could each hold a copy of the same cache line. If the cores close to those caches then each write to disjoint bytes within the line, then when the two modified copies were written back to memory, one of the writes will be lost. Figure 3.12 shows this happening; the write to the first element from core zero has been lost.

Throughout this section we use the term *cache line* to refer to a contiguous, fixed-length, naturally aligned, set of bytes in memory. When this is held in a cache, it will also have additional tags associated with it to represent the local state of the line. Precisely what those tags represent depends on the coherence protocol (and other architectural issues; for instance, architectures that have hardware support for transactional memory will have additional tag bits to hold the state required to implement it).

3.2.5 Cache coherence: the MESI protocol

For simplicity, we will initially describe the protocol as if there are only two cores and a single level of cache involved. Once we have that covered, we can move on to the more complicated cases of multiple cores and multiple levels of cache.

3.2.5.1 Protocol description

In the MESI protocol, a physical cache line can be in one of four states:

1. **Modified:** The line is only valid in this cache and contains modified data that has not yet been written back to memory.
2. **Exclusive:** The line is only present in this cache, but has not been modified.
3. **Shared:** The line is valid in this cache and in other caches that all hold the same value, which is also the value in memory.
4. **Invalid:** The physical line is free and is not holding data that represents any memory location. This is the initial state of a line, and the state it reverts to when it holds no useful data.

The state of a cache line can change as the result either of a local request from the associated core or as a result of a request transmitted over the *coherence fabric* from some other core. A local request can require the cache controller to interact over the coherence fabric with other caches and/or the memory controller before the request can be completed.

In each state in which a line is in the cache, there are five possible things that can happen:

1. **Local Read:** The core associated with the cache performs a read operation to data in the line.
2. **Remote Read:** Another core tries to read data in the line and its local cache does not have a copy, so it issues a read operation over the coherence fabric, which hits in the remote cache.
3. **Local Write:** The core associated with the cache performs a write operation to data in the line.
4. **Remote Write (Read for Ownership, RFO):** Another core wants to write data to the line, so it must ensure that it has an exclusive copy of the line.
5. **Local Flush:** The cache itself needs to evict the line to replace it with another or the local core has issued a cache-eviction instruction for that cache line.

Table 3.2 shows the impact of each action on the cache line and the requests it generates as a result. The same information can also be visualized as a state transition diagram, which is shown in Figure 3.13.

To show the protocol in action (in Figure 3.14), let's consider again our example above—in which we showed two cores writing to the same cache line and losing a write—to see how the MESI protocol avoids that problem.

We can see that because the protocol ensures that updates to a line can only happen when it is in Exclusive or Modified state (which ensure that it is only in a single cache), the multiple writes are correctly handled and no write is lost.

Table 3.2: State transitions of the MESI coherency protocol.

State	Request	External Operation	New State
Invalid	Local Read	Send read request	Shared if satisfied by another cache, otherwise Exclusive
Invalid	Local Write	Send read for ownership	Exclusive
Exclusive	Local Read	None	Exclusive
Exclusive	Local Write	None	Modified
Exclusive	Remote Read	Send data to requester	Shared
Exclusive	Read for Ownership	Send data to requester	Invalid
Exclusive	Local Flush	None	Invalid
Shared	Local Read	None	Shared
Shared	Remote Read	Send data to requester	Shared
Shared	Local Write	Send read for ownership	Exclusive
Shared	Read for Ownership	Send data to requester	Invalid
Shared	Local Flush	None	Invalid
Modified	Local Read	None	Modified
Modified	Local Write	None	Modified
Modified	Remote Read	Flush to outer cache/memory; then send data	Shared
Modified	Read for Ownership	Flush to outer cache/memory; then send data	Invalid
Modified	Local Flush	Flush to outer cache/memory	Invalid

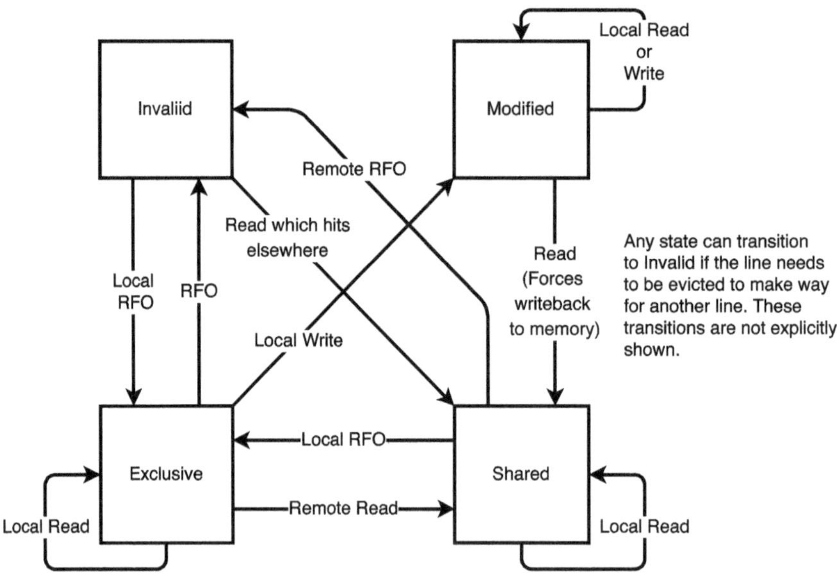

Figure 3.13: Cache line state transitions.

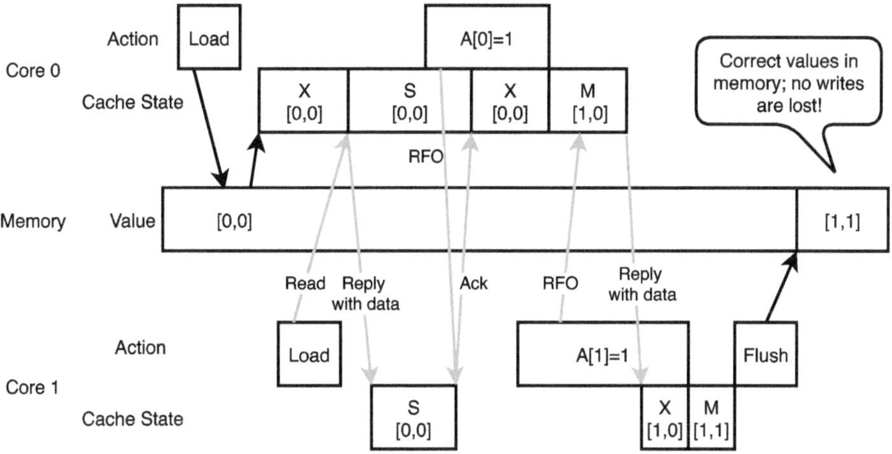

Figure 3.14: The MESI protocol in action.

3.2.5.2 Adding more cores

When we add more cores, we encounter a problem: how can the hardware find all of
the caches that have a copy of a shared line?

Historically, this problem was easy to solve, since all of the caches shared the
same single bus, and so all caches could see every transaction for a particular cache
line. However, in modern processors, that is no longer true, since physical signaling
issues mean that communication lines now have to be point-to-point connections at
the physical level.

One solution is to introduce another entity (often associated with a shared outer-
level cache), the *tag directory* (TD) or *cache home agent* (CHA). The CHA tracks both
the state of a cache line (one of the MESI states) and which caches in the processor
have copies of the cache line.

To ensure atomicity, each cache line in memory has a single CHA that is responsible for it. The mapping from the physical address of the cache line to the assigned CHA
is performed via an (unpublished) hash function from the physical address to one of
the CHAs.

Since the CHA has to ensure that it can move a cache line into an exclusive state,
it has to track the set of caches in which a line is present so that it can invalidate them.
For a line in the Exclusive or Modified state, there can only be one such cache. For a
cache line in the Shared state, there can be anywhere from two up to the number of
caches on the die, plus one for all off-die caches, if we're in a multi-die/socket system.
This can be a significant overhead (e. g., in a 32-core system with 64-byte cache lines,
it would lead to a 6 % overhead ($32/(8 * 64) = 1/16$)). Thus, the processor design may

assume that heavily shared lines are rare and switch to an additional "everywhere" state in the CHA when the number of caches in which a line is present is above a certain threshold. In that case, the protocol will require that all caches are tested when a line is in this "everywhere" state.

Another possibility (which is used in the Marvell[*] Arm processors) is to emulate a shared bus by broadcasting all transactions to all of the caches. In the Marvell implementations, this is achieved using bi-directional rings to implement the coherence fabric [124].

One peculiarity of the coherence protocol is *false sharing*. This occurs when two or more cores want to modify the same cache line, but at different offsets—that is, at two different memory locations that happen to be in this cache line. What happens is that one of the cores will "win" the cache line and move it into the Exclusive state to conduct the write operation. Once that write completes, the other core will get the cache line in the Exclusive state and also perform the write. Even though the two cores are not modifying the same memory, they are falsely sharing the same cache line just because they work on items that are so close to each other that they share the same cache line. While this not an issue of correctness, it may have a big impact on performance if many threads are falsely sharing a cache line (e. g., using the thread ID as an index into an array that is modified) or if this false sharing happens frequently in a loop. As we continually see, the time spent moving data is often the performance-limiting factor, so introducing this additional, unnecessary data movement can cause performance problems.

The detailed impact of this on core-to-core data transfers will depend on the design, since there are different options here. Current Intel processors implement a shared L3 cache in which, although a cache slice is co-located with a core, the data in that cache-slice could come from any core, whereas the AMD processor we are looking at has many L3 caches, each of which is shared by only four cores. The Marvell Arm processors follow the Intel style of having a shared distributed cache (though, as we saw above, their coherence fabric implementation differs).

While we could go into the details of each operation (and, if you want to understand the deep operation of caches, you should work this out), the details of the protocol are less important than the performance impact when we are trying to achieve the highest performance in our code, which has to move data between cores. The things we need to consider are:
1. **Line Placement**
2. **Core Placement**
3. **Degree of Sharing**

We will discuss each of these in the next section.

3.2.6 Performance implications

When we are attempting to optimize communication between cores (and, therefore, between caches), we want to ensure that the time taken to transfer data from one cache to another is as low as possible, and that our communication patterns are chosen to make that possible.

The actual topology of the on-die communication fabric affects the relative difference in performance that we may see. For interesting cases, it doesn't affect the fundamental issue, which is that the time to communicate between two cores will depend on how close they are to each other in the processor's communication fabric. (This would not hold true in a uni-directional ring, but since that avoids the inhomogeneity by making all cores a long distance away, it's not a common modern implementation!)

The three CPUs we are looking at here are:

- **AMD EPYC 7742 64-Core Processor:** This is a 64-core package built using "chiplets" that supports the x86_64 architecture [5]. Each core can support two SMT threads.
- **Marvell ThunderX2***: This is a monolithic 32-core die that supports the Arm V8.1-a architecture [87]. Each core can support four SMT threads.
- **Intel* Xeon* Platinum 8260L Processor:** This is a 24-core die that supports the x86_64 architecture [65]. Each core can support two SMT threads.

All of our machines are two-socket machines; more details of their configuration are given in Table 1.3, Table 1.2, and Table 1.1. In our measurements, we only use a single thread per core, since we are interested in the time taken for the communication between cores, not between threads that share all levels of cache. We have shown the results from these three rather different machines to demonstrate that the issues we are talking about here are not ISA or implementation specific, but show up more generally on all modern processors and constitute fundamental properties of processor architecture. Since we are not concerned with comparing the absolute performance of these machines, we do not show times, but rather per-machine ratios based on the best result we plot for each machine.

3.2.6.1 Line placement

At first glance, the address of the cache line we communicate with seems irrelevant; however, as we saw above, to maintain cache coherence, all communication between caches has to be controlled by the appropriate CHA, and since the hash to CHA is pseudo-random, it is entirely possible that the appropriate CHA for the line we are moving is on the opposite side of the die. Therefore, if we want to optimize communication, we would like to use a cache line whose CHA is co-located with one of the two cores that are communicating.

In Figure 3.15, we show the relative performance of communication between two arbitrarily selected cores using each of the cache lines in a single page for the commu-

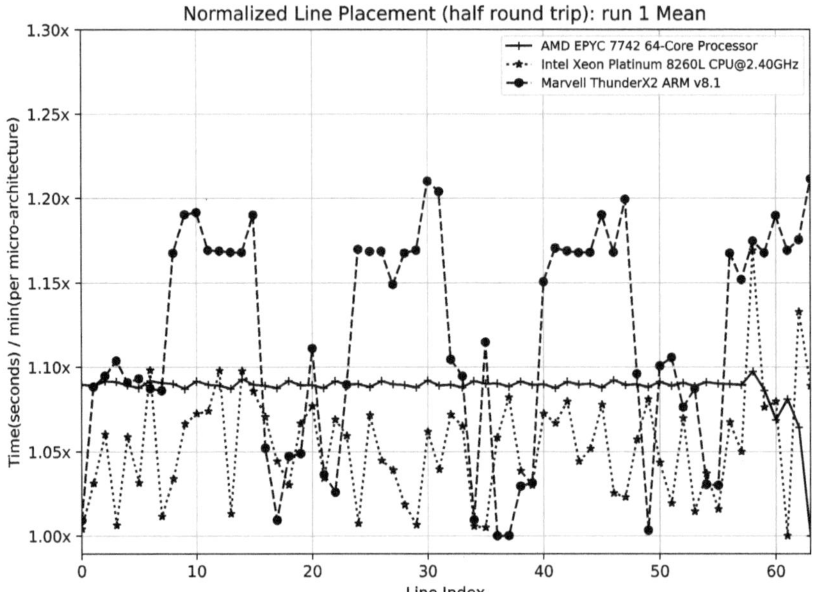

Figure 3.15: Effect of line placement on half round-trip time.

nication. We can see that on the Marvell ThunderX2 and Intel Xeon Platinum 8260L Processor, the worst line can take ~1.2 times the time of the best, whereas on the AMD machine, most lines are <1.1 times slower.

Although we only show one run (measuring 10,000 operations), we actually ran five separate repetitions of the experiment that show consistent results. Of course, if the code is re-run, the results for the specific lines will likely differ, since it is unlikely that the page that holds the communication channels will be placed in the same physical page.

Whether it is worth optimizing so that we can use well-chosen lines will depend on the target machine and also the complexity of finding a good line. Doing this would require locking the communication page down so that the OS does not choose to move it and then measuring the performance to find good cache line, for the specific point-to-point communication being optimized.

3.2.6.2 Core placement

For concreteness, consider 32 cores in a 4×8 grid. The longest possible distance between cores is then ten $((4 - 1) + (8 - 1))$ hops (between cores at opposite corners), so sending a message and receiving a reply has a cost of 20 hops. If the two cores were neighbors, the round-trip distance would only be two hops. Of course, this communication time may only be a small component of the overall time to satisfy the read or write request, since the system also has to perform the cache lookup and enqueue

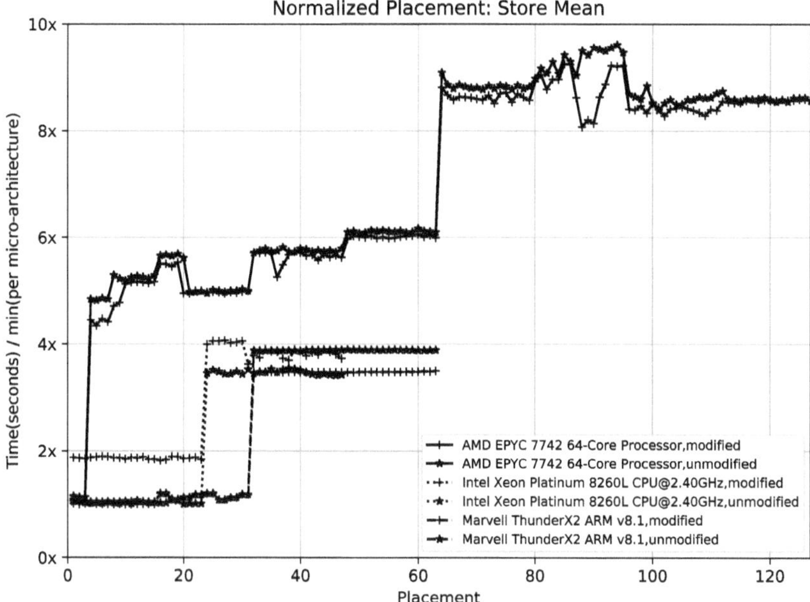

Figure 3.16: Effect of core placement on write time.

operations into the coherence fabric. However, if it is measurable and significant, we should try to choose a good mapping of the communication pattern we want onto the cores so that the communication is faster.

In Figure 3.16, we can see that within a single socket, the core placement makes little difference to the time to perform a store on the Intel Xeon Platinum 8260L Processor or the Marvell ThunderX2, whereas on the AMD EPYC 7742, the first four cores are much nearer to each other, since they share the same L3 cache. In that case, the caches are able to observe that the line is not shared with another L3 cache, so it does not need to look any further. As we should expect, all of the machines show a noticeable increase in latency when they have to handle data that was in a cache in a different socket.

3.2.6.3 Degree of sharing

We have seen that in a system with many caches, a single cache line can be present in many of them simultaneously. We have also seen that to perform a write operation, common coherence protocols require that the line is only in the cache of the core that is performing the write. Therefore, it is worth considering what the cost of moving a line from a heavily Shared state to the Exclusive state is, and how that changes as we alter the degree of sharing.

We can measure this, as shown in Figure 3.17. Since each group of four cores in the AMD machine share an L3 cache, we do not plot the data where we only have one

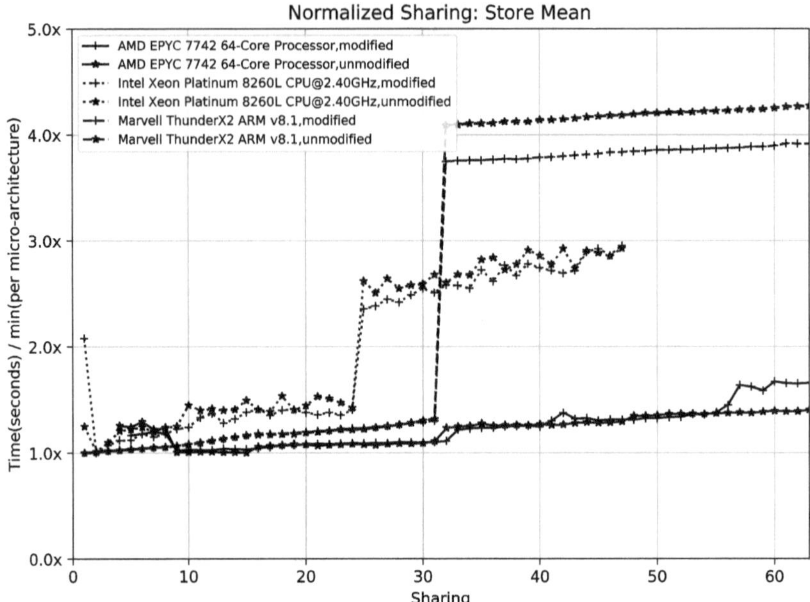

Figure 3.17: Effect of degree of sharing on store time.

through four threads since those cases are significantly faster and pervert the scale of the graph. We also restrict this to 64 cores (so that we can see the behavior on the Arm and Intel machines) so that the AMD data is all within a single socket. We can see that the effect we expected does exist, with a 1.3× difference in the write time inside a single socket on the Arm core, while the Intel shows ~1.5× inside a socket. The AMD machine also shows ~1.4×.

A related issue here is how long it takes for a write to become visible to all the other threads if they are all polling a single cache line. This seems a perverse thing to do, but is a pattern that occurs under contention in simple locks, when releasing threads in a hierarchical barrier, or when passing information to multiple threads from a single source. This pattern is typically called *broadcast*. While it may seem that all polling cores should see the updated cache line immediately, Figure 3.18 shows that they do not. Indeed, it shows a significant latency, with the time for the last thread to see a store increasing by more than 20× as we move from one polling thread to 63 in our Arm machine. Our Intel machine shows a smaller effect (~9×), while our AMD machine grows to ~2.5× between 4 and 10 pollers and is then quite flat. As in the previous graph, we do not show the second socket of the AMD machine, or the data for the case where all pollers are sharing the same L3 cache. If we were to show this, then the vertical scale would need to be much larger.

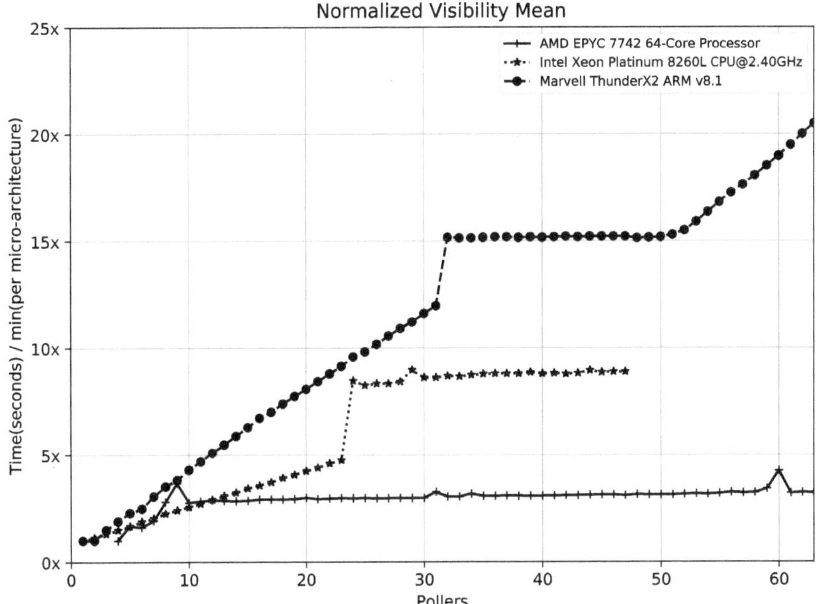

Figure 3.18: Time for last poller to see a store.

3.2.6.4 Performance conclusions

These graphs are all internally normalized (each implementation is normalized against its own best result), and therefore you cannot tell from this presentation which processor has the fastest communication. What you can see, though, is that the effects we are describing are common to all of these different implementations, despite them representing two different ISAs and three different design teams. These issues are therefore ones that we cannot ignore, since they are extremely likely to affect our code, whichever architecture and implementation we are targeting.

3.2.7 Non-uniform memory architectures

Our previous discussion has shown that latency for data access from a cache is affected by how physically close the core making the memory access is to both the cache in which the line is resident and to the CHA that has to service it. As well as the differences we have seen in the use of caches, systems often have multiple memory controllers that are not all equally close to any specific core. Thus, in a multi-socket system, each socket will likely have its own memory controller, and therefore the latency of a memory access will differ depending on whether the physical memory of the page containing the data is in a local memory system or a remote one.

Table 3.3 shows the normalized time required to complete a load or store operation in memory that is not initially present in a cache in three different two-socket

Table 3.3: Normalized memory access time.

Operation	AMD EPYC	Intel Xeon	Marvell TX2
Load	1.0	1.0	1.0
Store	1.2	2.2	0.8
Remote Load	1.5	1.0	2.0
Remote Store	1.5	5.3	1.6

machines. As usual, these are the AMD EPYC 7742, the Intel Xeon Platinum 8260L Processor, and the Marvell ThunderX2 Arm. Since the point here is not to compare the absolute performance of these machines, but rather to investigate whether the general properties are similar, the times have been normalized to that of a local load on the given machine.

Here, we can see that on each machine, loads from and stores to memory have different relative costs, though in the x86_64 machines, stores are more expensive than loads, while in the Arm machine, the reverse is true. This is likely because the x86_64 memory model is much stricter than the Arm one, and so requires the relevant cache line to be loaded before the store can complete, whereas in the more relaxed Arm model, the store can be committed earlier. In all of the machines, we see that the remote writes (to memory, in the other socket of these two-socket machines) are slower than to memory accessed via a local memory controller. In this test, the Intel machine seems to have the same cost for a remote as for a local load, though this may just be a result of its smart memory-access predictor defeating our attempts to make an unpredictable memory access pattern. Our critical point remains, though: these differences are important for performance tuning, since memory latency is frequently the performance-bounding factor. Although we have demonstrated NUMA effects resulting from accessing memory that requires inter-socket communication, in newer implementations that use "chiplets" within a single socket, or on Intel machines where SNC is enabled, NUMA effects may also likely occur within a single socket.

As the page tables allow any page to be mapped to any available physical memory, and, thus, to any memory controller, the performance of a code will depend on how that mapping has been performed by the operating system.

Many operating systems adopt a *first touch* policy, which means that a physical page is not allocated until a store to the page occurs, at which point the OS can see which core executed that write, so it can then allocate a physical page in memory that is near to that core. This has implications for how we write code; in particular, if possible, the initialization of arrays should be performed in parallel using the same distribution of array indices to threads as will be used inside the performance-critical parts of the code.

3.3 Conclusions

What should you carry forward from our brief overview of modern cores, caches, and memory systems? There are a couple of things that are really important to keep in mind when writing low-level, high-performance (runtime) code. Luckily, some of the conclusions are similar for the different platforms (that we looked at, but likely also others).

The hardware is much more complicated than you would reasonably expect. You've seen some of the tricks that the processor design teams implement to squeeze performance out of the cores of the processor. This is true for most processor architectures out there. While in some historic past, some processors (e. g., for embedded systems) were simple enough to reason about performance by looking at (assembly) code, modern processors with out-of-order engines, complex memory hierarchy, etc. are certainly much harder to understand. It may not only be hard to guess performance from looking at source code, it may even turn out to be impossible. When you attempt to do that though, you will likely draw conclusions that will misguide you in your implementation or optimization task.

Instead, we need to consider the real and measured performance of the hardware when writing our code. Our expectations and intuitions are often wrong. Looking at the numbers that we have presented in this chapter, you may even see that they do not match the given data sheet of the processor specifications, as these documents typically are geared to different metrics than those that we will need for a parallel runtime system. With his chapter in mind, we have the opportunity to write code that performs much better than the "obvious" options.

4 Compiler and runtime interaction

In this chapter, we will discuss the very basics of compilers, so that we gain a basic understanding of how a compiler works and how it processes the code written by the programmer. We also explain the implementation of a task-based programming model and how it interacts with a parallel runtime system. Here, we'll use Intel[*] Threading Building Block as a concrete example of such a system. To conclude the chapter, we will revisit compilers and show how a compiler for a parallel programming language can be implemented. For concreteness, the OpenMP[*] API will be our poster child for this discussion.

4.1 Compiler basics

Before we move on to explain how an OpenMP compiler (which we use to exemplify a compiler for a parallel programming language) generates parallel code, we must first understand what a compiler does and how it produces executable code out of human-readable sources.

Compilers are complex pieces of software. The `cloc` utility [27] counts over 1.7 million lines of non-comment source code in LLVM's `clang` sub-directories for `clang`, version 9.0.0; Table 4.1 shows a full breakdown of the languages used. The same line count for the whole LLVM version 9.0.0 compiler project (which includes runtime libraries, front-ends for other languages, tests and so on) is over 6.8 million lines. Thus, we can only scratch the surface of this topic in this book. If you want to read more about the theory behind compilers, we have to refer you to secondary literature. Although it is old, the "Dragon Book" [2] (famously named after the red dragon on the book's cover) is a good starting point. The series of "Tiger Books" (guess what...) written by Andrew W. Appel, e. g., "Modern Compiler Implementation in Java" [7], and the "Tree Book" by Grune et al. [52] are also good sources that cover more modern aspects of compiler principles.

Let's start our journey from human-readable source code to binary, executable code by looking at Figure 4.1. Historically, a compiler was split into two parts called the *front-end* and the *back-end*, each of which is responsible for certain kinds of code analysis and transformation. Some compilers added a *middle-end* with architecture-agnostic optimizations.

The front-end and the back-end are connected to each other through *intermediate code* or *intermediate representation* (IR)—that is, a representation of the source code that is close to executable code, yet abstract from the actual binary code that a processor can understand. This is an implementer's trick to reduce the complexity of writing compilers for n different source languages and m target architectures. Without intermediate code, we'd have to write $n \cdot m$ implementations, e. g., C to the Intel Architecture, C to the Arm[*] architecture, C++ to Intel Architecture, C++ to the Arm architecture,

https://doi.org/10.1515/9783110632729-004

Table 4.1: Lines of code in the LLVM `clang` compiler (version 9.0.0).

Language	Files	Blank	Comment	Code
C++	6041	192603	335685	1000946
C/C++ Header	2008	58861	96007	235133
C	3620	48423	218872	151927
XML	81	47	715	142803
Objective C	1568	17461	28756	57654
HTML	32	3643	309	30303
reStructuredText	68	14366	9111	29584
Objective C++	428	5063	5225	17073
Sum	14643	352085	710575	1716955

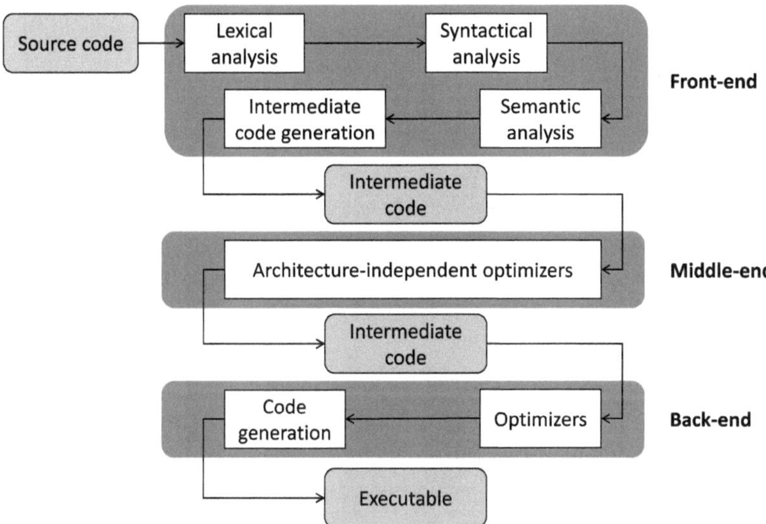

Figure 4.1: Stages of a compiler with front-end, middle-end, and back-end.

and so on. However, by using a common intermediate code, we can implement n compiler front-ends (C and C++ in our example) and m back-ends (e. g., Intel Architecture and Arm).

The very first step in producing executable code is to perform *lexical analysis*. In this phase of the compilation, the source code is read from the source file (or compilation unit) and is split into so-called *token*s. Most programming languages have certain classes of words, for instance, keywords (`int`, `subroutine`) or identifiers (`foo`, `printf`). Later compiler phases can be more efficient if they do not have to deal with sequences of characters, but rather tokens that can be represented by a unique integer ID. While the keywords of the language have their unique ID, the lexical analysis

```
for (i = 0; i < length; ++i)
  sum += compute(data[i], value);
```

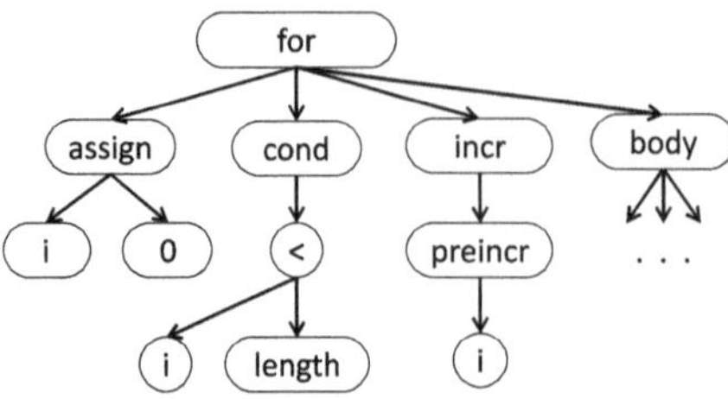

Figure 4.2: Example of a simplified abstract syntax tree.

inserts identifiers into a table and each identifier is replaced with a reference to the table. Two rather practical books about tools for this and the next phase are [83] (for lex and yacc) and [82] (for flex and bison).

The *syntax analysis* phase then receives the stream of token IDs from the lexical analysis and tries to match it with the formal syntax of the programming language for which the front-end is written. This phase knows about the syntactic structure of well-formed programs and rejects any stream of tokens that does not conform. If all the checks go well, the result is a tree representation of the code, which is called an *abstract syntax tree* (AST). Figure 4.2 shows a simplified AST for a small example of a for loop in the same figure.

The next step is to perform *semantic analysis*. This phase traverses the AST and makes sure that the program conforms to the specification of the programming language from a semantics perspective. For instance, the expression i = i + 1; is syntactically valid C/C++ code; however, if i had been declared as an enum type, then it would not be semantically valid.

Syntax checks include tests to ensure that all variables have been declared properly before they are used (if this is required by the language) and that they have the right type for their usage (this is what would catch our previous attempt to add one to an enum). The same happens for functions, procedures, classes, and other entities that are used by the code. For object-oriented languages, the compiler also validates the inheritance relationships of classes and objects, and checks the visibility of class members when they are used in the code. The compiler may also introduce implicit type conversations if required—for instance, when an integer value is assigned to a floating-point variable—or fill in default parameters to functions that do not have an explicit actual argument for the parameter at a specific call site.

The last step in the front-end is the generation of *intermediate code*. This compiler stage performs a final traversal of the AST and generates the intermediate code for the source code of the compilation unit (the source file). Depending on the design of this bridge between front-end and back-end, the intermediate code may still be a tree-like graph, a list of trees, or a plain sequence of statements similar to assembly code. Regardless of the actual structure, the intermediate code is an abstract representation of the program that still contains much more information than assembly code [92]. For the task of intermediate-code generation, the compiler defines a set of rules about how the respective nodes of the AST are processed and what sequence of IR instructions or trees should be emitted.

In classic (optimizing) compilers, at this point in the compilation, the compiler back-end takes over the intermediate code and begins code generation by feeding the intermediate code through various optimization passes and then into a machine-specific code generator.

In modern compilers, there are potentially many different optimization passes, each of which performs a transformation of the internal representation. Collectively, these may be known as the *middle-end*. The difference between middle-end and back-end passes is that code in the middle-end is architecture neutral, and applies optimizations that are portable and can be re-used without changes whatever the target machine is, whereas optimizations in the back-end are architecture specific (or even micro-architecture specific), since they consist of things like instruction scheduling for an out-of-order architecture, which will depend on the details of a specific processor implementation. Of course, there is no general way to decide where an optimization should be considered to happen in the chain. There are optimizations such as vectorization that have many features that are independent of the details of the target architecture, but at the same time, are clearly also target dependent, since there is no point in vectorizing code for a machine architecture that has no vector instructions. Some people even consider finding a good order of optimization passes in a compiler an art that is close to magic!

The optimization passes are often the part of the compiler that consumes most of the time needed for compilation. Depending on the quality of the compiler, several basic optimizations such as dead-code removal, common-subexpression elimination, constant folding, function inlining, and so on are commonly applied [92]. Compilers can also implement very sophisticated loop optimizations, e. g., loop tiling, loop skewing, loop unrolling [10], and polyhedral loop analysis [51]. The goal of these optimizations is twofold. First, the rule-based translation of the AST into intermediate code may create sub-optimal code patterns that the optimizers remove and replace, as they have a more extensive view of the generated code. Second, programmers want the compiler to optimize the code beyond what typical programmers want to do manually in their programs. This helps avoid some of the most error-prone and cumbersome source optimizations that might make the resulting code less readable and manageable.

The final stage in the compiler is *code generation*, which transforms the optimized intermediate code into actual binary code for the linker or assembly code for the assembler. At this stage, (in a register-based architecture) variables are assigned to the registers that are visible in the instruction set of the processor and the compiler selects the instructions to be used to execute the program. When compiling a program from a single source file, the compiler will also invoke a linker that produces the final executable image by binding all compilation units plus libraries into a single binary file that the operating system can load and execute. If you want to read more about how the linking stage works, we'd like to refer you to [81]. While the book is a bit dated, it nicely explains how the process of linking an executable and loading the resulting binary image to create a running process works.

Some modern compilers (such as LLVM [84]) can also be invoked at link time. Because all of the source files have been compiled at that point, the compiler can therefore see and analyze the whole program. This allows for additional optimizations, which were not possible due to the limited scope when compiling individual source files. This is known as link-time optimization (LTO) [70, 94].

4.2 Implementing a task-based parallel model

Now that we have an overview of how a compiler works for a base language like C++, we can use that compiler to implement a parallel programming model through a library. Intel Threading Building Blocks [151] (Intel TBB) is an example of such a task-based parallel programming model that is implemented as a library on top of C++. The C++ compiler does not provide any extra mechanisms beyond standard C++ language features, such as templates, lambda functions, etc. [123].

The basic API entry points for a task-based parallel programming model are quite simple. The first API routine needed is one that spawns a task and enqueues it for execution by the runtime. We will also need another API routine that awaits the completion of an enqueued task so that its parent can continue once the child task has finished execution. Of course, while this two-function interface may be sufficient for something simple, a real task-parallel programming model is more complex.

For instance, you could think of a runtime entry point that schedules a set of tasks for execution rather than doing it one by one, or an entry point that waits for the completion of more than one task, e. g., all tasks created by the same parent task. Both of these might improve performance, as there would be fewer interactions with the runtime and its task pools. A more advanced programming model might offer features beyond simple tasking (e. g., TBB's task graphs or the task dependences provided by other libraries). You could also think of adding other high-level "services" such as containers that are geared toward being used from multiple threads or tasks at the same time, or parallel algorithms such as the one offered by Parallel STL [71]. And, the whole topic of inter-task synchronization (such as mutual exclusion) requires additional entry points into the runtime system.

We will cover how to implement the internals of such a runtime in Chapter 9.

4.2.1 Lambda functions and closures

At this point, let's make a short digression and talk about lambdas and closures, because they are important concepts to understand and we will need them soon. A *lambda function* (sometimes also called a *lambda expression* or an *anonymous function*) is a function that does not have a name. This terminology stems from functional programming theory. Related to this is the notion of a *closure*. A closure holds both a function (e. g., a lambda function) and the data environment that carries all of the data that is needed to execute a specific runtime instance of the function. A key aspect of functional programming is that you can have function objects that have some arguments that are bound while others are not. Thus, one could have a general addition function that adds two numbers and then create an increment function by binding one of the arguments to the value one.

What we have just described constitutes the foundation of TBB. It makes heavy use of C++ templates—for instance, `tbb::parallel_for`—and C++ lambda functions to define parallel work for these templates and thus provide a high-level interface to the programmer. The earlier example in Listing 2.7 showed how lambdas are used to describe the body of a task. Effectively, we can view a task as a closure that consists of the outlined code of the lambda expression along with the captured values of all of the variables that the task will need when it is scheduled for execution.

Listing 4.1 shows a variation of the traditional "Hello World" example that uses the TBB library. Listing 4.2 shows the same code, but now with the lambda expression manually outlined, which is similar to what the compiler does automatically for us. The code of the tasks has been moved to a C++ class and into a function that implements the `operator()` method to turn the class into a functor class. Note that we are using `struct` for the class to have `public` visibility as the default for all definitions inside the class.

The values that the outlined code uses become arguments to the constructor of the newly created functor class. They are also private member variables of the class structure, so that the outlined code can access the variables through the `this` pointer. As you can see from the code pattern, a call-by-value argument becomes a regular class member of the same type (e. g., `i` and `value`). Call-by-reference arguments (e. g., `mtx`) are implemented as class members of a reference type to the base type of the saved variable (e. g., here `tbb::spin_mutex` for `mtx`).

At the original source location of the lambda function, the compiler removes outlined code, as we do in Listing 4.2. Instead of the lambda's code, it inserts code to construct a functor object of the new class, passing in all the variables that need to be captured for the closure. If the arguments are not given by name as in our example, but rather as a generic capture specification, the compiler needs to identify which variables to capture in the outlined code.

It may seem that a task-parallel model needs a very expressive programming language to implement tasking features. Certainly, C++ is a prime example of a base lan-

```cpp
#include <iostream>
#include <tbb/spin_mutex.h>
#include <tbb/task_group.h>

const size_t num_tasks = 8;

void answer() {
  using namespace std;
  tbb::task_group grp;
  tbb::spin_mutex mtx;
  float value = 21.0f;
  int factor = 2;
  for (auto i = 0; i < num_tasks; ++i) {
    grp.run([i, &mtx, value, factor]() {
      tbb::spin_mutex::scoped_lock lck(mtx);
      cout << "Task " << i << " says: the answer is "
          << (value * factor) << endl;
    });
  }
  grp.wait();
}
```

Listing 4.1: TBB example code before outlining for parallel execution.

guage that provides very expressive and sophisticated mechanisms that a task-parallel library can exploit. However, even less expressive languages like C are sufficient. At the most basic level, a function closure created from a lambda function boils down to two things: a pointer to a function and a pointer to a data environment. So, any programming language that has the data types and features to express these two concepts is powerful enough to create executable tasks and hand them over to the parallel library for execution. In the end, it's a matter of how much code the programmer has to write to mimic the concept of lambda functions and closures.

4.2.2 Enqueuing tasks in TBB

Now that we have a function object for a task in our hands, we need to hand it off to the runtime system to store it in the task pool for execution by one of the available worker threads. Listing 4.3 shows the starting point for the call to the run() method of the tbb::task_group class that we used in Listing 4.1. This code is from the TBB version

```cpp
#include <iostream>
#include <tbb/spin_mutex.h>
#include <tbb/task_group.h>

const size_t num_tasks = 8;

struct outlined_answer_task_0 {
  outlined_answer_task_0(size_t i_, tbb::spin_mutex &mtx_,
                         float value_, int factor_)
      : i(i_), mtx(mtx_), value(value_), factor(factor_) {}

  void operator()() const {
    tbb::spin_mutex::scoped_lock lck(this->mtx);
    std::cout << "Task " << this->i
              << " says: the answer is "
              << (this->value * this->factor) << std : endl;
  }

private:
  size_t i;
  tbb::spin_mutex &mtx;
  float value;
  int factor;
};

void answer() {
  using namespace std;
  tbb::task_group grp;
  tbb::spin_mutex mtx;
  float value = 21.0f;
  int factor = 2;
  for (auto i = 0; i < num_tasks; ++i) {
    grp.run(outlined_answer_task_0(i, mtx, value, factor));
  }
  grp.wait();
}
```

Listing 4.2: TBB example code after outlining for parallel execution.

```cpp
template <typename F>
void run(F && f) {
  internal_run<
  internal::function_task<typename internal::strip<F>::
      type>>(std::forward<F>(f));
}


template <typename Task, typename F>
void task_group::internal_run(__TBB_FORWARDING_REF(F) f) {
  owner().spawn(*new (owner().
                allocate_additional_child_of(*my_root))
                Task(internal::forward<F>(f)));
}


inline void interface5::internal::task_base::spawn(task & t)
{
  t.prefix().owner->spawn(t, t.prefix().next);
}


void tbb::internal::generic_scheduler::spawn(task & first,
                                             task *& next) {
  governor::local_scheduler()->local_spawn(&first, next);
}


void generic_scheduler::local_spawn(task * first,
                                    task *& next) {
  if (&first->prefix().next == &next) {
    // Spawn a single task
    size_t T = prepare_task_pool(1);
    my_arena_slot->task_pool_ptr[T] =
      prepare_for_spawning(first);
    commit_spawned_tasks(T + 1);
  }
  else {
    // Spawn multiple tasks in one go
  }
  // More code that we have omitted
}
```

Listing 4.3: TBB code fragments to store a task in the task pool.

that ships with the Intel Composer, version 2019.4.243, and the open source version of TBB [66].

The process begins with the `run()` method calling an internal implementation `internal_run()` (later versions of TBB make use of more modern C++ features to streamline this a bit and avoid the function call). In the `internal_run()` method, the code allocates memory for the task from the owner's memory pool and uses a placement `new` when calling the constructor of the `Task` object.

The `spawn()` method receives this new `Task` object and forwards it to the scheduler that is responsible for taking care of task execution. The interface is designed so that multiple task objects can be spawned by passing in an iterator-like `first` and `next` pair. When spawning a task, the scheduler implementation first allocates a slot `T` in the task pool and then puts the task into the reserved slot. Finally, the task is committed to the task pool.

In Chapter 9, we will cover in more depth what happens inside the runtime system, how the task pool is implemented, and how tasks are stored and retrieved from the task pool.

4.3 Compilers for parallel programming languages

The OpenMP API serves as our main example of how to implement a parallel runtime. It is now time to have a closer look at how a compiler implements the OpenMP API and how we can extend the sequential compiler to become a compiler for a parallel programming language.

Figure 4.3 shows what a parallel runtime system looks like for the OpenMP API. In comparison with Figure 1.1 in Chapter 1, you will notice that the application code now also makes use of the capabilities of the processor. The compiler is free to decide whether, for a particular feature of the language, it would like to invoke entry points of the runtime system or instead generate code to implement the feature directly. This is true for both sequential and parallel programming languages. For instance, the compiler can either implement atomic operations using whatever instructions are provided by the target architecture or it can call into the runtime library to perform the operation.

From a conceptual perspective, the compiler for a sequential language can be extended to support a parallel programming language. Figure 4.4 shows a high-level view of what the new compiler pipeline looks like. The major addition is a set of new passes in the middle-end and the back-end, and, possibly, an extension of the intermediate code to include information about parallelism. Before we discuss these, we will describe the modifications needed in the compiler front-end. Most parallel programming languages have additional syntax to help programmers express parallelism to the compiler; because the front-end is responsible for parsing and analyzing the

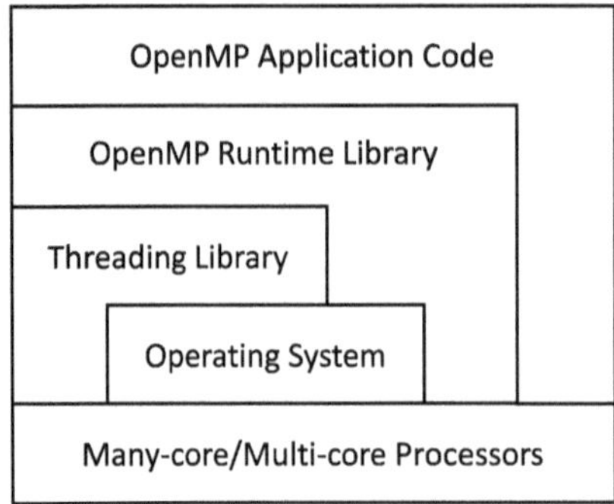

Figure 4.3: Layers of a typical OpenMP runtime library.

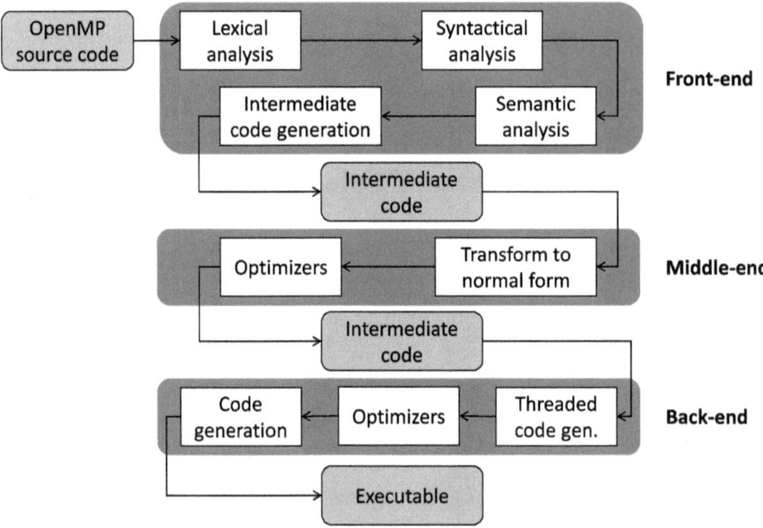

Figure 4.4: Compiler with a middle-end for parallel code transformation.

syntax of the source code, the front-end must be extended to parse the new parallel language correctly.

During lexical analysis, the compiler recognizes the keywords of the parallel programming language and emits corresponding tokens to the syntax analysis phase. There, these tokens are built into the abstract syntax tree as before, so that now the AST contains a full model of the parallelism described in the source code. OpenMP compilers parse the pragmas in C/C++ or directives in Fortran as additional syntax elements on top of the C, C++, and Fortran base languages. Figure 4.5 takes the example of Figure 4.2 and adds an OpenMP parallel loop to it. The figure also shows a potential

```
#pragma omp parallel for shared(data) reduction(+ : sum)
for (i = 0; i < length; ++i)
  sum += compute(data[i], value);
```

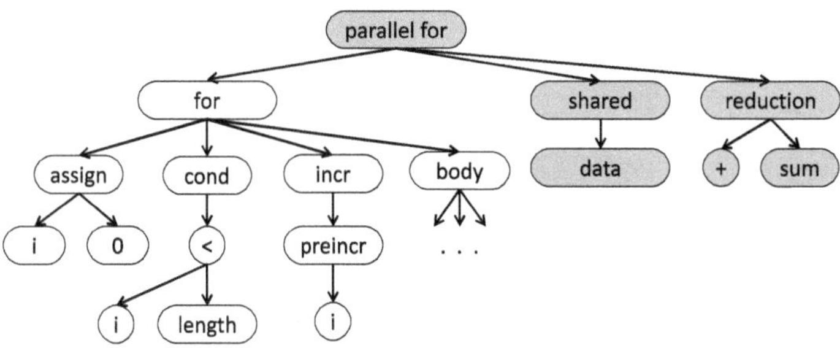

Figure 4.5: Example of the simplified AST for an OpenMP parallel loop.

representation of the AST with the OpenMP directive reflected as additional nodes in the AST.

Semantic analysis takes these additional pieces of information and checks the AST for incorrect usage of language features. For instance, OpenMP semantics prohibit using the same variable in conflicting ways in both a shared and private clause. Another example is the canonical loop form that the OpenMP specification requires the program to use for a parallel loop. The semantic analyzer must analyze the code, check that programmers have not attempted to use an illegal form, and reject the program if they have. In other parallel programming languages, there might be similar restrictions of how features can be composed. Semantic analysis will detect all these patterns and inform the programmer with a compiler diagnostic message.

The *middle-end* of the parallel compiler has to handle code transformations and optimizations that relate to the parallelism in the code. The intermediate representation that flows from the front-end to the middle-end still contains complete information about the parallel code structure. As in Section 4.1, the representation of parallelism in the intermediate code can be tree based or instruction based. This is a degree of freedom that the compiler developer has and is not only a matter of taste, but rather something that influences a lot of the compiler phases in the middle-end.

Note that in many compilers, the middle-end does not exist as a separate entity in the compiler tool chain. Some compilers merge the middle-end with the front-end and have corresponding AST transformations that translate parallel code into threaded code for the entry points of the runtime system. Other compilers merge the middle-end with the back-end and create threaded code as one of the optimization passes.

The first step in the middle-end is to transform the incoming intermediate code to *normal form*. This step is meant to reduce the complexity in later compiler stages by replacing complex parallel constructs with simpler parallel constructs by making

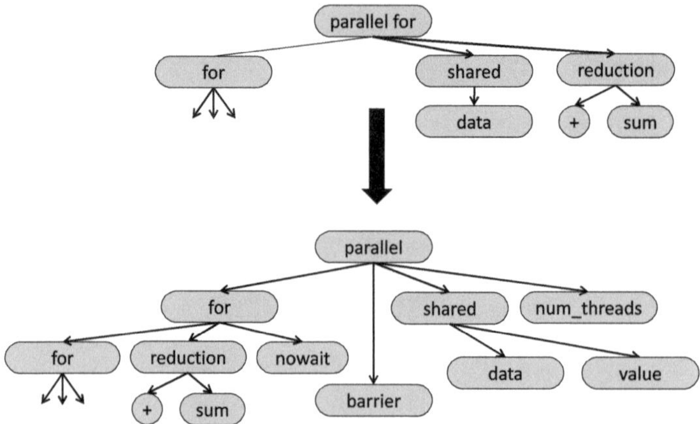

Figure 4.6: AST of Figure 4.5 with split parallel constructs and explicit clauses.

implicit constructs explicit and by reducing the number of parallel constructs that have to be supported in later code-generation phases.

Figure 4.6 shows an example transformation that might happen in an OpenMP compiler. The combined `parallel for` construct has been replaced by its two components, `parallel` and `for`. The simplification also added an explicit `num_threads` clause to the `parallel` AST node to explicitly determine the number of threads to be used while executing the resulting parallel region at runtime. The `for` construct received a `nowait` clause and the implicit barrier was converted to an explicit barrier at the end of the `for` construct.

The next stage in the middle-end implements specific optimizations that the compiler can make to exploit properties of the compiled parallel code. Some simple examples involve barrier fusion that merges consecutive barriers with no user code between them to avoid excessive thread synchronization. Parallel-region expansion [32] detects opportunities to fuse sequences of parallel and (maybe) sequential regions to reduce overhead that is associated with forking and joining parallel regions. Other optimizations may involve the analysis of task graphs resulting from the parallel region to create an improved execution pattern (or even deduce an ahead-of-time task schedule [112]), or for the analysis of the parallel code to perform the propagation of properties such as read-only access to variables [32].

Once the parallel phases of the middle-end are complete, the parallelization-specific information in the IR will be removed and, where necessary, calls to the parallel runtime will be inserted. The modified IR can then be optimized by other architecture-neutral middle-end passes before being passed to the back-end for code generation, as described in Section 4.1.

The choice of when to remove parallel information from the IR is still a matter of discussion among compiler writers. If the IR is flattened by removing explicit knowledge of parallelism and inserting runtime calls early, then it is easy to re-use existing optimization passes without change, since they see nothing new. That is clearly beneficial when initially implementing a parallel language, since it reduces the amount of work required. However, it also loses information that could have been useful for later compiler phases, which can result in poorer code being generated. Therefore, as parallel languages become more ubiquitous, there is more interest in passing information about parallelism deeper into the compiler even though this requires changing more code in the compiler.

4.4 Parallel code-generation patterns

Let's have a look at a few examples of code generation patterns that we need for a parallel language in general and specifically to support the OpenMP API. As you can imagine, there are countless code patterns that we could cover, so we will restrict ourselves to the ones that we believe are the most important: getting parallelism started and passing data into parallel regions, parallel loops and sequential regions within a parallel context, and finally, tasking constructs.

4.4.1 Code generation for parallel regions

The usual approach for code generation for multi-threading is to outline the parallel code into a so-called *thunk function* that are then handed to the threads for parallel execution. This is similar to what the C++ compiler does for TBB lambda expressions as described in Section 4.2.1, or the POSIX[*] Thread API that requires programmers to write run functions for the threads to execute [34]. You have seen an example of this in Listing 2.1. Listing 4.4 shows a short code example that a compiler transforms into the code of Listing 4.5.

The compiler takes the parallel region code and moves it to a new function called `__omp_thunk_answer_0()`. References to variables in the parallel region are changed to adhere to the semantics of data sharing in the OpenMP API. A `shared` variable (`value` in the example) becomes a pointer argument of the thunk function and points to the original storage location of the variable. The code generator moves variables that are marked as `private` into the thunk functions, which automatically allocates these variables in the threads' stacks and thus makes them private. A `firstprivate` variable is a hybrid of the two previous cases. The thunk function receives a pointer argument that refers to the original variable and then copies the value from the original storage location to the thread-local stack memory (here `tmp_factor`).

At the initial location of the parallel code, the compiler removes the outlined code and leaves invocations of the runtime system to fork parallel execution (`__omp_invoke_region()`) and to stop parallel execution (`__omp_end_region()`). Note

```c
#include <omp.h>
#include <stdio.h>

void answer(void) {
  float value = 21.0f;
  int factor = 2;
#pragma omp parallel shared(value) firstprivate(factor)
  {
    int thread_id = omp_get_thread_num();
    printf("Thread %d says: the answer is %f\n", thread_id,
           value * factor);
  }
}
```

Listing 4.4: Source code before outlining for parallel execution.

```c
#include <omp.h>
#include <stdio.h>

void __omp_thunk_answer_0(float * value, int * factor) {
  int tmp_factor = *factor;
  int thread_id = omp_get_thread_num();
  printf("Thread %d says: the answer is %f\n", thread_id,
         *value * tmp_factor);
}

void answer(void) {
  float value = 21.0f;
  int factor = 2;
  __omp_invoke_region(&__omp_thunk_answer_0, &value,
                      &factor);
  __omp_end_region();
}
```

Listing 4.5: Code of Listing 4.4 after outlining the code of the parallel region.

that the names `__omp_invoke_region()` and `__omp_end_region()` are placeholders for now. We will see real examples from production compilers later in this chapter.

To enter the parallel region, `__omp_invoke_region()` receives a set of pointers to the thunk function and all `shared` or `firstprivate` variables. It is up to the runtime library to manage threads and then to invoke the outlined thunk in each thread.

```
#include <omp.h>
#include <stdio.h>

void answer(void) {
  float value = 21.0f;
  int factor = 2;
  void __omp_thunk_answer_0(void) {
    int tmp_factor = factor;
    int thread_id = omp_get_thread_num();
    printf("Thread %d says: the answer is %f\n", thread_id,
           value * tmp_factor);
  }
  __omp_invoke_region(&__omp_thunk_answer_0);
  __omp_end_region();
}
```

Listing 4.6: Code of Listing 4.4 after outlining a nested thunk function.

For the sake of completeness, Listing 4.6 shows another way of creating an outlined thunk function. If the compiler supports nested function declarations (see also [42]) or even lambda expressions, then the thread-code generator can outline the thunk function as a nested function or as a lambda function. That way, the handling of data sharing becomes a lot easier for the code generator: instead of explicitly passing pointers, it can rely on the nested function implementation to access the right original variables from the outer scope, and some of the runtime complexity required to pass the arguments of the fork call to the threads can be removed.

4.4.2 Code generation for thread-parallel loops

Code generation for thread-parallel loops comes in two main flavors. First, we will look at how a compiler can emit code that assigns the loop iteration space of a loop to the worker threads of a parallel region. Second, most platforms can execute SIMD instructions, and so we will need to generate code for these units as well.

Let's start with multi-threaded loops. What we want to achieve here is to map the loop iteration space efficiently onto the available threads. Naturally, there will be several options for how the actual mapping can be implemented. This is called *loop scheduling* and will be described in more depth in Chapter 8, where we discuss different schedules and their implementation.

Let's take a simple example that performs an element-wise addition of two arrays a and b into array c (Listing 4.7). Since we already know how the parallel region is

```
void array_sum(double * c, double * a, double * b,
               size_t n) {
#pragma omp parallel shared(a, b, c) firstprivate(n)
#pragma omp for nowait
  for (size_t i = 0; i < n; ++i)
    c[i] = a[i] + b[i];
}
```

Listing 4.7: Element-wise array summation with a worksharing construct.

treated by the compiler from the previous section, we can ignore that aspect of the code-generation pattern.

From a high-level perspective, distributing a loop across a team of threads requires the following steps. First, the loop schedule needs to be set up in case some internal state to keep track of the loop execution must be initialized. Once that has happened, we need a way to hand out chunks of iterations to the worker threads. A chunk is the unit of work assigned to a thread and is the subset of the iterations of the original loop iteration space.

This chunk function is called whenever a thread has completed a chunk (or starts the execution of the loop) and will have to indicate to a thread that the last chunk of work has been finished and that the parallel loop is done.

Finally, a runtime entry point to clean up any internal state and to finalize the parallel loop would be good to have, so that it synchronizes thread execution with a barrier. Although that aspect may be part of the chunk function, it's usually a good idea to keep these things separate.

Conceptually, the resulting code for Listing 4.7 looks like that shown in Listing 4.8. In the `array_sum()` function, the only code that is left is the boilerplate code to invoke the parallel region that we saw in the previous section. In the thunk function, the compiler introduced the aforementioned functions to distribute the loop iteration space of the original loop across the available worker threads.

Everything starts off with the initialization of the loop schedule by calling `__omp_for_init()` to set up the work distribution for the worker threads. The function receives a set of arguments: the lower bound and upper bound of the loop iteration space (here: 0 and n), the stride of the loop iteration space (here, 1 for ++i), a pointer to a variable for the function to return the chunk size that will be used for the loop distribution, and the thread ID of the worker that is calling the function.

The original loop has been rewritten to process only the loop iterations of a loop chunk from the lower bound (lb) to the upper bound (ub) using the increment of the original loop. The lb and ub are set by calling the `__omp_for_get_chunk()` runtime function that returns these values via pointers to these variables. This function also receives the thread ID, as it will have to compute the correct lower and upper bounds

```
void __omp_thunk_array_sum_0(double * c, double * a,
                             double * b, size_t n) {
  size_t lb, ub;
  size_t chunksz;
  size_t tid = omp_get_thread_num();
  __omp_for_init(0, n, 1, &chunksz, tid);
  while (__omp_for_get_chunk(0, n, 1, chunksz, tid,
                             &lb, &ub))
    for (size_t i = lb; i < ub; i += incr)
      (*c)[i] = (*a)[i] + (*b)[i];
  __omp_for_fini(tid);
}

void array_sum(double * c, double * a, double * b,
               size_t n) {
  __omp_invoke_region(&__omp_thunk_array_sum_0,
                      &a, &b, &c, &n);
  __omp_end_region();
}
```

Listing 4.8: Compiler-generated loop-scheduling code for Listing 4.7.

for the calling thread. The function returns a `bool` to indicate whether the `while` loop
that surrounds the rewritten `for` should stop because no work for this thread remains,
and therefore the calling thread can cease to execute the parallel loop.

Finally, when the execution of the parallel loop has completed for the current
thread, it invokes the `__omp_for_fini()` function that cleans up the internal data
structures that may be needed for keeping information about the parallel execution of
the loop. Some compilers slightly adjust this code pattern so that they do not emit the
call to `__omp_for_fini()`, but rather finalize the loop execution in the last invocation
of the `__omp_for_get_chunk()` function when it hands out the last loop chunk.

Another way to assign loop iterations to threads is via tasks. As we know, this is
done via the `taskloop` construct in the OpenMP API, while the Intel TBB library approach provides a loop template `tbb::parallel_for`. Listing 4.9 shows the example
of Listing 4.7, modified to use the `taskloop` construct. Here, the implementation creates a task for each of the resulting loop chunks and adds the created task to the task
pool.

There are many implementation strategies. First, the compiler can treat the task
loop as syntactic sugar for a loop structure like the one that Listing 4.9 shows at the
bottom. The `taskloop` construct has been removed and has been replaced by code
to split the original iteration space of the task loop into loop chunks that run from

```c
// original code:
void array_sum(double * c, double * a, double * b,
               size_t n) {
#pragma omp parallel shared(a, b, c) firstprivate(n)
#pragma omp master
  {
#pragma omp taskloop
    for (size_t i = 0; i < n; ++i)
      c[i] = a[i] + b[i];
  }
}


// transformed code:
void array_sum(double * c, double * a, double * b,
               size_t n) {
#pragma omp parallel shared(a, b, c) firstprivate(n)
#pragma omp master
  {
    size_t grainsize = __omp_default_grainsize();
    size_t num_tasks = n / grainsize;
    size_t extra_iterations = n % grainsize;

    for (size_t tmp = 0; tmp < num_tasks; ++tmp) {
      size_t lb = tmp * grainsize;
      size_t ub = (tmp + 1) * grainsize;
#pragma omp task firstprivate(lb, ub) shared(a, b, c)
      for (size_t i = lb; i < ub; ++i)
        c[i] = a[i] + b[i];
    }
    if (extra_iterations) {
      size_t lb = n - extra_iterations;
      size_t ub = n;
#pragma omp task firstprivate(lb, ub) shared(a, b, c)
      for (size_t i = lb; i < ub; ++i)
        c[i] = a[i] + b[i];
    }
  }
}
```

Listing 4.9: Element-wise array summation using a task-loop construct.

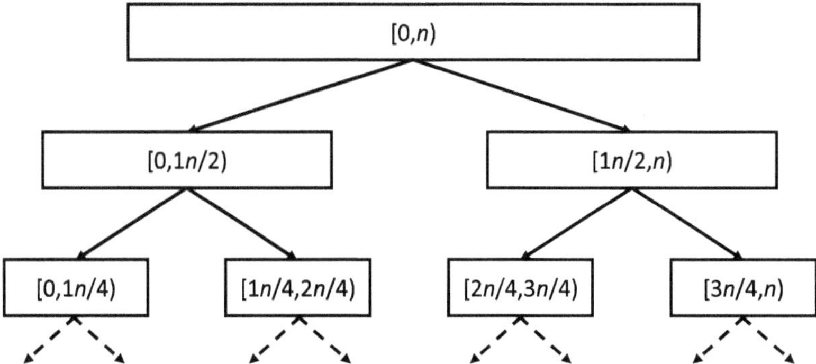

Figure 4.7: Recursive binary loop splitting to create tasks for parallel loop.

lb (lower bound) to ub (upper bound). The size of the chunks is determined by the grainsize variable that is set to some arbitrary default value. An outer loop creates loop chunk after loop chunk by computing the respective lb and ub values for the current loop chunk and creates a number of them determined by num_tasks. If the grain size does not evenly divide the number of loop iterations, a remainder task is created that processes the remaining loop iterations in a smaller chunk with between one and grainsize-1 iterations.

Another, slightly more complex strategy could save the creation of the remainder task by assigning one extra iteration to the first couple of tasks until all extra loop iterations have been distributed. Binary splitting is another popular one (also used by TBB for tbb::parallel_for). Assuming that the iteration space has $n = 2^m$ iterations, with m a positive integer number, the loop iteration space $[0, n)$ is split into two intervals of $[0, n/2)$ and $[n/2, n)$. The generated code then creates a task for each interval. The splitting and task-creation process continues until the interval is close to the desired grain size (see Figure 4.7). At this point, the implementation switches to loop tasks that execute their respective chunk of the loop iteration space. If the loop does not have $n = 2^m$ iterations, then some of the created chunks may be smaller on each of the recursion levels.

4.4.3 Code generation for SIMD-parallel loops

Let's switch subjects a bit and move from thread-parallel loops to SIMD-parallel loops. Code generation for SIMD units is usually done by what compiler writers call *unroll-and-jam* [4, 20]. As usual, there are other ways to do this, too. Because this is not a compiler book, we will keep things short and only discuss unroll-and-jam as one possible approach.

We'll use the example code in the upper part of Listing 4.10 as the example to explain the concept. This is the same element-wise sum that we have used for parallel

```
// original code:
void array_sum(double * c, double * a, double * b,
               size_t n) {
#pragma omp simd simdlen(4)
  for (size_t i = 0; i < n; ++i)
    c[i] = a[i] + b[i];
}


// unrolled version:
void array_sum(double * c, double * a, double * b,
               size_t n) {
  size_t ub = n - (n % 4);
  for (size_t i = 0; i < ub; i += 4) {
    c[i + 0] = a[i + 0] + b[i + 0];
    c[i + 1] = a[i + 1] + b[i + 1];
    c[i + 2] = a[i + 2] + b[i + 2];
    c[i + 3] = a[i + 3] + b[i + 3];
  }
  for (size_t i = ub; i < n; ++i)
    c[i] = a[i] + b[i];
}
```

Listing 4.10: Code for element-wise array summation targeting a SIMD machine.

loops. The simd directive instructs the compiler to generate a SIMD version of the loop for a machine that has four-wide SIMD registers.

The code at the bottom of the same listing shows the unrolled version of the array-sum example that is produced by the *unroll* phase of the compiler. In the example, we have unrolled the loop by four to target the Intel AVX2 instruction set architecture, which supports four double-precision values for each vector register. The main loop now runs with a stride of four to match the unroll factor. Before the loop, a new upper bound (ub) is computed so that the main loop's iteration space is a multiple of the unrolling factor. The remaining iterations are then processed in a second loop once the main loop has finished. This is called the *remainder loop*.

The body of the main loop is replicated several times so that each iteration of the unrolled loop processes four elements of the arrays. The code in the remainder loop is the same as before, as the remainder loop stays a non-SIMD loop in our simple example. Note that sophisticated compilers will also attempt to create a SIMD version of the remainder loop.

We now switch to assembly code in Listing 4.11, as we need to look at the *jam* phase of the code generation. For the sake of brevity, we only show parts of the assembly code that is generated for the main loop by clang, version 9.0.0.

```
# assembly code of the unrolled version:
array_sum:
  # function prologue omitted
.LBB0_2:
  vmovsd  ymm0, qword ptr [rsi + 8*rbx]
  vaddsd  ymm0, ymm0, qword ptr [rdx + 8*rbx]
  vmovsd  qword ptr [rdi + 8*rbx], ymm0

  vmovsd  ymm0, qword ptr [rsi + 8*rbx + 8]
  vaddsd  ymm0, ymm0, qword ptr [rdx + 8*rbx + 8]
  vmovsd  qword ptr [rdi + 8*rbx + 8], ymm0

  vmovsd  ymm0, qword ptr [rsi + 8*rbx + 16]
  vaddsd  ymm0, ymm0, qword ptr [rdx + 8*rbx + 16]
  vmovsd  qword ptr [rdi + 8*rbx + 16], ymm0

  vmovsd  ymm0, qword ptr [rsi + 8*rbx + 24]
  vaddsd  ymm0, ymm0, qword ptr [rdx + 8*rbx + 24]
  vmovsd  qword ptr [rdi + 8*rbx + 24], ymm0

  add       rbx, 4
  cmp       rbx, rax
  jb        .LBB0_2
.LBB0_3:
  # remainder loop & function epilogue
```

Listing 4.11: Assembly code of the array code of Listing 4.10.

The assembly code shows four blocks, each of which corresponds to one line of the unrolled body of the main loop. The first block computes $c[i + 0]$ by first loading $a[i + 0]$ through the movsd instruction that loads (mov) a scalar value (s) of double-precision type (d). It then uses the addsd instruction to also load $b[i + 0]$ as a memory operand and add it to the value in register ymm0. Finally, the block ends with a store operation of ymm0 back to memory at the address of $c[i + 0]$. The same block repeats with different increments of eight bytes for the offset to memory to process $c[i + 1]$, $c[i + 2]$, and $c[i + 3]$.

The addq, cmpq, and jne instructions implement the increment of the loop counter and the test of the loop condition, including the branch to iterate if the counter has not yet reached the last iteration.

In Listing 4.12, the code remains unchanged but the assembly instructions have been sorted so that similar operations on the arrays are now placed next to each other.

```
# assembly code of the unrolled version, sorted:
array_sum:
  # function prologue omitted
.LBB0_2:
  vmovsd   ymm0, qword ptr [rsi + 8*rbx]
  vmovsd   ymm0, qword ptr [rsi + 8*rbx + 8]
  vmovsd   ymm0, qword ptr [rsi + 8*rbx + 16]
  vmovsd   ymm0, qword ptr [rsi + 8*rbx + 24]

  vaddsd   ymm0, ymm0, qword ptr [rdx + 8*rbx]
  vaddsd   ymm0, ymm0, qword ptr [rdx + 8*rbx + 8]
  vaddsd   ymm0, ymm0, qword ptr [rdx + 8*rbx + 16]
  vaddsd   ymm0, ymm0, qword ptr [rdx + 8*rbx + 24]

  vmovsd   qword ptr [rdi + 8*rbx], ymm0
  vmovsd   qword ptr [rdi + 8*rbx + 8], ymm0
  vmovsd   qword ptr [rdi + 8*rbx + 16], ymm0
  vmovsd   qword ptr [rdi + 8*rbx + 24], ymm0

  add      rbx, 4
  cmp      rbx, rax
  jb       .LBB0_2
.LBB0_3:
  # remainder loop & function epilogue
```

Listing 4.12: Sorted assembly code of the unrolled code of Listing 4.11.

The first group of instructions loads the four elements of the a array. The next group performs the addition operations and the last group stores the results to the c array.

The next step is to jam the instruction groups, such that the first four load instructions are merged into a single SIMD load instruction (movpd, p for a packed four-element SIMD vector). The same happens for the next group, in which the addsd instructions are replaced with a single addpd instruction that performs an element-wise addition of two four-wide SIMD registers. The final movpd writes back four elements of the c array in one go.

4.4.4 Code generation for sequential constructs

Now that we have been going parallel for a while, let's go sequential again. Some parallel programming languages offer features that enable you to have a short sequential

```
# assembly code of the unrolled version:
array_sum:
  # function prologue omitted
.LBB0_2:
  vmovupd ymm0, ymmword ptr [rsi + 8*rbx]
  vaddpd  ymm0, ymm0, ymmword ptr [rdx + 8*rbx]
  vmovupd ymmword ptr [rdi + 8*rbx], ymm0

  add       rbx, 4
  cmp       rbx, rax
  jb        .LBB0_2
.LBB0_3:
  # remainder loop and function epilogue
```

Listing 4.13: Jammed SIMD code of the unrolled code of Listing 4.12.

region within a parallel section of code. There are usually two flavors of this: execute on the main thread or execute on any one thread. In OpenMP code, this is handled by the master (or the masked construct) and single constructs, respectively.

The master construct is syntactic sugar for comparing the encountering thread's ID returned by omp_get_thread_num() with that of the main thread. It is up to the implementation to determine how the main thread is identified. The slightly more flexible masked construct with the filter clause is similar, but requires us to compare the thread ID with the result of the evaluation of the expression in the filter clause.

Listing 4.14 shows the code pattern used to implement the OpenMP master construct (code fragment at the top). It is replaced by an if statement that determines the thread ID and only executes the code of the master construct if the ID is zero (code in the middle).

Some implementations choose to have a slightly more complex code pattern that is shown as the third code fragment in Listing 4.14. The compiler now emits calls into the parallel runtime library to enter (__omp_enter_master()) and leave (__omp_leave_master()) the code region of the master construct. The first function checks the thread ID within the runtime system and returns a bool that informs the thread whether it may enter the code region (value is true) or skip it when the return value is false. The call to __omp_leave_master() then finishes the execution of the region.

The generated code for the single construct executes a code region with any thread (but only once) is quite similar (see Listing 4.15). As with the master construct, the generated code calls a runtime function (__omp_enter_single()) to determine whether the current thread is allowed to enter the code region. When the thread finishes the execution of the code region, the region is completed by calling the runtime entry point __omp_leave_single().

```c
// original code:
void main_thread_only() {
#pragma omp master
  {
    printf("This code only runs in the main thread!\n");
  }
}

// translated code (simple):
void main_thread_only() {
  if (omp_get_thread_num() == 0)
    printf("This code only runs in the main thread!\n");
}

// translated code (more complex):
void main_thread_only() {
  bool ok_to_enter = __omp_enter_master();
  if (ok_to_enter) {
    printf("This code only runs in the main thread!\n");
  }
  __omp_leave_master();
}
```

Listing 4.14: Implementing the master construct.

```c
// original code:
void single_construct() {
#pragma omp single
  {
    printf("This code only runs on one of the threads!\n");
  }
}

// translated code:
void single_construct() {
  bool ok_to_enter = __omp_enter_single();
  if (ok_to_enter) {
    printf("This code only runs on one of the threads!\n");
  }
  __omp_leave_single();
}
```

Listing 4.15: Implementing the single construct.

The more complex pattern of a pair of *enter* and *leave* API routines is useful for an implementation if additional bookkeeping is required when entering and leaving such constructs, or if profiling hooks are required. The implementation of the interfaces for master and single are shown in Section 6.9.

4.4.5 Code generation for static tasking

The OpenMP API offers a feature that is best described as *static tasking*—that is, a set of concurrent tasks that are known at compile time. The OpenMP sections construct introduces a set of statically described tasks, each delimited by a section construct (cf. Section 2.8.1 in [100]) that marks the individual static tasks. See Listing 4.16 for an example of three static tasks. If the parallel region is executed with fewer threads than the number of section constructs, then some of the constructs will be serialized. If more threads than section constructs are available, then some of the threads will continue past the sections construct and either wait at the implicit barrier for all of the threads that are executing sections to finish, or, if a sections construct with the nowait clause was used, start to execute later code.

4.4.6 Code generation for dynamic tasking

TBB handles task creation through the native capabilities of C++ lambda functions and template interfaces to the TBB library. We will now describe how tasking can be implemented in a compiler for a parallel programming language.

We have seen in Section 4.2.1 how a compiler can handle the compilation of lambda functions. If the parallel language offers lambda functions, then the code transformation for tasks is rather obvious. The task region is transformed into a lambda expression that is then passed to a runtime API entry point to schedule the lambda function for the execution as a task. Listing 4.17 shows how this can be done. The code fragment at the top shows the original code for an OpenMP task region with shared, firstprivate, and private variables.

In the lower code fragment, the lambda transformation has been applied. We left the parallel and master constructs for readability. See Section 4.4.1 and Section 4.4.4 for the respective code-generation patterns.

The transformation rules for data-sharing clauses are:

- **shared:** A variable that is declared shared is captured as a reference to the lambda expression.
- **firstprivate:** A firstprivate variable becomes a call-by-value argument of the lambda expression.
- **private:** A private variable is not passed to the lambda function, but a local variable in the body of the lambda function is created.

```
// original code:
void parallel_sections() {
#pragma omp parallel
#pragma omp sections
#pragma omp section
  { // section 1
    code_for_section_1();
  }
#pragma omp section
  { // section 2
    code_for_section_2();
  }
#pragma omp section
  { // section 3
    code_for_section_3();
  }
}


// translated code:
void parallel_sections() {
#pragma omp parallel
#pragma omp for schedule(static, 1)
  for (auto tmp = 0; tmp < 3; ++tmp)
    switch (tmp) {
    case 0:
      code_for_section_1();
      break;
    case 1:
      code_for_section_2();
      break;
    case 2:
      code_for_section_3();
      break;
    }
}
```

Listing 4.16: OpenMP static tasks and corresponding code generation pattern.

Some base languages (or your parallel programming language) may not support lambda functions (e. g., neither C nor Fortran has such support at the time of writing). This makes the situation slightly more complex, and a compiler must emulate the concepts of lambda expressions. Ultimately, the code transformation is very similar, albeit involving a bit more low-level code to deal with moving data around.

```cpp
#include <iostream>
#include <omp.h>

// before task outlining:
void answer() {
  float value = 21.0f;
  int factor = 2;
  int thread_id = -1;
#pragma omp parallel
#pragma omp master
  {
#pragma omp task shared(value) firstprivate(factor) \
                 private(thread_id)
    {
      thread_id = omp_get_thread_num();
      std::cout << "Thread " << thread_id
                << " says: the answer is "
                << (value * factor) << std::endl;
    }
  }
}

// after task outlining:
void answer() {
  float value = 21.0f;
  int factor = 2;
  int thread_id = -1;
#pragma omp parallel
#pragma omp master
  {
    auto __omp_thunk_answer_0 = [&value, factor]() mutable {
      int thread_id = omp_get_thread_num();
      std::cout << "Thread " << thread_id
                << " says: the answer is "
                << (value * factor) << std::endl;
    };
    __omp_create_task(__omp_thunk_answer_0);
  }
}
```

Listing 4.17: Transforming a task construct into a lambda expression.

```cpp
int32_t __omp_answer_thunk_0(char * data) {
  float * value;
  int factor;
  int thread_id;
  memcpy(&value, data, sizeof(float *));
  memcpy(&factor, data + sizeof(float *), sizeof(int));

  thread_id = omp_get_thread_num();
  std::cout << "Thread " << thread_id
            << " says: the answer is "
            << (*value * factor) << std::endl;
  std::flush(std::cout);

  return 0;
}

void answer() {
  float value = 21.0f;
  int factor = 2;
  int thread_id = -1;
#pragma omp parallel
#pragma omp master
  {
    char * data =
        __omp_data_alloc_task(sizeof(float *) + sizeof(int));
    float * value_ptr = &value;
    memcpy(data, &value_ptr, sizeof(float *));
    memcpy(data + sizeof(float *), &factor, sizeof(int));
    __omp_task_create(__omp_answer_thunk_0, data);
  }
}
```

Listing 4.18: Transforming a task construct into low-level runtime entry points.

For the `answer()` function in Listing 4.17, an OpenMP compiler might generate the code pattern of Listing 4.18. The code of the task has been moved to a thunk function, which is then called indirectly by calling the runtime system's API. In our example, the compiler generated two calls to the runtime system. First, `__omp_data_alloc_task()` to allocate memory to store data that the task needs to receive. The argument indicates how much memory needs to be allocated. Second, a call to `__omp_task_create()` to enqueue the task in the task pool. This call passes the pointer to the allocated memory and a pointer to the thunk function to the runtime system so that it can ultimately invoke the outlined function with the correct arguments.

We have to slightly change the transformation rules for data-sharing clauses now so that they are as follows:
- **shared:** For a variable that is declared shared, the compiler generates code to take the address of that variable and copy the pointer to the task's data portion.
- **firstprivate:** A firstprivate variable is copied into the data portion of the task.
- **private:** A private variable is not passed to the task via the data portion, but a local variable in the body of the thunk is created.

According to these rules, the compiler takes the address of the shared variable value in value_ptr and then copies the pointer to the task's data portion. The factor variable is copied by value, as it is marked firstprivate. As before, the thread_id becomes a variable that is declared in the thunk function as a local variable because it is a private variable. When storing a variable into the data area of the task, the compiler assigns a slot for each variable. It will respect the natural alignment of the data type stored, e. g., eight bytes for the type double in C/C++. Each slot is offset by the accumulated sizes of the preceding variables in the data area. In our example, this is an offset of sizeof(float *) bytes (usually eight on a 64-bit architecture) for the second variable factor.

In the example of Listing 4.18, we use memcpy() to indicate the movement of data from the original variables to the created task's data portion. This is to make the example easier to follow; most compilers will, of course, generate appropriate intermediate code and then assembler code to perform the copy inline. Smart compilers may recognize our uses of memcpy() as a built-in function and replace them with load/store operations for primitive types like int, or even for simple structured types.

In Section 2.3.3, we saw mechanisms to synchronize task execution by making tasks await the completion of other tasks. We will now describe the taskwait and taskgroup compiler transformations.

Generating code for an OpenMP taskwait construct (or a similar construct in other parallel programming languages) is quite simple. The construct

```
#pragma omp taskwait
```

translates to the invocation of a runtime function to wait for the completion of all child tasks of the current task:

```
__omp_taskwait();
```

The taskgroup construct is slightly more complex, as it is a scoped synchronization construct. At the end of the construct, the encountering task must wait for the completion of all descendant tasks that have been created within the scope of the taskgroup construct. As a consequence, the compiler must generate an enter call to inform the runtime about the start of a taskgroup region and a leave call to inform the runtime that this special region has ended and it needs to wait for the completion of tasks. In that regard, the pattern is similar to the one we saw for the single construct.

A code fragment like

```
#pragma omp taskgroup
  {
    some_code_that_creates_tasks();
  }
```

is transformed by the compiler into something like this:

```
__omp_enter_taskgroup();
  some_code_that_creates_tasks();
__omp_leave_taskgroup();
```

Note that all the magic for keeping track of the created descendant tasks and the implementation of the waiting mechanism is part of the runtime system and so the compiler does not need to generate any extra code for this purpose.

4.5 Example OpenMP implementations

In the previous section, we discussed how to implement some of the features of the OpenMP language in a rather abstract way by using C examples and giving a high-level overview of the code generation patterns of the compiler. Now that we have an idea of how things will work, we can look in more detail at three actual implementations: the GNU Compiler Collection [40], the Intel compiler [61], and the LLVM compiler [84].

4.5.1 GNU compiler collection

Unfortunately, GCC's implementation of the OpenMP extensions in the compiler is not very well documented, except in the code itself. Some may say that this is documentation enough, but it makes it more difficult to extract the relevant pieces for a project like this book.

Each of the language front-ends (for C, C++, and Fortran) parses the OpenMP parallelization directives and augments the abstract syntax tree with them. As part of what GCC calls *gimplification*, the AST is translated into a higher-level intermediate code called *GIMPLE*, which the compiler then feeds into the middle-end. It also performs architecture-independent optimizations before handing the code off to the back-end for architecture-specific optimizations and code generation.

During the gimplification process, GCC lowers the AST representation of parallelism, discovers what variables are used from parallel regions, and explicitly adds data-sharing clauses for those variables, as we discussed in Section 4.3. Listing 4.19 shows the GIMPLE code after the AST has been transformed and the compiler has made the data-sharing clauses explicit. We have slightly reformatted the code. If you

```
answer ()
{
  float value; int factor;
  value = 2.1e+1; factor = 2;
  {
    .omp_data_o.1.factor = factor;
    .omp_data_o.1.value = value;
    #pragma omp parallel firstprivate(factor) \
                          firstprivate(value) ...
      {
        .omp_data_i = (struct .omp_data_s.0 &...)
          &.omp_data_o.1;
        factor = .omp_data_i->factor;
        value = .omp_data_i->value;
        {
          int thread_id;
          thread_id = omp_get_thread_num ();
          D.2703 = (float) factor;
          D.2704 = D.2703 * value;
          D.2705 = (double) D.2704;
          printf ("Thread %d says: the answer is %f\n",
                  thread_id, D.2705);
        }
        #pragma omp return
      }
    .omp_data_o.1 = {CLOBBER};
  }
}
```

Listing 4.19: GCC high-level intermediate code for Listing 4.4.

want to reproduce the output for your code, you can do so with the `-fdump-tree-all` compiler switch.

There are a few things that are worth looking at in the generated code of Listing 4.19. First, the compiler introduced a `struct` to copy the values of the `value` and `factor` variables into the parallel region. In the parallel region, the values are then copied to their respective thread-local variables. Second, the compiler made an analysis and optimization that detected that the `value` could be changed to a `firstprivate` variable instead of a `shared` variable.

The next step outlines the code and creates code to invoke GCC's runtime for multi-threading (*libgomp*, see [41] for the actual runtime interface). Listing 4.20 shows the

```
answer._omp_fn.0 (struct .omp_data_s.0 & restrict
                   .omp_data_i)
{
  double D.2720;
  float D.2719;
  float D.2718;
  int thread_id;
  float value;
  int factor;
  ...
  factor = .omp_data_i ->factor;
  value = .omp_data_i ->value;
  thread_id = omp_get_thread_num ();
  D.2718 = (float) factor;
  D.2719 = D.2718 * value;
  D.2720 = (double) D.2719;
  printf ("Thread %d says: the answer is %f\n", thread_id,
         D.2720);
  return;

}

answer ()
{
  int thread_id;
  int factor;
  float value;
  struct .omp_data_s.0 .omp_data_o.1;
  ...
  value = 2.1e+1;
  factor = 2;
  .omp_data_o.1.factor = factor;
  .omp_data_o.1.value = value;
  __builtin_GOMP_parallel (answer._omp_fn.0, &.omp_data_o.1,
                           0, 0);
  .omp_data_o.1 = {CLOBBER};
  return;
}
```

Listing 4.20: GCC low-level intermediate code for Listing 4.4.

final output of these stages. We have again reformatted the code and shortened it slightly (indicated by . . .) to fit the format of the book.

As Listing 4.20 shows, the compiler has outlined the code into a thunk function (`answer._omp_fn.0()`), which receives a pointer to the `struct` that Listing 4.19 already showed. In the original code location, the compiler generates code to allocate the `struct` on the stack of the main thread, fill in the slots for the arguments of the parallel region, and then pass it to `__builtin_GOMP_parallel()`. This function is a placeholder name for `GOMP_parallel_start()` from the *libgomp* library and will be replaced with the actual function call when the back-end of GCC performs the final code generation to produce assembly code.

4.5.2 Intel compilers and the LLVM compiler

We will discuss the Intel and LLVM compilers in one section because they are similar, since they target a common OpenMP runtime system and thus produce comparable code that invokes the same entry points for parallel execution. However, they differ wildly in the internal design of their respective compiler stages.

At the time of writing, the Intel compiler uses an intermediate representation (called *ILO*) that supports explicit parallelism in the intermediate code. Future versions of the Intel compilers will be based on LLVM, but retain a similar late processing of parallel constructs. The Intel compiler has no middle-end, but instead performs the lowering from OpenMP directives to the parallel ILO code in the front-end [135]. The parallel intermediate code is then given to the back-end for optimization [136] before producing multi-threaded code for the remaining passes in the back-end.

Because the Intel compiler's intermediate code is aware of the parallelism in the code, some of the early optimization passes before the parallelism is finally removed can perform optimization on the parallel code and maintain parallel semantics. These optimizations include inlining, code restructuring, constant propagation, and profile-guided optimization.

While working on the parallel code, the compiler builds a hierarchical graph with what Intel calls *thread entry* and *thread exit* nodes while building the graphs to analyze the intermediate code. With these nodes added, later optimization passes do not only see function calls to outlined code, but still have all information about parallel execution in structures like the control flow graph. This helps the compiler conduct advanced analysis passes to detect parallel loop structures, cache `threadprivate` variables, inter-procedural optimizations, and code generation for SIMD instructions [135].

In contrast, as of version 9.0.0, the `clang` compiler removes OpenMP directives before the intermediate code is generated. Similar to the Intel compiler, the `clang`/LLVM compiler does not have an explicit middle-end to support parallel languages. The front-end performs the lowering of the parallel code and produces LLVM byte code

```
TranslationUnitDecl 0x15ae438 ...
| ...
`-FunctionDecl 0x16db0e8 ... answer 'void (void)'
  `-CompoundStmt 0x16dbdc0 ...
    |-DeclStmt 0x16db228 ...
    | `-VarDecl 0x16db1a0 ...
    |   `-FloatingLiteral 0x16db208 ... 'float' 2.100000e+01
    |-DeclStmt 0x16db2e0 ...
    | `-VarDecl 0x16db258 ...
    |   `-IntegerLiteral 0x16db2c0 ... 'int' 2
    `-OMPParallelDirective 0x16dbd80 ...
      |-OMPSharedClause 0x16db318 ...
      | `-DeclRefExpr 0x16db2f8 ... 'value' 'float'
      |-OMPFirstprivateClause 0x16db4f8 ...
      | `-DeclRefExpr 0x16db338 ... 'factor' 'int'
      `-CapturedStmt 0x16dbce8 ...
        |-CapturedDecl 0x16db620 ... nothrow
        | |-CompoundStmt 0x16dbb78
        |     ... openmp_structured_block
        | | ...
        | |-ImplicitParamDecl 0x16db690
        |     ... implicit .global_tid. ...
        | |-ImplicitParamDecl 0x16db6f8
        |     ... implicit .bound_tid. ...
        | |-ImplicitParamDecl 0x16db788
        |     ... implicit __context ...
        | ...
```

Listing 4.21: The clang compiler's simplified AST for the code in Listing 4.4.

(LLVM's intermediate representation) that invokes the parallel runtime system. This intermediate code is then given to the back-end for further processing and optimization. However, the LLVM project is undertaking an effort to change the LLVM intermediate representation to retain more information about parallelism and to move the processing of parallel constructs in the AST to an intermediate step similar to a middleend, so in future versions information about parallelism will propagate further into LLVM, allowing more parallel-specific optimizations, such as those described in [38].

Listing 4.21 shows the abstract syntax tree that clang creates for the code in Listing 4.4. For brevity, we have removed a few parts of the structure (marked by . . .). You can easily spot the nodes in the tree that correspond to the OpenMP parallelization directives in the code: OMPParallelDirective for parallel and OMPSharedClause and OMPFirstprivateClause for shared and firstprivate, respectively.

```
define dso_local void @answer() #0 {
  ...
  store float 2.100000e+01, float* %value, align 4
  store i32 2, i32* %factor, align 4
  %0 = load i32, i32* %factor, align 4
  %conv = bitcast i64* %factor.casted to i32*
  store i32 %0, i32* %conv, align 4
  %1 = load i64, i64* %factor.casted, align 8
  call void (%struct.ident_t*, i32,
            void (i32*, i32*, ...)*, ...)
      @__kmpc_fork_call(... @.omp_outlined. ...,
      float* %value, i64 %1)
  ret void
}

define internal void @.omp_outlined.(...,
        float* dereferenceable(4) %value,
        i64 %factor) #1 {
  ...
  %value.addr = alloca float*, align 8
  %factor.addr = alloca i64, align 8
  %thread_id = alloca i32, align 4
  ...
  store float* %value, float** %value.addr, align 8
  store i64 %factor, i64* %factor.addr, align 8
  ...
  %call = call i32 @omp_get_thread_num()
  store i32 %call, i32* %thread_id, align 4
  ...
  %conv1 = sitofp i32 %3 to float
  %mul = fmul float %2, %conv1
  %conv2 = fpext float %mul to double
  %call3 = call i32 (i8*, ...) @printf(...)
  ret void
}

...
```

Listing 4.22: LLVM byte code as generated by the clang compiler for Listing 4.4.

After all semantic checks have passed, the `clang` compiler produces LLVM byte code from the AST representation. At this stage, the compiler also removes all OpenMP constructs from the code and produces byte code that invokes the runtime system to create parallelism when the program is executed. Listing 4.22 shows the resulting LLVM byte code for Listing 4.4. Again, we have removed a few bits here and there (marked with ...) to reduce the complexity of the presented code.

As you can see from Listing 4.22, the `clang` compiler has created an outlined function `.omp_outlined.` that contains the code of the parallel region. You can still spot the call to `printf` in there. At the origin of that outlined code, the byte code now contains a call to `__kmpc_fork_call()`, which is the runtime's entry point to create a team of threads for parallel execution. In contrast to GCC's implementation, the call is similar to Listing 4.5 and passes `shared` and `firstprivate` variables as arguments to the runtime API using the calling convention of the base language.

4.6 Conclusions

This chapter has scratched the surface of how to implement the application-facing layers of a parallel programming model. We wish we could have covered all compiler-related topics to the extent and at the level of detail that this interesting topic deserves. That said, there are tons of research papers about library and compiler approaches to parallelism, so one can easily spend a lifetime trying to read up on everything.

In the next chapter, we change subjects and dive deeper into the software stack and the implementation of a parallel runtime. We are going to talk about memory management, which is a very important topic not only for application programmers, but also for runtime implementers. Memory is undoubtedly a precious (and sometimes scarce) resource. Being smart about how to manage allocation and deallocation in a parallel runtime while trying to reduce overhead in both the memory footprint and overhead will be key for a fast parallel runtime.

5 Fundamental parallel runtime mechanisms

Underlying most parallel runtimes, there are some fundamental mechanisms on which the functionality required by the programming language is implemented. These are components that can be used in different places and will be called by the higher-level interfaces that are directly targeted by the compiler, and they will form the basis of the remainder of this book. We will cover these fundamental topics in this chapter.

There are three topics that this chapter will explore. First, how to manage the parallelism that is exposed by the parallel programming model and the programmer. Efficiently spawning parallelism is key to performance, especially in the case of fork/join programming models. Second, the runtime system needs to lay out the parallelism and factor in the structure of the machine. Most notably, NUMA-aware parallelism and stopping the OS from arbitrarily moving threads around is an important concept. Third, and lastly, the runtime system needs to maintain memory to keep important information about its state. Memory management and memory layout are thus some very important topics in this chapter.

5.1 Managing parallelism

As we saw in Chapter 2, parallel programming models not only create parallelism, but also offer many ways of synchronizing threads or tasks with other threads or tasks, and offering constructs to reduce parallelism for a short period. Thus, a parallel runtime system has to support these patterns and provide efficient implementations for them.

5.1.1 Spawning parallelism

In Section 2.1.1, we saw code that implemented fork/join parallelism by creating and destroying system-level threads. While this is an obvious implementation, for most parallel applications, it is overly costly, since making operating system calls to create threads and determine that they have finished execution is expensive (~28 µs per thread on our Marvell[*] Arm[*] machine).

The OpenMP[*] API also has requirements about the persistence of per-thread state—also called *thread-local storage* (TLS)—that are hard to meet if operating system threads are constantly being created and destroyed. While it would be possible to have the OpenMP runtime library provide its own implementation of thread-local storage, it is much more efficient to use the mechanisms that are provided by modern compilers and operating systems, since these entirely remove the need for any calls into a runtime library.

If we have a two-socket machine with 32 cores per socket and start one thread per core (remembering that one thread must already be running), then that will take 63

https://doi.org/10.1515/9783110632729-005

times 28μs or about 1.8 ms. This is clearly untenable from a performance point of view, since many parallel regions will run for significantly less time than this. Therefore, we are driven to maintain a *thread pool*, in which we create threads once and then keep them around, even during the periods when the language semantics only require a single thread to be executing.

As we will discuss in Chapter 7, we may be able to significantly reduce the time to enter and leave a parallel region when we do not have to invoke the operating system to create and destroy threads, reducing it to ~ 1 μs (i. e., more than a 1500× faster) at 64 threads on the same machine.

However, if we maintain a thread pool, then all of our threads remain in existence even when the semantics of the program only require a single thread. We must therefore consider how to park these threads when they have no work to do in a way that releases their hardware resources at a low level (e. g., SMT resources on architectures with SMT support), and, ideally, also lets the operating system know that they have no work to do so that it can move the hardware to a lower-energy state, or use the logical core to perform other useful work. This is more important in a shared environment than in one where a single job owns a whole slice of a machine, since when the resources are not shared, it is less likely that there's other work that could be executed.

This leads us to our next topic, which is how a thread should wait.

5.1.2 Waiting

At first glance, waiting seems a really simple thing to do. Overall, it just means doing nothing, right? However, somehow a thread needs to be notified when it is time to wake up and resume useful work. One way to implement this could be to send a thread into a tight loop polling a single cache line until a flag is modified and the thread can continue. Job done.

Unfortunately, things are not that simple because we do not necessarily want a waiting thread to be consuming precious resources that could be more useful to solve the computational task at hand, or not even be used at all. Such resources are things like those inside the processor cores, which we saw in Chapter 3, e. g., ALUs, ROBs, load and store buffers, and the physical register file. Another one is energy (for which someone will have to pay, and whose use has a CO_2 cost). Of course, the hardware resources inside the single core may not be reusable if no other SMT threads can run on the same core. However, if we allow our waiting threads to consume energy, then they will generate heat that can raise the temperature of a whole die/chiplet and cause clock-throttling, which in turn may slow down threads, even those that are running on other cores.

If code is operating in an environment in which other applications are running and sharing the hardware, then releasing logical cores back to the operating system is even more important, since the OS may be able to use the logical core to execute other code.

So, there's an obvious conflict here. On the one hand, we want our thread to be woken with the smallest possible latency as soon as the condition on which it is waiting is satisfied, but on the other hand, we want to release resources as soon as we can so that they can either be shut down to save energy or be used by a thread that has useful work to do.

If we really believe that the wake-up latency of a single thread is critical, then polling in a tight loop is unbeatable, whereas if we believe that we must release resources to the operating system as soon as we can, then the runtime should send the thread to sleep. If the cost of the system call to wake a thread is small, then we would obviously always do that. Alas, this is not the case. We measure the time to wake a thread that is waiting via a Linux[*] futex() system call [130] as ~1.2 µs on our Arm machine. Since the time to wake up a polling thread directly is near to 100 ns, we can see that this has non-trivial overhead.

We can conclude that waiting is more than just making a thread sit somewhere and twiddle its virtual thumbs. Thus, a parallel runtime needs to find a proper trade-off between spin waiting and signal-like waiting.

Some people differentiate spin waiting, or *busy waiting*, (by which they mean testing a condition in a tight loop with no delay) from polling (by which they mean testing a condition in a loop with a delay). We do not make this distinction, since spin waiting is merely the extreme case of polling with no, or minimal, delay. The critical distinction we do make is between polling, where a thread continues to execute and tests the condition itself, and waiting in the kernel where the thread is suspended until woken by another thread which knows that the condition is satisfied.

Some parallel implementations offer a choice to the user by exposing environment variables or interfaces to specify a waiting time. Until the waiting time is over, a thread enters a spin-wait loop and goes to sleep after the waiting time has passed. Another approach taken by parallel runtimes is to have a *throughput* mode and a *performance* mode. In throughput mode, the runtime frees up resources when a thread has to wait so that they can be consumed by other threads (i. e., the thread sleeps in the kernel immediately). In performance mode, the library will keep the threads active to make them as responsive as possible when they have to resume working. The OpenMP API offers the OMP_WAIT_POLICY environment variable to control this behavior.

Section 6.7 in Chapter 6 goes into more detail on the issues involved in implementing waiting.

5.2 Management of parallelism and hardware structure

With the aforementioned capabilities of efficiently managing parallelism in general, a parallel runtime system also needs to take care to map the parallelism onto the available hardware as well as possible.

As we have seen in Chapter 3, hardware has a certain structure in which the components have been composed. There are the physical cores that may have several hardware threads and that may share parts of the cache hierarchy. Then there's the NUMA structure of the system, with memory that has a lower access latency and a higher memory bandwidth.

5.2.1 Detecting the hardware structure

A typical operating system offers interfaces to query the structure of the hardware. Some OSes, like Linux, provide information to the user through files. The Linux kernel exposes the hardware structure and its capabilities (e. g., instruction set architecture) via the `/proc/cpuinfo` file that one can read and parse. However, it is usually much easier and more portable to let other people worry about the gory details of how each operating system describes the hardware on which it runs, and to use a library that someone else has already written to access this information in an architecture-independent and OS-independent way.

The *hwloc* library [99] provides low-level information about the system structure, and also shell commands, which allow you to look at this structure from the command line. In the previous chapter, Figure 3.11 already showed the output of the `hwloc-ls` command. But hwloc also provides a library interface that allows a program to determine the properties of the machine on which it is running without having to understand how each hardware platform and OS encodes that information.

Using a library like this makes it possible to deal sensibly with both NUMA information and information about the mapping of logical cores to physical cores, and cores to the higher levels of the hardware hierarchy (dice, tiles, or packages). Since each physical break introduces extra latency in communication (as we saw in Figure 3.16), it is important to know where individual threads are executing to obtain the best possible performance.

If all that is required is NUMA information and you are prepared to restrict your code to running on Linux, then the *libnuma* library [133] provides convenient access to that information. Listing 5.1 shows a function that prints the NUMA layout of a machine.

The code starts by asking libnuma to return the number of configured NUMA CPUs (in this book's terminology: logical cores) and the number of configured NUMA domains, which libnuma calls *nodes*. The code then constructs a bit mask that will have a bit for each core of the system, which will be set after calling `numa_node_to_cpus()`

```c
#include <stdio.h>
#include <numa.h>

int print_numa_domains(FILE * stream) {
  int ncpus = numa_num_configured_cpus();
  int nnuma = numa_num_configured_nodes();
  struct bitmask * mask = numa_bitmask_alloc(ncpus);

  if (!mask)
    return -1;

  fprintf(stream,
          "This system has %d core%s in"
          "%d NUMA domain%s\n\n",
          ncpus, (ncpus != 1) ? "s" : "", nnuma,
          (nnuma != 1) ? "s" : "");

  for (int n = 0; n < nnuma; ++n) {
    fprintf(stream, "Cores in NUMA domain %d:\n", n);

    if (numa_node_to_cpus(n, mask)) {
      numa_bitmask_free(mask);
      return -1;
    }

    for (int c = 0; c < ncpus; ++c)
      if (numa_bitmask_isbitset(mask, c))
        fprintf(stream, "%d ", c);

    fprintf(stream, "\n");
  }
  fprintf(stream, "\n");

  numa_bitmask_free(mask);

  return 0;
}
```

Listing 5.1: Detecting NUMA structure of a system using libnuma.

if the corresponding core belongs to the queried NUMA domain. The code then checks the bit and prints the core's number if the bit was set. For our Intel[*] Xeon[*] Platinum 8260L Processor that we use throughout this book, the output of the function will look like this (we use ... to indicate that we have shortened the output):

```
This system has 96 cores in 4 NUMA domains

Cores in NUMA domain 0:
0 1 2 3 7 8 12 13 14 18 19 20 48 49 50 51 55 56 60 61 62 ...
Cores in NUMA domain 1:
4 5 6 9 10 11 15 16 17 21 22 23 52 53 54 57 58 59 63 64 ...
Cores in NUMA domain 2:
24 25 26 27 31 32 33 37 38 39 43 44 72 73 74 75 79 80 81 ...
Cores in NUMA domain 3:
28 29 30 34 35 36 40 41 42 45 46 47 76 77 78 82 83 84 88 ...
```

Once the runtime system has queried the system and understands the structure of the machine, it can take this data into account when creating the threads for execution, scheduling work, or allocating memory in the NUMA domains to keep it as local as possible for the intended usage.

5.2.2 Thread pinning

One such optimization that a runtime can apply either automatically or per user request is thread pinning. Even though the OS will typically try to schedule processes or threads on the same cores, from time to time, it may happen that the OS scheduler decides to move a thread around between cores. This can be for wear-leveling purposes, so that all cores are used roughly equally to spread the heat across the processors and thus make sure that they degrade at approximately the same rate. It can also simply happen as a sleeping process or thread wakes up, consumes a few time slices, and then goes back to sleep. A running thread then needs to move while that other process is working.

We have seen that spatial and temporal locality is hugely important to achieving performance, since access to data from a local cache is significantly faster than from one elsewhere in the system. We also saw that remote accesses in a NUMA system can be costly in terms of increased access latency, and reduced memory bandwidth.

For instance, in the OpenMP API, every thread has an integer thread number, or thread ID (returned by `omp_get_thread_num()`), which is in the range of $[0, N_{\text{Threads}})$. It is this value that is used when determining how iterations are mapped to threads in a statically scheduled loop. However, on its own, this index tells us nothing about where in the machine the thread may run, and therefore, without specifying additional information, you cannot assume that thread 0 and 1 are near to each other in

```c
#define _GNU_SOURCE
#include <sched.h>
#include <pthread.h>

static pthread_mutex_t mtx = PTHREAD_MUTEX_INITIALIZER;

static void * run(void * data) {
  int thread_id = * (int *) data;
  unsigned core;

  cpu_set_t cpuset;
  CPU_ZERO(&cpuset);
  CPU_SET(thread_id, &cpuset);
  pthread_setaffinity_np(pthread_self(), sizeof(cpu_set_t),
                         &cpuset);

  pthread_mutex_lock(&mtx);
  printf("Hello World from thread %d on core %d!\n",
         thread_id, sched_getcpu());
  pthread_mutex_unlock(&mtx);
  return NULL;
}
```

Listing 5.2: Pinning threads to cores using the POSIX[*] thread API.

the machine, or even that they are not free to move around the machine at the whim of the operating system.

Therefore, for the highest performance, the runtime can (and should!) take the information it has about the system and use it move the threads to the right places in the system and then keep them there by preventing the OS from moving them around. This is called *thread pinning*. If there is an interface for the user also to control the placement of threads (e. g., the OpenMP `proc_bind` and `OMP_PLACES` features), then the compiler and runtime will have to offer the corresponding interfaces.

Again, the interfaces to instruct the OS to pin threads in the system are specific to the operating system. The libnuma library provides calls, such as `numa_sched_setaffinity()`, that take a bit mask that has a bit set for each of the cores on which a thread may be scheduled by the operating system. Another option is the pthread-specific `pthread_setaffinity_np()` function that is illustrated in List-ing 5.2, in which we're showing a slightly extended `run()` function that we used in Listing 2.1.

The code in Listing 5.2 takes a thread ID that is passed via the `data` argument of the `run()` function (the main thread has to create these IDs upfront when the threads are created). The code then constructs a cpuset of type cpu_set_t, similar to the bit mask in Listing 5.1. In this mask, a set bit indicates that the OS can schedule the thread on that logical core; if the bit is unset, then that logical core is not available for scheduling the thread. To simplify things, we assume that we can pick any core of the available system and we use the thread ID to pin the thread to a core with that same number. Finally, the printf() shows the thread ID and asks the Linux kernel about the currently active core for the thread by calling the sched_getcpu() system call.

5.3 Memory management in parallel runtime systems

Memory management in a parallel runtime system is a very important topic that can quite dramatically influence the performance when a parallel program is being executed. In this section, we will explore how to deal with this fact and how to efficiently allocate memory for all sorts of data structures that a parallel runtime system will need.

5.3.1 Memory efficiency and cache usage

Every (parallel) runtime system usually needs some memory for its own consumption. It needs to keep an internal state, maintain arrays and lists of things (e. g., the threads it is using for execution), hash tables for rapidly looking up stuff, and various other things that we will cover in the remaining chapters of this book.

From Chapter 3, we know that modern processors typically have multiple levels of caches to reduce memory latency. As the internal data that the runtime system needs is also flowing through the caches of the processors, the runtime system needs to be very aware of its own memory footprint and how it interacts with the cache. The goal is to reduce capacity misses (or even conflict misses) in the caches, which will evict data used by the application and thus impact the application code's performance. Thus, keeping unnecessary data or touching data that is not strictly needed for the functioning of the runtime system will cause performance issues.

Since caches are organized in cache lines, data structures that contain data that is accessed from multiple cores need to be placed such that they are in different cache lines. This will help to avoid false sharing, when they are accessed from different threads on different cores. Proper alignment of data to minimize cache use and avoid false sharing requires that the layout of the individual data elements within a class or struct is considered, as well as the alignment of the whole object itself. At the same time, data that is needed on the same (physical) core should be represented as densely as possible to exploit spatial locality in the caches and avoid unnecessary cache-line motion by making sure that a single line contains related fields that will be accessed together.

If the system has multiple NUMA domains, then the runtime system should take data placement into consideration so that what is needed on one processor is placed in the NUMA domain(s) that are associated with that processor package. With the first-touch policy in mind, a runtime system must initialize data in parallel so that threads that will later access the data on one processor have touched the data on the very same processor during the initialization phase of the runtime. For instance, the thread objects to keep track of a thread should be initialized by the thread which it describes, rather than the main thread that started initialization of the runtime.

5.3.2 Single-threaded memory allocators

After these rather general remarks, let's have a look at how to do memory management. The task of memory management is to provide memory to the runtime system and the application code. Memory management can either be done manually via the new and delete operators in C++, the malloc() and free() functions in C, or automatically via garbage collectors that save programmers from having to free memory that has been allocated. In any case, memory management involves handing out memory when requested and taking it back to the pool of available memory when it is no longer needed. The component that takes care of this is the *memory allocator*.

While we are interested in parallel runtime systems, it is still instructive to look at memory allocators for single-threaded cases, as they are the foundation upon which memory allocators for a parallel system are implemented.

The main task of the memory allocator is to hand out, or *allocate*, memory for the requester and take it back, or *deallocate*, when the memory is no longer needed. Thus, the memory allocator needs to keep track of the memory that is currently allocated so that it can find unallocated chunks when a new request for an allocation comes in.

The memory that belongs to a process is usually divided into at least two memory segments [81, 126]:

- **Text Segment:** Sometimes also referred to as the *code segment*, this segment contains the executable code of the process and is typically read-only.
- **Data Segment:** This is the segment that contains the data that the program can work with and is readable/writable. This part of the memory is divided into the *heap* and the *stack*.

On some operating systems, the data segment is sub-divided into the *BSS* segment (historically, this means "Block Started by Symbol") and contains data that is initialized to zero via the operating system and a data segment that contains initialized data [126].

A useful convention on, for instance, Linux is that the heap grows "upwards" from lower virtual addresses to the higher ones (starting at the end of the text segment or the BSS segment), while the stack grows "downward" from a high address to the lower

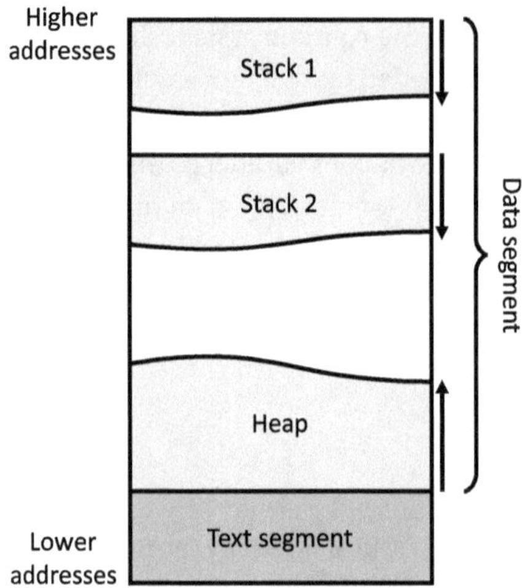

Figure 5.1: Memory layout of a multi-threaded process.

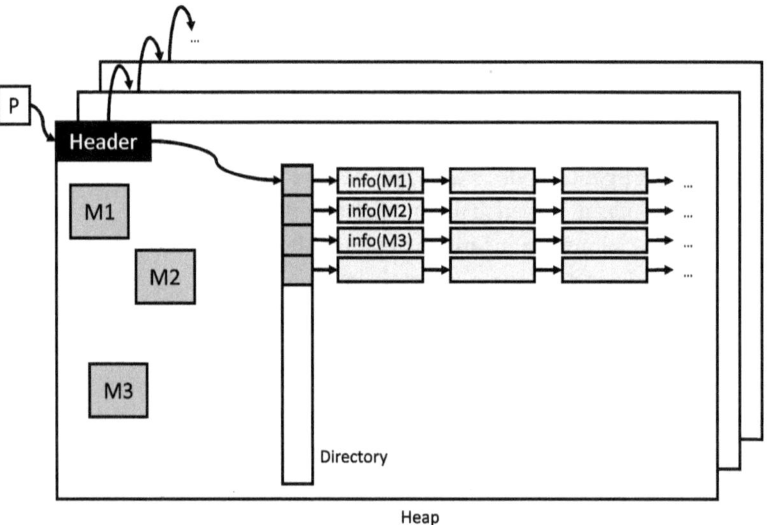

Figure 5.2: Conceptual structure of a memory allocator (from [37]).

ones. If multiple threads exist in the process, then for each thread, a stack is reserved that starts at a high address and grows downwards (see Figure 5.1).

Figure 5.2 shows the conceptual structure of a memory allocator. The allocator requests a *memory slab* from the operating system through one of the system calls. Such a slab is a larger amount of memory that will be used to serve the allocation requests or *chunks* (*M1* and *M2* in the figure). Since a slab has a fixed size, it can be exhausted when all of its memory has been allocated, or it may not be able to serve a memory

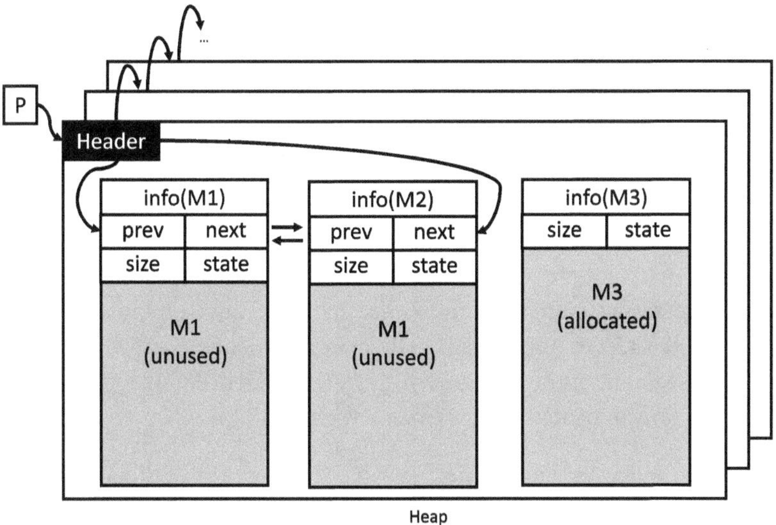

Figure 5.3: Storing the allocation directory in memory chunks.

request because of fragmentation if there are many free, yet small, memory chunks none of which is large enough to satisfy the request. If a slab runs out of memory, then the allocator will request a new slab of suitable size from the operating system and add it to the existing slabs.

At the beginning of each slab, the allocator reserves some memory that contains a header to hold information it requires, such as where to find the *directory* and a pointer to the next slab (or a set of slabs). The pointer P holds the address of the header of the first slab in memory so that the allocator can access it and from there find all of the other information in its database.

The directory is the central place to store information about current allocations: their address in memory (or offset within the slab), their size, and other status information. When memory is returned to the allocator, it must find the relevant information in the directory and mark the allocation as free. It may then choose either to leave the allocation as is, or try to find adjacent, free chunks that can be merged to create larger blocks of free memory to tame fragmentation.

Some allocators (e. g., `malloc()` in glibc [44]) store the directory with the allocated and unused memory chunks by reserving some extra space in front of each allocation, which is then used to store the directory information (see Figure 5.3). In the figure, memory chunks *M1* and *M2* are currently unused, so the directory information is stored in the memory chunk. Each of the free chunks is connected via a doubly-linked list, so that the memory allocator can navigate through them when it tries to find a matching memory chunk for an allocation request.

Chunk *M3* is currently allocated, and so most of its memory space contains payload data from the requester. At the start of the chunk, but before the address that was given to the requester, the allocator keeps a small header that holds the size of the memory chunk and, perhaps, some status flags. When the chunk is given back to the memory allocator, the allocator finds the first (or last) in the chain of unused chunks, reads the size of the released chunk, and creates the header to add the chunk into the linked list.

One final thing that's missing from our description is how to actually request memory from the operating system. Each OS has its own way to handle this. On UNIX[*]-like systems like Linux, the system calls `brk()` and `sbrk()` grow or shrink the size of the data segment of the process [74, 129]. On modern Linux systems, the memory slabs are usually allocated via the `mmap()` system call [74, 131]:

```
new_slab = mmap(NULL, slab_size, PROT_READ | PROT_WRITE,
                MAP_ANONYMOUS | MAP_PRIVATE, -1, 0);
```

With this system call, the Linux kernel allocates enough pages for `slab_size` bytes that will be accessible to the process for both reading and writing (so regular memory). The `MAP_ANONYMOUS` tells the kernel that the memory is not backed by a file on disk and should be initialized to zero. The `MAP_PRIVATE` mode restricts visibility of the memory to the current process (in contrast to the shared segment of Section 2.1). If the `mmap()` call is successful, it returns the pointer to the beginning of the slab it has allocated.

5.3.3 Multi-threaded memory allocators

With this basic understanding of how memory allocators work, we can now turn toward how to allocate memory in a parallel program and a parallel runtime system.

Because the data segment is part of a single process, there is only one such segment that the memory allocator can allocate memory from. While we can assume that operating system calls are thread safe (so two threads each calling `mmap()` to map additional memory will perform safely), in some cases there is clearly danger if more than one thread makes a conflicting call at the same time. For instance, `sbrk()` is not inherently thread-safe, as two threads can call it with conflicting requests. One thread may ask to decrease the amount of memory, while another thread would like to increase the memory space. How should this race be handled? In effect the two threads are each attempting to update a single shared variable (the position of the "break") without synchronization. The fact that the variable is hidden inside the OS does not avoid the inherent race condition.

At the same time, the memory allocator needs to keep track of the memory given out to the application code. When multiple threads concurrently request and release memory, the structures in Figure 5.2 and Figure 5.3 have to be modified concurrently.

This is another race condition that needs to be handled so that these key data structures are not in an inconsistent state at any given time.

A typical solution is to use a global lock that protects the memory allocator from concurrent accesses. However, this means that only one thread can enter the internals of the memory allocator and modify the state of the allocation directories or request new memory from the operating system. As you can imagine, this solution is not scalable. The single-threaded memory allocator quickly becomes the bottleneck for programs (or runtimes!) that request and release memory frequently.

So, let's get back to the drawing board. What if we could apply the same trick that we used with the threads' stacks so that each thread has its own slab of memory to allocate from? Each thread would then also have its own directory to keep track of memory chunks. When a thread requests memory, the memory allocator will look at the directory for the requesting thread, determine whether a new memory slab needs to be created, and return the appropriate memory chunk. As long as the OS uses system calls like `mmap()` to create new memory slabs, no locking will be needed. If an old-style `sbrk()` equivalent must be used, then this needs to be protected by a lock. But, that lock would only be required to ensure mutual exclusion for when a new slab is added. As long as the allocator is not doing this, it can pretty much operate without mutual exclusion.

Alas, things are not that easy! If there's a guarantee that a thread that requested memory will also always release it, then this approach will work. However, things become complex (and maybe even complicated) when threads can pass pointers between themselves. If one thread allocates memory and another thread releases it, the releasing thread would have to modify the allocation directory that is owned by the initial requester. So, we are back at square one, and have to apply mutual exclusion to protect each thread's directory from concurrent access by other threads.

The need for mutual exclusion does not seem to have improved the situation a lot, although replacing the global directory, and therefore the global lock, with localized, per-thread locks is useful. If threads mostly request and release memory for their own usage and if the crossover of memory between threads does not happen frequently, then the per-thread locks will mostly be acquired and released by the owning thread. In this case, multiple threads can run concurrently. In the other extreme, when all allocations are released by other threads, the implementation will get closer to the behavior of a memory allocator with a global lock.

However, we can do better than this. If one abandons the idea of a centralized directory (even if it's per thread), but distributes the information about allocations with the allocations themselves (similar to what glibc's `malloc()` does), then we can optimize the behavior and reduce the need for locking even further. We do not necessarily need to return a memory chunk to the thread that initially requested it. The memory allocator may simply keep a released chunk of memory from a thread and use it to serve allocation requests coming from that same thread. Effectively, the thread that releases the memory chunk will become the new owner of that chunk.

With this modification, the memory allocator does not have to protect allocation requests (unless the OS has to supply a new memory slab) and it does not need to protect the release of memory (unless it is returned to the OS). The two operations become updates that are local for each of the threads, and thus can be performed without any locking.

Memory allocators like glibc's `malloc()` may also implement additional smartness, like special lists of returned allocations to find the best match for a new allocation request, heuristics to determine if and when returned memory chunks should be coalesced to reduce memory fragmentation, or may also implement additional heuristics to determine the sizes of newly requested memory slabs and, even, when to switch to a global memory allocation algorithm. There's a vast implementation space and some interesting memory allocators to read about beyond what we are able to present here. Some good examples are FreeBSD* jemalloc [36], Google* TCMalloc [43], and the Intel TBB `tbb::scalable_allocator` [151].

5.3.4 Specialized memory allocators for parallel runtime systems

In the context of a parallel runtime system, there's more we can exploit when it comes to how to deal with memory allocations. While a memory allocator for a (parallel) application needs to serve any allocation request for any sizes of memory chunks (barring out-of-memory issues, of course), a parallel runtime system is something that you, as the implementer, can control. This means that the runtime system can use memory allocators that specialize in the task of serving the requests for memory coming from the runtime system. They do not have to be generic in the sense that they would have to serve any request coming from elsewhere.

We will see throughout this book that a parallel runtime system uses internal data structures that are very predictable in their size and their alignment needs in memory. Lock objects (see Chapter 6) to implement mutual exclusion are typically of the same size (for the same lock implementation). Linked-list elements or buckets in a hash table can be of the same size. Task descriptors (see Chapter 9) may also require the same amount of memory to store them. The implementer of the runtime also knows (well, should know) exactly when and where those objects are created and when the runtime needs to allocate them. Thus, we can exploit this knowledge to optimize the task of memory management to the needs of the runtime system's implementation.

Let's assume that the runtime's memory allocator can be specialized to provide three classes of memory allocations: for 32 bytes, 128 bytes, and arbitrary sizes. What the allocator can do is to pre-allocate enough entries for the 32-byte and 128-byte cases. Any time the runtime requests an allocation of either size, the memory allocator returns a 32-byte chunk from the memory slab for that size or a 128-byte chunk for a 128-byte request. Since all allocations in one of the classes will be of the same size, no memory fragmentation will occur. For the arbitrary-size allocation, the implementation would call a multi-threaded memory allocator to request whatever size the runtime is asking for.

5.3.5 Thread-local storage

Thread-local storage is an important concept and can be divided into two different categories:

1. **Local or Private Variables:** These are allocated on the thread's stack (see Section 4.4.1), and so are naturally private to a single thread without requiring any special addressing or other handling. Examples are the OpenMP `private` and `firstprivate` clauses.
2. **Static/Global Variables:** In serial code, static or global variables are allocated once (by the linker), and they can be accessed directly, either by their absolute address or via the *global offset table* (GOT), which is used to handle such data in position independent, relocatable shared libraries. An example is the `threadprivate` directive in the OpenMP API or the `thread_local` specifier in C++11.

Because each thread has its own stack, stack variables fall out for free. When we talk about TLS variables, we are normally considering the second case of static or global-scope variables. Although the use of static and global variables is frowned upon by programming stylists (for very good reasons), such variables still occur in user code— for instance, as `COMMON` blocks in older Fortran codes, where the variables often need to be shared between functions and subroutines in a code but also need to be replicated in each thread to enable parallelization of the code.

Modern compilers and operating systems have realized that support for TLS is important, and therefore, in many architectures they maintain a specific register that points to the base of the area of the TLS for a thread. On x86_64, one of the (otherwise unused) segment-base registers is used (either `FSbase` or `GSbase`, depending on the OS). Linux on Arm keeps the TLS address in the `tpidr_el0` register. This allows the linker to manage the allocation of space inside the TLS area, and the compiler to generate efficient code to access TLS variables.

For instance, code that uses the `_Thread_local` specifier of C11 to create some thread-local space:

```
int getTLSValue () {
  extern _Thread_local int value;
  return value;
}
```

generates the assembly code shown below when compiled by clang 9.0.0 for Linux on an x86_64 machine:

```
getTLSValue:
  mov   rax, qword ptr [rip + value@GOTTPOFF]
  mov   eax, dword ptr fs:[rax]
  ret
```

Here, you can see that a value is loaded from the GOT (which is accessed relative to the program counter). The value that was loaded is then used as an offset into the *F* segment (by using the fs: syntax) so that the computed address is the value of the FSbase register plus the offset.

The implementation of a parallel runtime will almost certainly need TLS state, since it will need to identify threads. We will also see that the per-thread state needs to be maintained by the runtime and TLS will be used to associate these structures with the native OS thread. For instance, in an OpenMP runtime, thread identities are visible to the user code (via a call to omp_get_thread_num()). Another example is pointers to shared data, such as those that are used to implement barriers, dynamic loop scheduling, and task pools.

Of course, if a thread can obtain its identity, it can then proceed from there to provide its own TLS implementation by using its identity to index into an array (or std::unordered_map) that holds the value of a given TLS variable for each thread. While this does provide an OS-agnostic way to write a TLS implementation, it increases the latency to find the pointer to the thread-local storage area of a particular thread.

Historically, the interface to the LLVM OpenMP runtime is designed along those lines. Back in the day, when the first OpenMP implementations were written, TLS was not well supported or implemented in the then available operating systems. So, the LLVM OpenMP interface requires that the compiled code passes a special gtid argument to almost all of its functions (remember this when you see an anonymous int32_t argument in the code listings throughout the remainder of this book!). The argument gtid is short for global thread ID. The assumption is that finding the gtid is expensive (for instance, it might require a comparison with each thread's stack range to determine which thread's local memory was to be accessed). The compiler optimized this by calling a runtime function to compute the global thread ID and cache this value in a local variable on the stack. This is an optimization whose day has now (long) passed, and loading and passing the additional argument is now merely additional overhead, since loading a TLS variable is fast.

The OpenMP specification has strict rules about the preservation of TLS state (except in very specific circumstances), which mean that code like that shown in Listing 5.3 must not trigger the assert() function.

```
void test_for_tls() {
  static int myId;
#pragma omp threadprivate(myId)
#pragma omp parallel
  {
    myId = omp_get_thread_num();
  }
  // Thread executing the serial region must have
  // been thread zero inside the parallel region.
  assert (myId == 0);
#pragma omp parallel
  {
    // My TLS must be preserved.
    assert (myid == omp_get_thread_num());
  }
  // And the initial thread must still see zero after
  // parallel region.
  assert (myId == 0);
}
```

Listing 5.3: Required preservation of thread-local state in OpenMP code.

5.3.6 Data layout of the thread objects

When dealing with data structures inside the runtime library, we have the same issues to contend with as does multi-threaded user code. In particular, we want to ensure that we avoid false sharing, so we need to separate data that will be accessed by a single thread from that which is genuinely shared and place the two types of data in separate cache lines. When choosing how to lay out the data in data structures, the aim is to avoid unnecessary data movement, which normally means separating data that is widely shared from that which is accessed only by a single thread.

C++ supports the `alignas(size)` attribute so that we can request the compiler to align appropriately, provided that we know the cache-line size we require. In general, this works, though the C++ standard allows implementations to silently ignore alignment requests for alignments greater than `alignof(std::max_align_t)`, which may be as low as 8 or 16 bytes. It's therefore worth checking whether the specific compiler you intend to use respects alignment requests for higher (e. g., 64 B) sizes.

As an example, here is the data layout part of the class declaration of a naive broadcast class (see Section 7.3.2 for details of the operation this class will perform):

```cpp
class NaiveBroadcast : public
    alignedAllocators<CACHELINE_SIZE> {
  // Put the payload and the flag into the same cache line.
  CACHE_ALIGNED std::atomic<uint32_t> Flag;
  InvocationInfo const * OutlinedBody;

  // And the per-thread state into a cache line/thread.
  AlignedUint32 * const NextValues;

public:
  // ...
};
```

You can see that:

1. The whole class object is aligned to CACHELINE_SIZE and uses a specific cache-aligned allocator to ensure that (C++17 would make this easier; however, this code only assumes that we are using C++14).
2. The flag is aligned, while the pointer to the OutlinedBody is not aligned, and so will, intentionally, be in the same cache line as the Flag. This means that when a thread reads Flag, the value of OutlinedBody will also be fetched into the cache and will be ready to be used because of the spatial locality of the fields.
3. The per-thread counters pointed at by NextValues are each placed in a cache line to avoid false sharing, but we allow the pointer itself to be in the same line as the flag. Since many threads will be polling, the line will be in a shared state before the flag is set anyway, so there's no point in placing this in a separate line.

If the compiler ignores the alignment requests, one can introduce manual padding to align members of a struct:

```cpp
struct aligned_data {
  union {
    int x;
    char padding1[8];
  };
  union {
    int y;
    char padding2[16];
  };
  char padding3[LINE_SIZE - 32];
  int z;
};
```

The code fragment uses a mix of `union` and `struct` types to construct a certain alignment of data. Component `x` is aligned to the alignment of type `struct aligned_data`. On an x86_64 machine and Linux, this is likely to be a 16 B alignment. The union types that wrap `x` and `y` are to change the default alignment of the `int` type in C by introducing padding. While it is usually 4 B-aligned, the field `padding1` changes `y` to be 8 B-aligned. Finally, `padding3` pushes `z` to the next cache line.

When doing this kind of low-level data alignment, you will have to make sure that you take into account what the alignment properties of the respective primitive data types are according to the language specification of C or C++. A memory allocator can take the alignment needs of structured types into account when returning a pointer from a pool of memory (see Section 5.3.4). For instance, for lock objects, the runtime's memory allocator can ensure alignment by only handing out memory that is properly aligned to the boundaries of cache lines.

5.4 Conclusions

This chapter was all about the miscellaneous, cross-cutting concepts and implementation strategies that a parallel runtime system needs as the foundation for all of the things that we are going to present in the following chapters.

The parallel runtime can now understand the specific structure of the machine that it is supposed to run on. We know how to pin a thread to a particular processor core so that the OS does not interfere with cache locality for that core. We have looked at some aspects of memory management in the runtime system so that the runtime's effect on the execution of the application code is reduced. We have also worked through the high-level concepts of memory allocation in a parallel system and learned how a parallel runtime system can make use of knowledge of the context in which allocations happen for optimizations.

6 Mutual exclusion and atomicity

As soon as we have multiple threads executing in the same address space, we have the possibility that they will interfere with each other, since updates made by one thread can potentially occur in places which another thread does not expect. This is called a race condition and needs to be avoided to ensure that a parallel program computes the correct result.

In this chapter, we'll start off with an analysis of the problem and describe ways to avoid such race conditions. We'll quickly explain what options hardware can provide to help solve this problem. We will then show how to use these hardware capabilities in a parallel runtime system to implement locks for mutual exclusion as well as speculation. Of course, we will have an in-depth look the implementation options at hand and discuss how they compare performance wise. So, let's go!

6.1 The mutual exclusion problem

Let's start with a piece of code that can go terribly wrong. Consider a simple example like the one shown in Listing 6.1. Consider also what can happen if two threads are each attempting to push entries into the queue implemented as a linked list at the same time. Although this is only a few lines of code it can fail if execution occurs in the order shown in Table 6.1.

What has happened is that, because of the race conditions, one of the two newly created entries has been lost. It is not linked into the list because the second thread overwrote the pointer to it in the Front field with a pointer to a different entry that points to the previous Front entry.

There are similar issues in pop(). Consider what would happen if:
1. One thread removes the last item and stores 0 into Front after another thread has tested isEmpty(), but before it has read the Front and Value fields.
2. One thread reads the Front field after another thread has deleted it.

Clearly, we need to somehow ensure that multiple threads do not step on each other's toes while accessing a shared data structure.

There are two fundamental approaches to achieving this:
1. **Mutual Exclusion:** Ensure that only a single thread at a time can access the relevant data structure and make updates by restricting the execution of a block of code to a single thread at a time.
2. **Speculation:** Allow multiple threads to look at the data structure, but ensure that they can detect any updates that will cause problems, abort, and re-execute if necessary.

https://doi.org/10.1515/9783110632729-006

```cpp
class LIFOQueue {
  struct LIFOEntry {
    LIFOEntry *Next;
    int Value;

    LIFOEntry(LIFOEntry *N, int V) : Next(N), Value(V) {}
    ~LIFOEntry() {}
  };
  LIFOEntry *Front;

public:
  LIFOQueue() : Front(0) {}
  ~LIFOQueue() {}
  bool isEmpty() const {
    return Front == 0;
  }

  void push(int V) {
    Front = new LIFOEntry(Front, V);
  }

  int pop() {
    if (isEmpty())
      return 0;

    int V = Front->Value;        /* Remember the value */
    LIFOEntry *N = Front->Next;  /* and next pointer */
    delete Front;                /* Delete front element */
    Front = N;                   /* Update to point to next */
    return V;                    /* Return the value */
  }
};
```

Listing 6.1: Implementation of a simple LIFO queue.

Table 6.1: Failing interleaving for the linked-list example of Listing 6.1.

Thread Zero	Thread One
Read Front	
	Read Front
Create new entry	Create new entry
Store into Front	
	Store into Front

Exclusion is implemented using critical sections and locks (*mutexes*), speculation using either hardware speculative execution ("transactional memory") that can check for interfering reads and writes, or hardware instructions that allow for the detection of conflicts at a smaller granularity, e. g., *load linked* and *store conditional* (LL-SC) or atomic compare-and-swap (CAS) instructions that atomically check that a value is as expected before updating it.

In some architectures, the hardware only provides speculative operations, so other atomic operations such as claiming a lock have to be built on top of these.

6.1.1 Hardware support for locks: atomic instructions

To perform a single operation atomically, the hardware has to encode an operation that loads from memory, operates on the value read, and stores it back to memory, while ensuring that no other update to the memory location(s) being updated happens between the read and the write. In a CISCy architecture, instructions that both read and write memory are common, since memory can be used as an operand in most instructions. Therefore, adding atomic operations like this is architecturally relatively easy. However, an operation that both reads and writes memory does not fit naturally into a RISC instruction set, where memory values are loaded into registers, updated there, and then stored back to memory from the registers. This has led many RISC machines to separate the operations using a specific load (the load linked) to mark the start of the operation and a corresponding store (the store conditional) to mark the end.

The precise details of hardware support depend on the instruction set architecture and the revision of that architecture. However, either CAS or LL-SC is a minimum foundation on which everything else can be built. We'll show here how to do this, but one should use the higher-level instructions if it is possible and if they are available, since they have better performance, as we can see in Figure 6.1.

The figure shows the total machine throughput for the atomic operation of incrementing a `uint32_t` when that is implemented on the same Arm[*] processor either using the LL-SC operations or the newer single atomic instruction. You can see that the implementation using the single atomic instruction has about a six times higher performance than that of the LL-SC loop. Luckily for us, modern compilers for the Arm architecture know this, and will generate the fast code for this operation if we allow the compiler to target a more modern instruction set.

6.1.1.1 Atomic instruction implementation

To ensure that an atomic instruction can perform atomically, it must guarantee that no other store operation can occur between the time when it loads the value and the

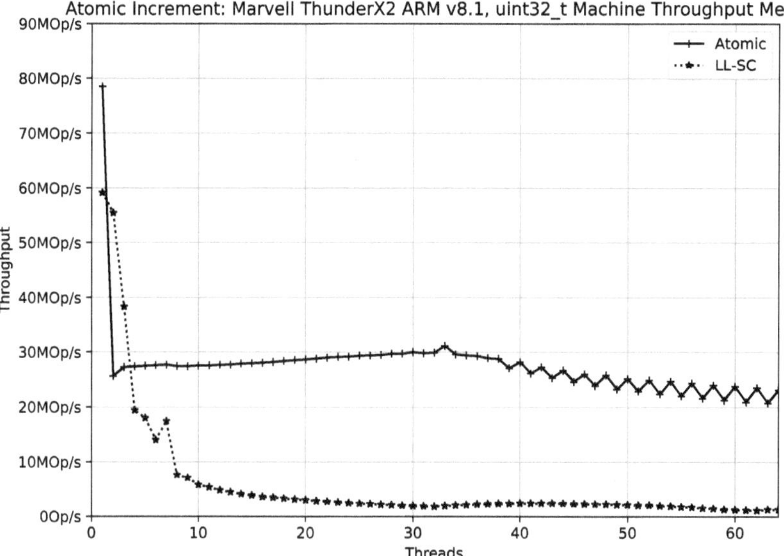

Figure 6.1: LL-SC performance vs. performance of native atomic operations.

time when it stores the updated value. If we look back at our description of the cache protocol in Section 3.2.5, then we can see that the core can achieve this if

1. the instruction is treated as a store, so the cache line is moved into an exclusive state before the load operation occurs, and
2. the cache does not respond to external requests until after the store operation has occurred.

This implementation ensures that if a line can be acquired in an exclusive state, the operation will complete, so forward progress is guaranteed.

An important side-effect here is that in a CAS operation, the line will be moved to an exclusive state even if the compare fails and no update is performed. This has non-trivial performance implications, since it requires that all other copies of the line are invalidated, even though the line may end up unchanged.

6.1.1.2 LL-SC instruction implementation

In the LL-SC implementation, the processor uses a different approach, in which the load and store operations are separated and do not occur in a single instruction. This fits much more easily into RISC architectures that normally avoid read-modify-write operations in a single instruction.

LL-SC instructions can be implemented in different ways: either in the cache (as in the open-source RISC-V[*] [110] or DEC[*] ALPHA[*] [118] architectures), or by having separate hardware that explicitly handles the global state (as in the Arm architecture).

In the cache-based implementation, the locking load still places the line in the exclusive state. At the associated store, the state is checked, and, if the line is no longer exclusive (i. e., it has been snooped by some other core or evicted from the cache for some reason), then the store fails, and that failure is reported as one of the results of the operation, allowing the code to retry it.

In the Arm architecture specification, the locking load sends a message to a centralized hardware resource that maintains information about which memory regions are being monitored; similarly the conditional store checks that no other operation has altered the monitored state by checking with the central monitor. From a programmer's point of view, this is merely an alternative implementation that has effectively identical semantics to the cache-local implementation, though, possibly, with different performance.

The problem with both of these implementations is that they do not trivially guarantee forward progress, since two threads can interleave their accesses and each can repeatedly shoot down the other without either of them successfully storing a result. To overcome this, architectures introduce additional rules restricting the size of the region between the LL and SC, and which instructions can be executed there. With these restrictions, the implementation can ensure that a thread has a window in which it can make an update without the line being snooped away or the monitor being dropped.

For instance, the RISC-V Instruction Set Manual [110] includes this text (as well as some other more architecture-specific restrictions) in Section 8.3; ("LR" is the RISC-V "load reserved" mnemonic, which is the equivalent of the "load linked" we have been referring to, while "SC" is "store conditional"):

> The standard A extension defines *constrained LR/SC loops*, which have the following properties:
> - The loop comprises only an LR/SC sequence and code to retry the sequence in the case of failure, and must comprise at most 16 instructions placed sequentially in memory.
> - An LR/SC sequence begins with an LR instruction and ends with an SC instruction. The dynamic code executed between the LR and SC instructions can only contain instructions from the base "I" instruction set, excluding loads, stores, backward jumps, taken backward branches, JALR, FENCE, FENCE.I, and SYSTEM instructions. ...
> - The code to retry a failing LR/SC sequence can contain backwards jumps and/or branches to repeat the LR/SC sequence, but otherwise has the same constraint as the code between the LR and SC.

Similar restrictions apply in other architectures that implement LL-SC.

6.1.1.3 Difference between LL-SC and CAS

At first glance LL-SC and CAS appear to be very similar. Each is speculative, since there is no guarantee that any given "load, operate, CAS" or "LL, operate, SC" sequence will succeed, so the work represented by "operate" may be discarded. However, the CAS sequence is more powerful than the LL-SC one, because it has a global forward-progress guarantee. The CAS can only fail if some other thread has successfully performed an update and, therefore, the whole computation is progressing. On the other hand, the SC operation can potentially fail even if all that another thread did was load the relevant cache line, an operation that does not, itself, ensure forward progress.

6.1.1.4 Transactional memory

Various processors (at least some from Intel[*] [68], IBM [69, 79], and, soon, Arm architecture ones [8]) provide limited support for hardware transactional memory operations. This is similar in concept to the speculative execution we have been discussing in LL-SC, but with the extension that instead of explicitly tagging a single load, the processor is placed into a transactional, speculative execution mode in which all loads are monitored and all stores are kept private until the whole transaction is committed, at which point they all become visible simultaneously. Hardware detects read and write conflicts and aborts the transaction if any are detected.

As with LL-SC, there are no forward-progress guarantees when using this hardware transactional memory; the code that uses it has to detect failure and back off to some other implementation (such as taking a normal lock) when failures occur.

6.1.2 The ABA problem

If we are using CAS to build atomic operations, we must detect competing changes in the value we have read so that we can abort and retry. For simple arithmetical instructions, such as *atomic add*, this is fine: each useful update will change the value, and the updated value will be directly dependent on the value that was read. However, when CAS is being used to manage a data structure, we have the problem that the value being checked is frequently that of a pointer, when our dependence is really on values that are in the data structure that is being pointed to. This opens the possibility that although the pointer we are checking has remained the same, the values on which we really depend have changed and we will not be able to detect this.

This is known as the *ABA problem* because the way that the changes can happen is that one thread sees the pointer with value A and then stalls; meanwhile, another thread alters the value to B, potentially releasing the memory pointed to by pointer value A; then that memory is reused, and the newly allocated object is put back into the shared data structure. Now our initial thread will see the value that it is checking is still that which it expects (i. e., A), whereas the data structure has actually changed and the values it read are invalid. Unpleasant consequences ensue!

Note that the ABA problem does not occur with LL-SC or transactional memory, since both of those implementations will see the store to the location that is monitored and will let the operation fail. This is effectively the mirror image of the forward-progress problem; here, the eagerness to abort gives the hardware-monitoring implementations an advantage, whereas in the other case, it was a problem since it removed the forward-progress guarantee.

6.2 Should we be writing locking code?

Linus Torvalds argues persuasively that we should not [137]. He says:

> "I repeat: **do not use spinlocks in user space, unless you actually know what you're doing.** And be aware that the likelihood that you know what you are doing is basically nil...
>
> You should **never ever** think that you're clever enough to write your own locking routines. Because the likelihood is that you aren't (and by that "you" I very much include myself—we've tweaked all the in-kernel locking over decades, and gone through the simple test-and-set to ticket locks to cache-line efficient queuing locks, and even people who know what they are doing tend to get it wrong several times)."

Given that, why are we bothering to explain how to do something that you shouldn't do? The answer is that even if we're not going to write our own locking scheme it is useful to understand both the issues and caveats that are involved in evaluating the performance of a lock implementation and the constraints under which lock implementations are working. This is useful even if we do follow Linus' advice:

> "So you might want to look into not the standard library implementation, but specific locking implementations for your particular needs. Which is admittedly very annoying indeed. But don't write your own. Find somebody else who wrote one, and spent the decades actually tuning it and making it work."

The other reason why this chapter is worthwhile is that we also discuss more than just locking, as we are covering how to write atomic operations that do not require locks to achieve mutually exclusive updates to memory.

As with any optimization, we should consider the circumstances under which we are hoping that our own locks can outperform pre-existing implementations. If a critical section takes a long time to execute (say, 1 ms), and the lock code adds 10 µs, then that reflects only 1 % overhead, so even if we could halve the lock overhead, we can only expect an improvement from 1.01 ms to 1.005 ms, or approximately 0.5 %. Therefore, this doesn't seem an area where we can usefully spend our time for optimization. The more fundamental problem here is that the large critical section is likely to be limiting the available parallelism. Fixing this will require algorithmic changes in the user code to reduce the time spent in the critical section, or, ideally, eliminate it completely.

This is a place where profiling can be misleading. For a heavily contended lock, a profiler that has no lock awareness will show a lot of time spent in the lock-acquisition code, which makes it look as if that is something that should be optimized. However, this really just reflects the higher-level problem that the lock is contended, and therefore, the thread has to wait somewhere before it can acquire the lock. This is not an issue that can be fixed by improving the lock-acquisition code. While profiling is critical to direct tuning efforts, we must always consider how to interpret what the profiler is telling us.

At the other extreme, we have very small critical sections. Here, the lock overhead can be larger than the time in the critical section, which seems worrying. However, from an overall performance point of view, the application will need to be executing huge numbers of these critical sections (since, by definition, they are extremely short) if they are to be significant in the overall runtime. So, again, we may be optimizing the wrong thing, or at the wrong level. Maybe the operation could be expressed directly as an atomic instruction rather than as a critical section, or perhaps the operation is some form of reduction that could be better implemented by using a per-thread accumulator and combining each thread's results at some suitable point.

Travis Downs discusses similar issues in his "Performance Matters" blog [33].

6.3 Classes of locks

In this chapter, we only discuss the simplest kind of lock: one which can be acquired by a single thread and later released by the same thread, and can only be owned by a single thread at a time.

There are other classes of lock with slightly different semantics, such as:

- **Recursive Lock:** A recursive lock can be claimed many times by a single thread and then has to be released as many times as it was claimed. It can clearly be implemented relatively easily as a wrapper around any of the basic locks we discuss.
- **Reader/Writer Lock:** A reader/writer lock can be used to guard data structures such as hash-tables in which many threads may be performing lookups concurrently, while any update must occur on its own. It allows many concurrent readers to hold a read lock simultaneously, while only a single writer is permitted which holds off all other (reader or writer) accesses.

Although these locks are important we do not discuss them because the underlying principles we explain here also apply to these more complex locks. Therefore if you understand the issues we expand on below you should be able to handle the more complex cases, too.

6.4 Properties of lock algorithms

Before we discuss specific lock implementations, we need to consider the properties that we want them to have and the circumstances in which they are used. This will affect which implementation choice is optimal.

As above, the two extreme utilization cases are:
1. **Uncontended:** Lock conflicts are very rare.
2. **Heavily contended:** It is very likely that more than one thread is trying to execute inside the critical section at the same time.

When there is contention, an important property of a lock is whether it is fair or not. In other words, whether the lock is handed out to threads in the same order as they attempted to claim it. Implementing this property has a cost, and the property is irrelevant for uncontended locks, since, if there is no contention, then no thread is ever waiting for the lock, so discussing the order in which waiting threads receive the lock is pointless.

Similarly, we must consider whether we expect the code to be operating in a controlled environment, in which the machine is being used to execute only the code we know about, or whether it is being run in a "noisy" environment, in which other processes and threads will also be running or many interrupts will be occurring.

In the noisy case, the behavior of the operating system scheduler will be important, and we must ensure that the operating system is aware of when and how our threads are polling so that it can make good thread-scheduling choices.

6.4.1 Lock-performance metrics

If we are to evaluate lock implementations, we have to consider what aspects of the lock implementation affect the performance of the lock, how those code paths map to the underlying hardware operations, and how the lock behavior affects other threads.

For an uncontended lock, all of the execution time spent in acquiring and releasing the lock is additional overhead, since the thread executing that code could be executing useful application-code instructions; however, as this is a small code path, the possibility of interference with other threads is low.

For contended locks, things are much more complicated, since threads that are trying to claim the lock have to wait until it becomes free, and how the threads wait can have a significant impact. Also, the critical path for the contended lock does not include the time entering the wait, since the thread would have to be waiting anyway. Rather, the critical path here is the time between the holder of the lock releasing it and another thread claiming it. In effect, this time is added to the duration of the critical section.

We therefore need to measure a variety of properties of a lock under different levels of contention:
- **Overhead:** The time to claim and release an uncontended lock.
- **Exclusive time or throughput:** The time during which the lock is held after one thread has released it, but before another has acquired it. Under contention, this is the real overhead of the lock, though it is not a time between operations in the

same thread. The reciprocal of the exclusive time gives the machine throughput (i. e., the number of empty critical sections that the given number of threads can execute in one second). For a perfect lock, we would want this time to be zero (and the throughput to be infinite), though in reality, this is obviously unattainable.
– **Interference:** How much impact the use of the lock has on other threads that are not waiting.

To understand the difference between the first two metrics, look at Figure 6.2. In the uncontended case, the lock overhead is obvious; it is all of the heavily shaded time, which is the time it takes to execute the lock-acquisition code and the lock-release code. This is relatively easy to measure, since it can be measured inside a single thread. To avoid simply measuring the performance when the lock is in the local cache, our benchmark randomly picks a lock from an array of 4,096 locks. Even with 64 threads, this gives a low probability of conflict, while reducing the chance that the lock is still resident in the L1 cache of the processor claiming the lock.

However, for the contended lock, things are more subtle. Here, we can see that the time spent inside the lock functions includes time when the thread is waiting because of the lock contention, which is caused by the application code, not the lock implementation. Instead, the time that reflects real lock-implementation overhead is the lock-induced exclusive time—that is, the time between one thread starting to release the lock and the next successfully acquiring it and beginning to execute the code inside the user's critical section. This time is unrelated to the time it takes to start to claim the lock and deciding to wait, since the application has to wait anyway.

Instead, it is made up of overhead in the release operation before it marks the lock as free, the time it takes for other threads to notice that change, and whatever other overhead there is in the claim operation after it has claimed the lock, but before it returns to the user code. This total time can be either longer or shorter than the time measured for the lock release in the releasing thread. This time is labeled explicitly in Figure 6.2.

To measure it, we measure the overall machine throughput executing empty critical sections as we add threads, all of which are contending. The critical time is then related to the machine throughput like this (the reciprocal that we have mentioned before):

$$T_{\text{Critical}} = \frac{N_{\text{Threads}}}{\text{Throughput}(N_{\text{Threads}})}.$$

Although we measure T_{Critical}, we show our results in terms of throughput, since small differences in T_{Critical} can be hard to see, whereas total machine throughput (which would ideally be constant as we add threads under full contention) is easier to look at and understand.

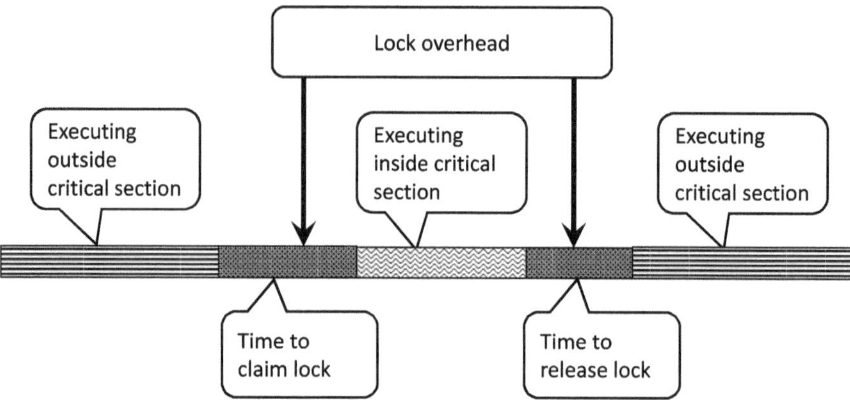

Uncontended lock timeline

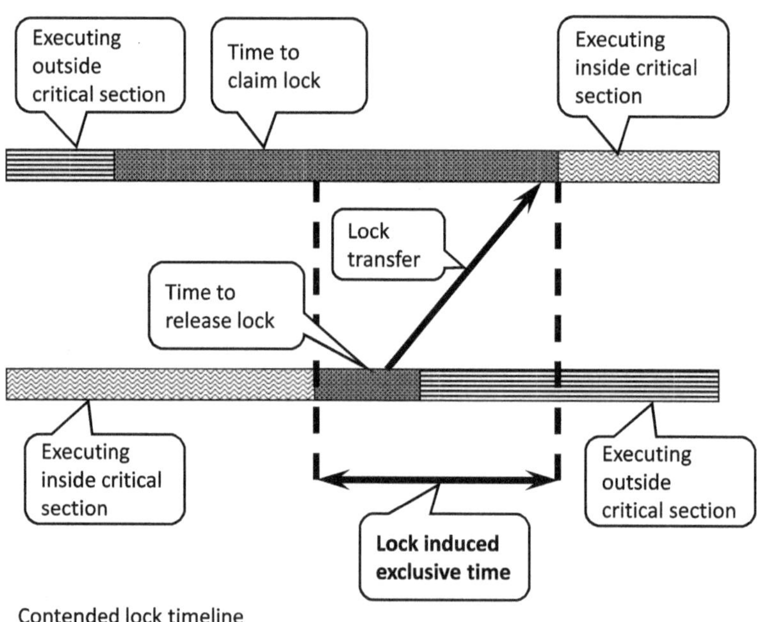

Contended lock timeline

Figure 6.2: Uncontended and contended lock timelines.

Interference is a less obvious issue. The problem here is that when we have many threads waiting for a lock, we need to ensure that they are not consuming machine resources that could be better used by threads that are performing useful work.

6.4.1.1 Measuring lock performance

Although we know what we want to measure, we have to be careful how we set up our experiments so that we really do measure what we think we are. The obvious way to measure the contended-lock performance is to have each thread execute in a tight loop claiming and releasing a single lock. However, this can give us a very misleading result for unfair locks, since the thread that has just released the lock can get around the loop fast enough to reclaim it immediately (as the lock is in its cache) before any other thread has managed to acquire the lock. This leads to an execution pattern in which each thread executes a large clump of iterations consecutively, and, therefore, we rarely see the true lock-transfer time.

One can convincingly argue that this "clumping" is a good thing, since it reduces the lock overhead and therefore achieves higher machine throughput. However, that comes at the cost of significantly larger unfairness; instead of having to wait for at most $N_{\text{Threads}} - 1$ critical-section times, a thread may have to wait arbitrarily long. This might be alright for some codes, but it is certainly unexpected, and will cause problems with others.

To demonstrate this, we'll measure throughput with more than one lock, as well as with only one, and we will also show the percentage of lock reclaims.

6.5 Lock algorithms

In this section, we will show a number of simple lock implementations that one might consider, along with a discussion of their performance.

In each case, we implement the lock on top of a simple abstract base class, since this simplifies all of our code that is measuring lock performance, as it can be passed an instance of the base class and can then measure its performance without needing any implementation details. This does add a small amount of overhead, since calls to the lock operations are now all indirect. However, modern processors can predict such calls well, and something similar is likely to be required in a real runtime implementation anyways.

In Listing 6.2, we show the abstract lock base class. You can see that it is an interface that is almost identical to the interface of the `std::mutex` class in C++, and meets the C++ standard's `BasicLockable` requirement. We use the terms *acquire* and *release* as synonyms for the C++ terms *lock* and *unlock*, respectively.

```
// A base class so that we can have a simple interface to
// our timing operations and pass in a specific lock type
// to use. This does mean that we're doing an indirect call
// for each operation, but that is the same for all
// cases, and should be well predicted and cached.
class abstractLock {
public:
  abstractLock() {}
  virtual ~abstractLock() {}
  virtual void lock() = 0;
  virtual void unlock() = 0;
  virtual char const *name() const = 0;
};
```

Listing 6.2: Simple abstract lock base class.

6.5.1 Test-and-Set locks

One of the simplest, most obvious, locks is a Test-and-Set lock. Listing 6.3 shows one possible implementation of such a lock.

Here, we use the `std::atomic<>::compare_exchange_weak` function to test whether `locked` is `false` and, if so, store the value `true` into it. If the CAS operation succeeded, then we have acquired the lock. If it failed, then some other thread is holding the lock and the acquiring thread must wait until it releases it. Instead of the CAS operation, we could use a simple exchange operation by exchanging `true` and then checking the old value returned by the exchange to see whether it was `false` (in which case the lock was previously free); this is the kind of small optimization rabbit hole that it is easy to run down.

Note that we explicitly request the memory-model semantics that we require, which are `std::memory_order_acquire` when taking the lock, to ensure that none of our loads from inside the critical section float above the point where we claim the lock, and `std::memory_order_release` in the lock release, so that all of the stores that were executed inside the critical section are globally visible before the lock-releasing store, thus preventing another thread from seeing an inconsistent view of the memory we updated inside the critical section.

At first glance, it's hard to see how we could do anything faster than this simple lock operation; the acquire operation is a single atomic load, and the release a single store (with appropriate memory-fencing operations). What could be faster than these single instructions? However, if we remember Figure 3.17, we can see that this may be a poor way to perform the release operation, since the store can be slow when there are many threads polling.

```
// A Test-and-Set lock
class TASLock : public abstractLock {
  CACHE_ALIGNED std::atomic<bool> locked;

public:
  TASLock() : locked(false) {}
  ~TASLock() {}
  void lock() {
    bool expected = false;
    while (!locked.compare_exchange_weak(expected, true,
                                 std::memory_order_acquire)) {
      expected = false;
      architecturalYield();
    }
  }
  void unlock() {
    locked.store(false, std::memory_order_release);
  }
  // Name and factory omitted.
};
```

Listing 6.3: Implementation of a Test-and-Set lock.

Another significant problem with this lock implementation is that the spin-wait loop includes an atomic operation. In some implementations, atomic operations pull the cache line into an exclusive state, even if the operation ultimately does not perform a write (as shown here, where the CAS will fail to match under contention). This means that each polling thread is generating huge amounts of cache-coherence fabric traffic that can interfere with other operations. When we measure that interference (by measuring the bandwidth achieved by a thread that is simply performing memcpy() operations while executing in parallel with N threads that are attempting to claim a single lock), we can see that this is, indeed, a big problem as soon as one of the polling threads is in a different socket.

In Figure 6.3, we show the performance of the memcpy() operation as we add polling threads normalized to its maximum performance. We can see that the memcpy() performance drops to $< \frac{1}{80}$ of the maximum on the Intel[*] Xeon[*] Platinum 8260L Processor when we have TAS lock-polling threads in the other socket from the thread executing the memcpy(). With a TTAS lock (see next section), the performance still drops slightly when we have any poller in the second socket on the Intel machine,

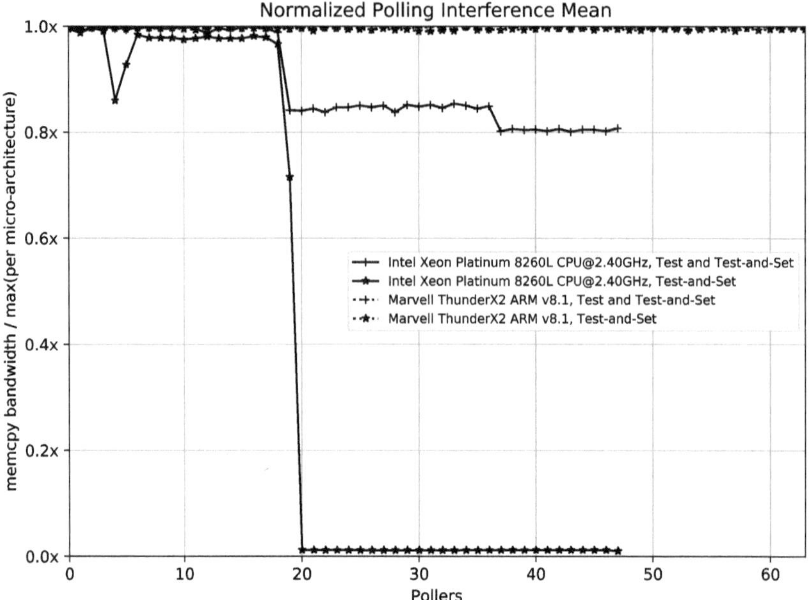

Figure 6.3: The effect of polling interference.

but only to 80 %, not to 1.25 %. With the Marvell[*] Arm processor, there is negligible impact from either lock. It may be that the Arm architecture's more relaxed memory model and different inter-socket coherence protocols remove this problem.

Since it has this unpleasant property, which we can easily fix, we won't consider the TAS lock further.

6.5.2 Test and Test-and-Set locks

The Test and Test-and-Set lock [113] is a small improvement over the TAS lock. As we saw, the TAS lock can have unpleasant interference properties when threads are polling, since they are all performing atomic operations. We can overcome this by checking whether the lock is busy before attempting to acquire it when we are polling. Since we're not (pretending) to update the lock value, the cache line in which it resides can be shared, allowing each polling thread to have its own local copy. Therefore, the polling now generates no coherence traffic until the lock is released (or, in our code, until another thread arrives and makes an initial attempt to claim the lock, when it will need to pull a local copy to its cache). The only changes from the TAS lock are in the lock() function, which is shown in Listing 6.4. With these small changes, the interference with other threads that we were seeing is removed, as can be seen in Fig-

```cpp
void TTASLock::lock() {
  for (;;) {
    bool expected = false;
    // Try an atomic before anything else so that we don't
    // move the line into a shared state then immediately
    // to exclusive if it is unlocked.
    if (locked.compare_exchange_strong(expected, true,
                                std::memory_order_acquire))
      return;
    // But if we see it locked, wait until it is unlocked
    // before trying to claim it again.  Here we're polling
    // something in our cache.
    while (locked) {
      architecturalYield();
    }
  }
}
```

Listing 6.4: The Test and Test-and-Set lock's `lock()` method.

ure 6.3 where the achieved `memcpy()` performance on the Intel Xeon Platinum 8260L Processor is no longer destroyed by polling threads in the second socket.

6.5.3 Ticket lock

A *ticket* (or *bakery*) lock is another simple lock that looks initially attractive, since it is simple to understand, requires little space, and is fair. Our implementation is shown in Listing 6.5.

The idea here is that when trying to acquire the lock, a thread claims the next ticket by atomically incrementing the `next` field, and then waits until the value of `serving` matches that of its ticket. This is how the queue at the cheese counter in our local supermarket works; you pull a ticket from the roll on the counter and wait until the "serving" display shows your number.

The problem with this lock is that under contention, many threads are polling the same cache line, and therefore, we are a long way to the right in the "visibility" graph (see Figure 3.18), so the time before another core sees the store can be significant. Since the ticket lock is fair, there is only one thread that can execute next, so we expect the average delay here to be $\sim \frac{1}{2}$ of the time we measured for the visibility.

As the Test and Test-and-Set lock also has many threads polling a single cache line, you might expect it to suffer for the same reason. However, since it is unfair, any

```cpp
class ticketLock : public abstractLock {
  // Place serving and next in different cache lines.
  // Entering the lock and updating next doesn't need to
  // disturb those waiting for serving to update.
  CACHE_ALIGNED std::atomic<uint32_t> serving;
  CACHE_ALIGNED std::atomic<uint32_t> next;

public:
  ticketLock() : serving(0), next(0) {}
  ~ticketLock() {}

  void lock() {
    uint32_t myTicket = next++;
    while (myTicket !=
            serving.load(std::memory_order_acquire)) {
      architecturalYield();
    }
  }
  void unlock() {
    serving.fetch_add(1, std::memory_order_release);
  }
  // Name and factory omitted.
};
```

Listing 6.5: Implementation of a Ticket lock.

of the polling threads can claim the lock and the relevant lock-transfer time will not be the time until a specific thread sees the value, but rather the time until the first one does.

6.5.4 Queuing locks

To overcome the problems we have seen with the previous locks, we can use a queuing lock. The archetypal example is the Mellor-Crummey & Scott (MCS) lock [89].

The idea here is that we maintain an ordered queue of waiting threads, in which each thread is polling its own go flag. Since only one thread is polling each of these cache lines, each can be in the cache of the spin-waiting thread, so polling will generate no coherence traffic. Similarly, since only a single thread is polling each line, the write to wake the waiting thread is, in effect, a point-to-point communication, and we are at the very left of our visibility plot (see Figure 3.18) where the transfer time is the lowest.

The code for an MCS lock is shown in Listing 6.6. You can see that this is slightly more complicated than the previous locks, and has additional infrastructure requirements, such as the need for a thread to have an identity (which, here, we get from `omp_get_thread_num()`, since that is an easy way to do it in our implementation). You can also see that each thread requires its own lock entry (in this code, an instance of the `MCSLockEntry` class), which is the data structure from which the queue of waiting threads is built.

The `tail` member points to the tail of a singly-linked list of threads whose head is the thread that currently owns the lock. To acquire the lock a thread exchanges a pointer to its `MCSLockEntry` with the tail; if the tail was previously zero (we use zero as a synonym for the C++ `nullptr` value here), then no thread holds the lock, so this thread now does so and it can proceed. If the tail was not previously zero, it needs to finish linking itself into the list, which it does by storing a pointer to itself into the `next` field of the previous tail (which it knows, because that is the value it got back when it exchanged itself into the global tail pointer). It then has to wait until it is woken, by spinning on its own go flag.

On release, the releasing thread attempts to CAS zero into the tail pointer, using its own `MCSLockEntry` pointer as the expected value. If no other thread has joined the queue, then `tail` will still be pointing at this thread, which means that there will be no queue of waiting threads, and therefore releasing the lock completely is appropriate. However, if the CAS fails, then the releasing thread knows that there is another thread waiting that it must wake up. It needs to tread carefully, though. Although another thread has updated the `tail` pointer, it may not yet have stored its pointer into the releasing thread's `next` field. Therefore, the releasing thread has to wait until that update has happened (by polling its `next` field until it is non-null). Once it has that pointer, it can set the go flag in its successor to allow it to enter the critical section, reset its own state, and finish.

One implementation detail of the MCS lock is that it requires a slot in the queue for each of the threads (see `MAX_THREADS` in Listing 6.6). There are several ways to handle this. For Listing 6.6, we have chosen to go the easy way and statically define an upper bound for the number of threads. A more dynamic approach might be needed, which will make the implementation more complex, as now queue slots have to be added when the number of threads outgrows the pre-allocated queue slots, when the lock is initialized.

6.6 Actual code performance

As we now have a reasonable lock zoo, it's time to measure their performance on the metrics we previously discussed. Since we have already shown that the TAS lock has unacceptable interference properties, we won't include it, and as the other locks all behave reasonably on that metric, we won't show the interference data for them.

```cpp
class MCSLock : public abstractLock {
  class MCSLockEntry {
  public:
    CACHE_ALIGNED std::atomic<MCSLockEntry *> next;
    std::atomic<bool> go;

    MCSLockEntry() : go(false), next(0) {}
    ~MCSLockEntry() {}
  };

  CACHE_ALIGNED std::atomic<MCSLockEntry *> tail;
  MCSLockEntry entries[MAX_THREADS];

public:
  MCSLock() : tail(0) {}
  ~MCSLock() {}

  void lock() {
    int myId = omp_get_thread_num();
    MCSLockEntry *me = &entries[myId];
    MCSLockEntry *t =
                tail.exchange(me,std::memory_order_acquire);

    // Anyone in the queue?
    if (t == 0)
      return; // No. So I now own the lock.
    // Yes: we must link ourself into the previous tail.
    t->next = me;

    // Then wait to be woken
    while (!me->go) {
      architecturalYield();
    }
  }

  // to be continued
```

Listing 6.6: Implementation of the Mellor-Crummey & Scott lock.

```cpp
// continued

void unlock() {
  int myId = omp_get_thread_num();
  MCSLockEntry *me = &entries[myId];

  // Need a sacrificial copy, since compare_exchange
  // always switches in the value it reads
  MCSLockEntry *expected = me;

  // Am I still the tail? If so no-one else is waiting.
  if (!tail.compare_exchange_strong(
          expected, 0,
          std::memory_order_release)) {
    // Someone is waiting, but they may not yet have
    // updated the pointer in our entry, even though
    // they've swapped in the global tail pointer, so
    // we may need to wait until they have.
    MCSLockEntry *nextInLine = me->next;
    for (; nextInLine == 0; nextInLine = me->next) {
      architecturalYield();
    }

    // Now we know who they are we can release them.
    nextInLine->go = true;
  }

  // Reset our state ready for the next acquire.
  me->go = false;
  me->next = 0;
}

// Name and factory omitted.
};
```

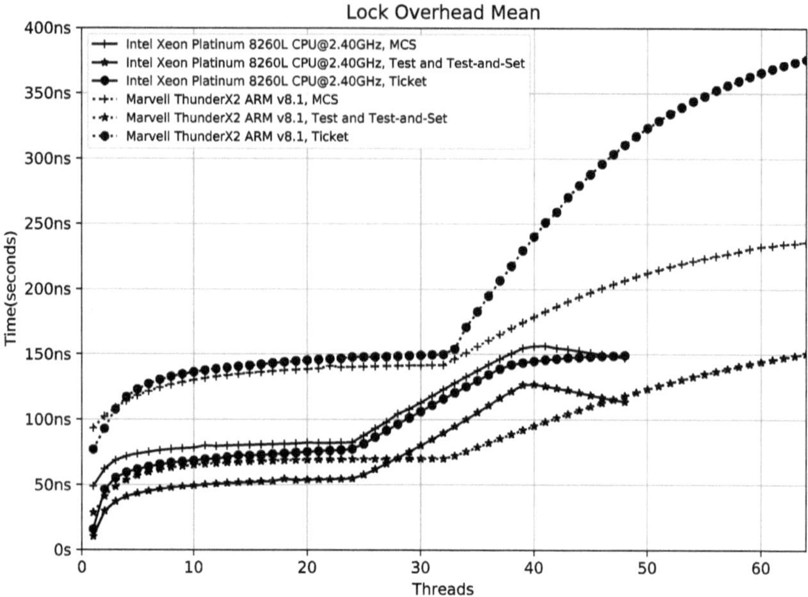

Figure 6.4: Lock overhead (sum of lock and unlock time in thread).

6.6.1 Uncontended lock overhead

In Figure 6.4, we show the time measured in a thread to acquire and release a lock. Since this metric is of interest for uncontended locks, our benchmark code selects a random lock from an array of 4,096 locks. Since there are at most 64 threads, the chance of contention is low, but, by doing this, we ensure that we're not measuring the performance when each thread has its own lock that is in its L1 cache.

We can see that the lowest overhead lock here is the TTAS lock, which wins on all of our machines, while the highest overhead is the MCS lock on the Intel Xeon Platinum 8260L Processor, and the Ticket lock on the Arm processor.

However, before we decide that the TTAS lock is the winner and the one that we should use, we should also take a look at how the locks perform, when there's some more contention.

6.6.2 Contended lock throughput

In Figure 6.5 and Figure 6.6, we show the overall machine throughput for each type of lock when all threads are attempting to execute an empty critical section. This measures the lock-induced exclusive time, which we discussed in Section 6.4.1. Under this high contention, the best we can hope for is that the throughput remains constant as we add threads. Since we are measuring machine throughput, higher is better on

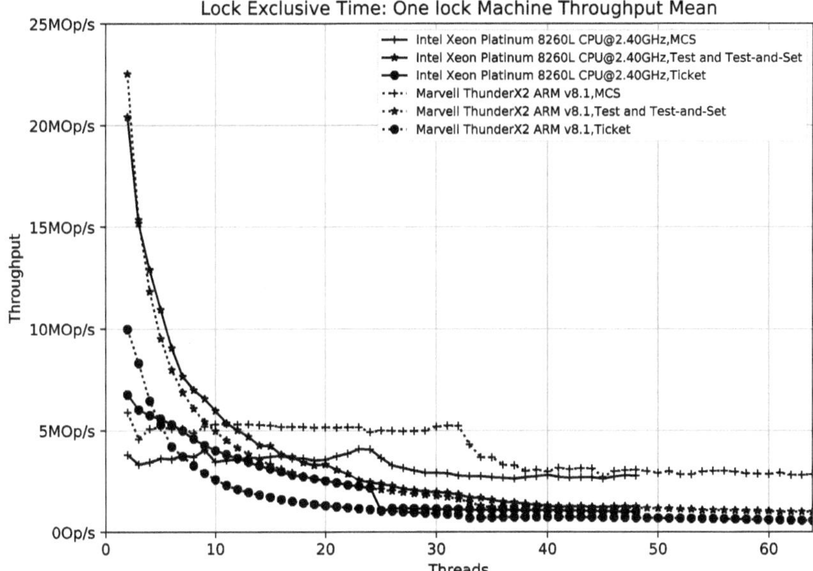

Figure 6.5: Contended lock total machine throughput (single lock).

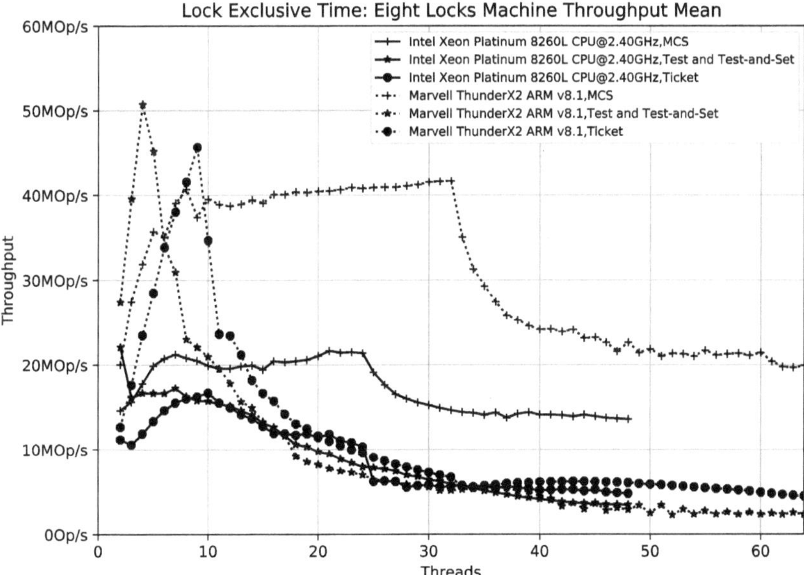

Figure 6.6: Contended lock total machine throughput (eight locks).

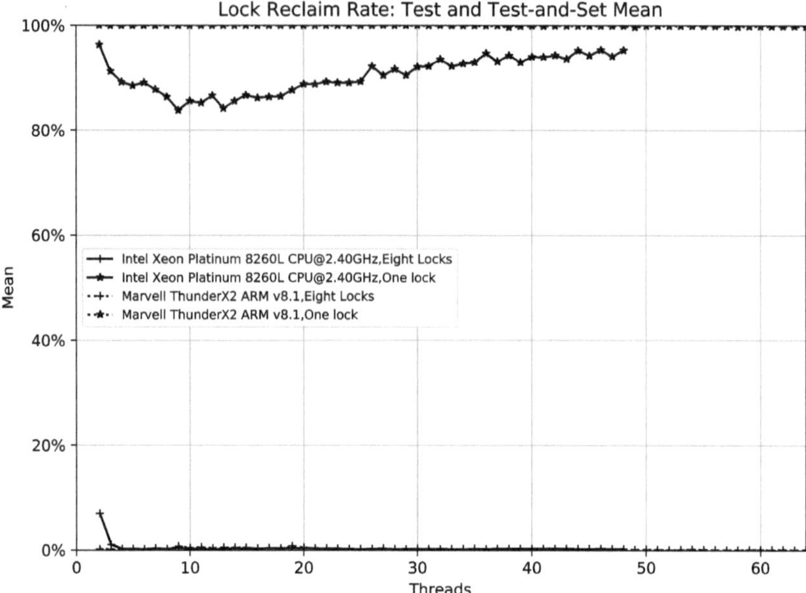

Figure 6.7: TTAS lock-reclaim rate.

these graphs, whereas in our previous measurement of overhead lower was better. We omit the single thread case in these plots, since that is clearly uncontended, and as it has very good performance, it affects the scale and makes it hard to read the more interesting data.

Figure 6.5 shows that data. Here we can see that the TTAS lock performs well on both machines at low thread counts, but drops off in performance on the Arm machine where MCS outperforms it above 11 threads, and on the Intel Xeon Platinum 8260L Processor, MCS outperforms it above 17 threads.

Figure 6.6 shows what happens if we use eight locks, where each thread is iterating over each of the locks (so with fewer than eight threads we are measuring only a partially contended case). Under these circumstances, the MCS lock is better on all of the machines, while the TTAS lock is poor on all of them, and is even out-performed by the Ticket lock at high thread counts.

In Figure 6.7, we can see the reason for this performance difference. When contending for a single lock, the TTAS lock has a close to 100 % reclaim rate on the Arm machine, and >80 % on the Intel processor. However, when we iterate over eight locks, the rate drops effectively to 0 % on all of the machines. There is then no clumping, and we're paying the full cost of lock transfer. Since they are fair, the MCS and ticket locks can never show clumping under contention.

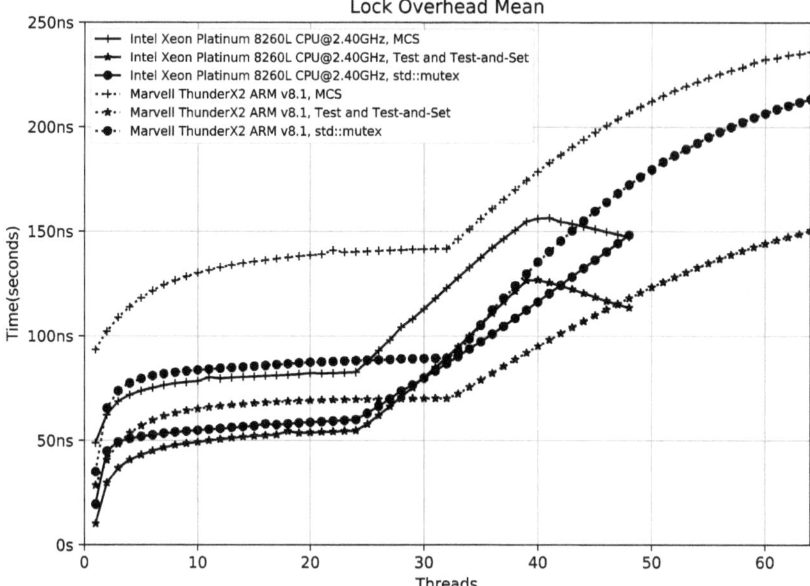

Figure 6.8: Uncontended lock overhead (including `std::mutex`).

6.6.3 Performance conclusions

The Ticket lock is consistently out-performed by other locks in either of our test cases, so we will not consider it further.

This seems to leave our unfair TTAS lock and our fair MCS lock as the most interesting locks; however, although we have been comparing our lock implementations amongst themselves, we haven't yet looked at other, already existing lock implementations. Therefore, we will now add in the `std::mutex` class of C++ and compare that with our MCS and TTAS locks.

First we look at the lock overhead. Figure 6.8 shows the lock overhead of our MCS and TTAS locks as well as `std::mutex`. The `std::mutex` has an overhead that is less than our MCS lock, but higher than the TTAS lock.

Figure 6.9 shows the throughput for our TTAS and MCS lock compared with that of `std::mutex` when contending for a single lock. Here we can see that the `std::mutex` provides the highest throughput on both of the tested machines above 12 threads, beating the MCS lock by 2× when we're contending across sockets.

Compressing the way we display the data risks removing important aspects of the actual performance of each lock, and may focus too much on our specific test cases. However, if we are prepared to do that, we can summarize that data in a table by taking the geometric mean over all thread counts, and then expressing that as a ratio to the best performing lock. That gives us Table 6.2, where we can see that on all of our

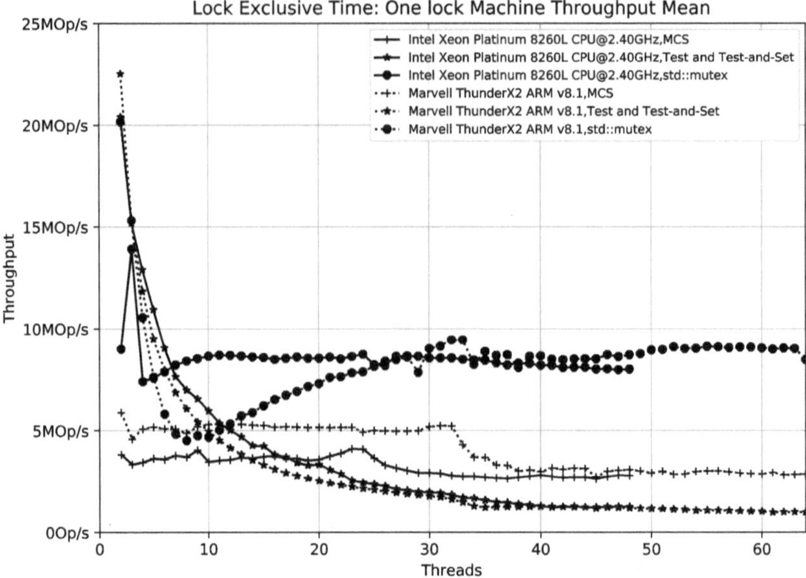

Figure 6.9: Contended lock total machine throughput (one lock).

Table 6.2: Geometric mean lock performance of one and eight locks.

	Single Lock			Eight Locks		
Architecture	MCS	TTAS	std::mutex	MCS	TTAS	std::mutex
Marvell TX2	49%	25%	100%	100%	21%	56%
Platinum 8260L	38%	34%	100%	100%	50%	65%
Geometric mean	43%	29%	100%	100%	33%	61%

machines the `std::mutex` has the best overall performance when dealing with a single lock, beating our best lock (TTAS on the Intel machine, MCS on the Arm processor) by more than a factor of two in this completely contended case. When we have eight locks, our MCS queueing lock is the best on all machines, but the `std::mutex` achieves 61% of its performance. Figure 6.10 shows the same measurement when iterating over eight locks.

Although this shows one of our locks beating the `std:mutex` in one configuration, it does not invalidate Linus' position, since it is only one test case, and his argument is that optimizing for a specific benchmark is suboptimal in general. On that basis, aside from now understanding the issues better, we seem to have been wasting our time, and should just use `std::mutex`. However, we have not yet looked into how we should be waiting, and that can also affect lock performance.

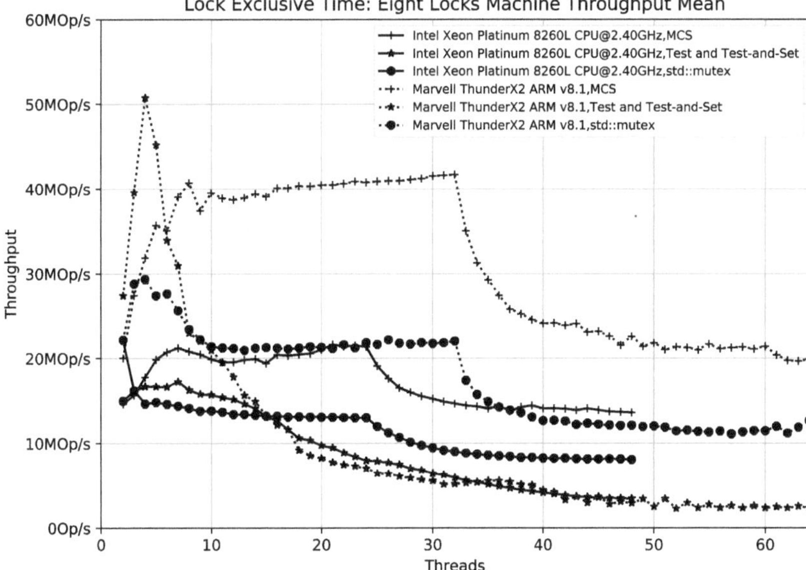

Figure 6.10: Contended lock total machine throughput (eight locks).

6.7 How to wait

We saw above that our locks have significant issues when we have contention and, therefore, threads have to wait. This is a more general issue than simply being a problem for locks, since it occurs anywhere a thread has to wait until a condition is satisfied. That could be at a barrier, where a thread is waiting for all threads to complete the work required before any can proceed, at the end of a serial region, or, as in this case, where a thread is waiting for a lock to be freed.

As usual, there are many different ways and metrics that we can use to measure and assess the performance of this wait operation, and they represent conflicting desires:

- **Wakeup Latency:** From a selfish point of view, we want the waiting thread to wake up as soon as possible after the condition on which it is waiting is satisfied. This can normally be achieved by a very tight polling loop. However, this has unpleasant side effects, since it consumes CPU time and energy both of which could, potentially, be used more productively for the progress of the whole program if they were used by other threads. Since spin-wait loops block the CPU, if there is over-subscription it will also hurt us, as the polling thread could be preventing the thread that could be doing useful work (in the case of a lock, the thread that holds the lock) from running.

  In machines with simultaneous multi-threading (i. e., where the physical cores are shared between multiple hardware threads at the instruction level), this can also prevent useful work from occurring in another thread which is sharing the same CPU hardware, even though that thread has been scheduled by the operating system and is running.

- **Resources Consumed:** From a higher level, we want to minimize the resources consumed by the waiting thread. Ideally we want the operating system to know immediately that the thread has no useful work to perform so that it can either use the hardware on which the thread was running for some other thread, or put the hardware into a low power state.

  However, this will certainly increase the latency between the condition on which the thread is waiting being satisfied, and the thread resuming execution, since if the thread is suspended in the kernel, then the thread which satisfies the condition must make a system call, and the kernel must then cause the woken thread to return from the call it made to suspend itself. Both of those operations take significantly more time than a simple load or store.

In all of our lock code above, we have prioritized wakeup latency by polling in tight loops (though with an architecture-specific SMT yield), but the results show that this is a poor choice, since the `std::mutex` is providing higher performance in many cases.

6.7.1 Backoff strategies

We have seen that our active polling can have an unpleasant performance impact on unrelated code, and will always be consuming more CPU time and energy that could be used elsewhere, and therefore we shouldn't be doing it. But what should we do?

The first, and simplest, thing to do on machines which have simultaneous multithreading is to tell the CPU hardware that the executing thread can give up resources to other SMT threads which share the same core. That is achieved by executing an instruction (on x86: `pause`; on an Arm processor: `yield`) that passes that information straight to the hardware. Smart out-of-order cores can then attempt to give more of the shared resources to other threads which can benefit from them, or, if all threads are in this state, potentially lower the clock frequency and reduce energy consumption.

The second thing to do in cases where the act of polling can itself cause any form of interference is to reduce the rate at which we poll—for instance, by a random exponential backoff between the points when we check the condition. This cannot be used usefully for fair locks, since it is very likely that a thread which has been waiting for a long time will have moved to polling rarely, yet if it has waited a long time its turn is likely to arrive soon.

The third thing which we can potentially do is to tell the hardware explicitly that the executing thread is waiting for a specific cache line to be modified. The x86 architecture has had support for doing that in kernel code for a long time, but, at least in Intel's implementations, the relevant instructions (`monitor` and `mwait`) could not be executed outside the operating-system kernel. However, future Intel processors will implement new instructions in the "WAITPKG", which provides `umonitor`, `umwait` and `tpause` instructions [68]. The `tpause` instruction enhances the `pause` instruction to add an elapsed time for which the pause should wait, and, can also be executed inside a speculative region, whereas `pause` would always cause a speculative abort. This allows the construction of code which can wait for a change to any one of a set of cache lines, whereas the basic `umonitor` and `umwait` instructions can only wait on a single line. To achieve this, start speculative execution, read each of the lines of interest to get them into the transaction's read set, then `tpause`. If any of the lines is externally modified the transaction will abort because of the write-after-read conflict, so the thread will be woken from the `tpause`. Unfortunately, we do not have support for WAITPKG instructions on our machines so we can't measure the performance, but it should be useful in the future.

The most drastic way of waiting is to inform the operating system that the thread is waiting for a lock. At this point, we're at the mercy of the operating system's lock-related primitives. Since we cannot look at all OSes, we'll just take a quick look at Linux,[*] as it is the primary OS used in HPC environments, and is representative of the sort of functionality any OS must provide,

Linux implements a system call named `futex()` [130], which is designed to be used to handle waiting for a resource. In its simplest form, threads wait on a futex, and another thread wakes some number of them (normally either one, or all, but the system call allows any number to be woken).

Figure 6.11 shows the time between the root thread making the futex call to release the waiting threads, and the last thread to leave on our Arm and Intel machines. We can see that this is large when we have many threads waiting on the same futex (>100 µs at 32 threads in the same socket, and >500 µs to wake 63 other threads spread across the two sockets on the Marvell Arm machine and ~170 µs to wake 47 threads on the Intel Xeon Platinum 8260L Processor). It is also obvious that the time is increasing super-linearly on the Arm machine. Given these timings, we should clearly not be using a futex to implement a broadcast operation in which many threads are waiting on the same futex, but rather limit the number waiting on a single futex, and then either have the root thread iterate over all of them, or use a tree so that we can have many futex calls in flight in different threads at the same time.

What this is telling us is not that we should not use a futex, but rather that we need to be careful where and how we use it. As ever, we need not worry hugely about a 5 µs cost if we have already been waiting for 250 µs, since it is only 2 % overhead which is likely to be immeasurable, and informing the OS that threads are idle has benefits.

Another idea which looks initially attractive is to use a call (such as Linux' `sched_yield()`) to attempt to allow another thread to execute, but not mark the

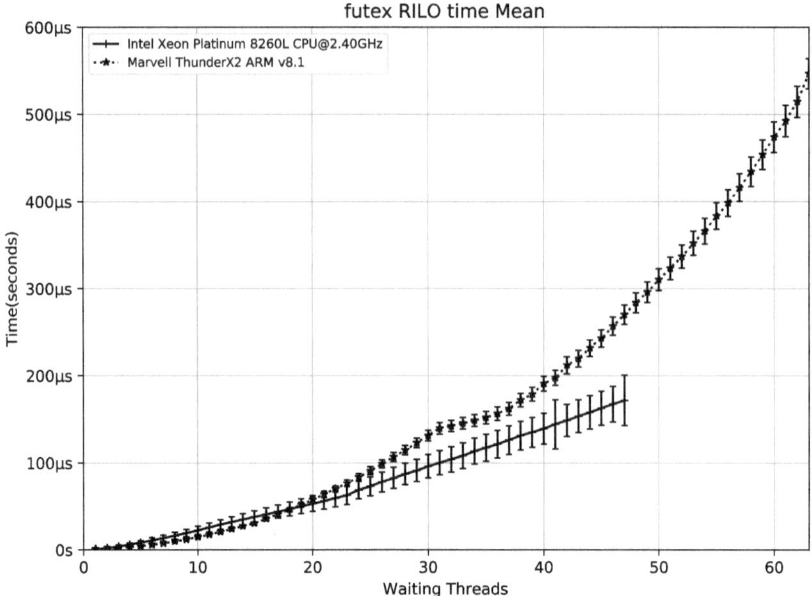

Figure 6.11: Futex RILO (Root In, Last Out) time.

current thread as suspended, so the OS will run it again at some future point. Unfortunately, this is based on a scheduling model that no longer exists, and will very likely lead to this thread being immediately re-scheduled, so the effect is merely that of an expensive polling loop in which the delay is implemented by entering the kernel and leaving it.

If we avoid locks which perform an atomic operation unguardedly on every polling operation (such as the poorly performing TAS lock), but rather use locks which always read before attempting to update (such as the TTAS lock), or are merely reading (such as the ticket lock or any queuing lock like MCS), then reducing the frequency with which we poll while waiting will only increase the latency for a specific thread to succeed in claiming the lock without reducing impact elsewhere, since the polling loads will all be to a shared copy of the cache line of the lock in L1 cache. However, by implementing a backoff, what we achieve under high contention is clumping (increasing the unfairness of the lock). We make it much more likely that the thread which released the lock will be the one which immediately claims it back. That can improve the overall machine throughput, since it means that many of the transfers of the lock's cache line between threads do not happen, so our average lock exclusive time drops.

The exponential backoff is implemented in the class shown in Listing 6.7. It calls a cheap maximum-length feedback-shift register random-number generator (initialized from the stack address, so it will be different in each thread) to generate a random number that is then masked by a mask that represents $2^n - 1$. This value is then scaled

```cpp
// We use the CPU "cycle" count timer to provide
// delay between around 100ns and 25us.
class randomExponentialBackoff {
  const float smallestTime = 100.e-9; // 100ns
  // Multiplier used to convert our units into timer ticks
  static uint32_t timeFactor;
  mlfsr32 random;
  uint32_t mask;             // Limits current delay
  enum { maxMask = 255 }; // 256*100ns = 25.6us
  uint32_t sleepCount;       // How many times has sleep been
                             // called
  uint32_t delayCount;       // Only needed for stats

  enum {
    initialMask = 1,
    delayMask = 1, // Do two delays at each exponential value
  };

public:
  randomExponentialBackoff(): mask(1), sleepCount(0),
                              delayCount(0) {
    // Racy; doesn't matter since everyone will set the
    // same value.
    if (timeFactor == 0)
      timeFactor = smallestTime /
                   tsc_tick_count::getTickTime();
  }
  void sleep() {
    uint32_t count = 1 + (random.getNext() & mask);
    delayCount += count;
    tsc_tick_count end =
        tsc_tick_count::now().getValue() + delayCount *
        timeFactor;
    // Up to next power of two if it's time to ramp.
    if ((++sleepCount & delayMask) == 0)
      mask = ((mask << 1) | 1) & maxMask;
    // Wait until after the time we decided.
    while (tsc_tick_count::now().before(end))
      architecturalYield();
  }
  uint32_t getDelayCount() const { return delayCount; }
};
```

Listing 6.7: Random exponential backoff.

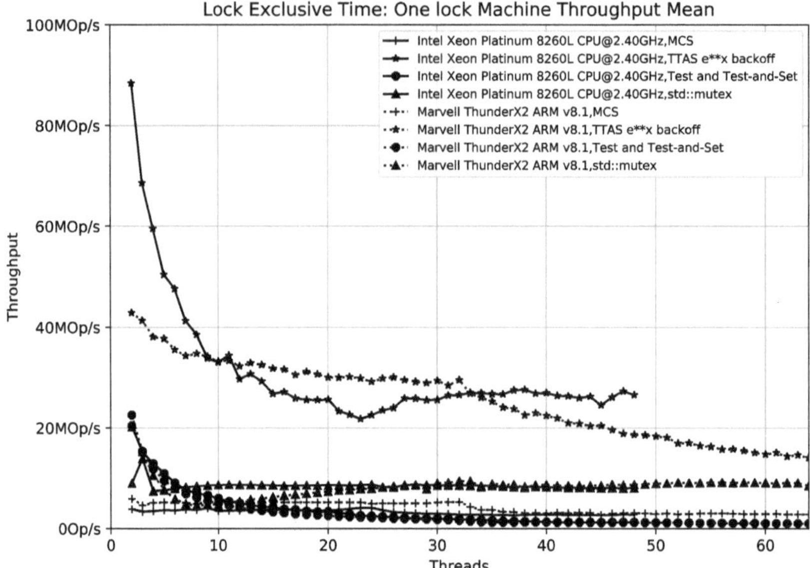

Figure 6.12: Contended lock machine throughput with backoff (one lock).

Table 6.3: Geometric mean performance of one and eight locks with TTAS e^x.

	Single Lock			Eight Locks		
Architecture	MCS	TTAS e^x	std::mutex	MCS	TTAS e^x	std::mutex
Marvell TX2	16 %	100 %	33 %	100 %	32 %	56 %
Platinum 8260L	11 %	100 %	29 %	100 %	70 %	65 %
Geometric Mean	13 %	100 %	31 %	100 %	47 %	61 %

to give a number of high-resolution timer ticks and a target time is computed. The code then waits (yielding to any other SMT threads) until the time is after the target time before polling again.

We can see the performance improvement if we apply this random exponential backoff to our TTAS lock, as shown in Figure 6.12 and Figure 6.13. Adding this data to our summary table in Table 6.3, we can see that the TTAS with exponential backoff is now the best-performing lock in the single-lock case, beating the `std::mutex`. Once again, this is the result of achieving even higher clumping. Whether the clumping is acceptable will depend on the specific use case, since this is amplifying the unfairness of the lock. (It also suggests some of the other more complicated approaches to exclusivity, such as "delegation", in which a single thread is used for the critical updates, as described in [91].)

Despite all this, the `std::mutex` remains a reasonable overall choice, especially when we remember the maxim that "the best code is the code I do not have to write."

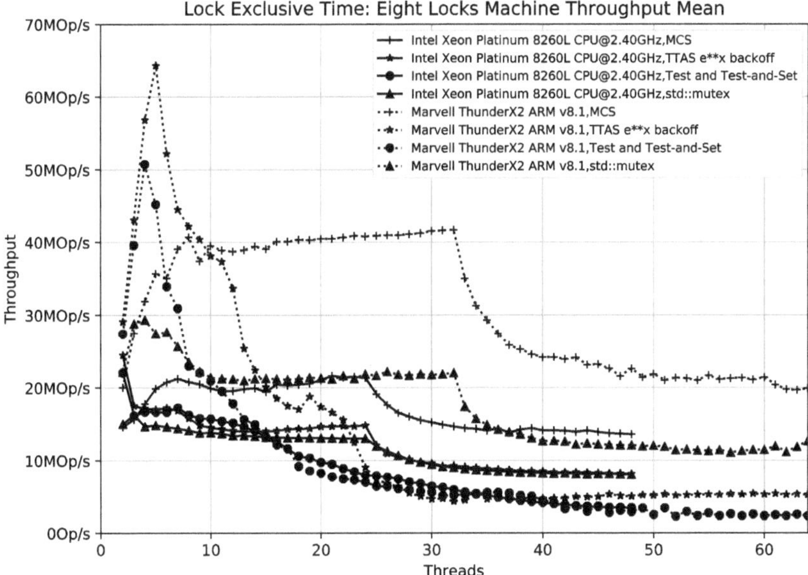

Figure 6.13: Contended lock machine throughput with backoff (eight locks).

6.8 Transactional synchronization

Various modern processors provide some form of transactional memory support (at least some from IBM and Intel do, while the Arm architecture has a specification for "transactional memory extensions").

As we saw before, our real problem is that of interference between concurrently executing threads causing one of them to see an inconsistent state of the data on which they are operating. Locks solve that problem by enforcing serialization of specific sections of the code and thus preventing interference, provided that the programmer has moved the relevant operations inside that critical section. Transactional synchronization takes the opposite approach, allowing many threads to execute speculatively, but detecting potential inconsistencies and aborting speculation should a race condition be observed. Under some circumstances, that can provide higher performance for less programming effort because it maintains parallelism which would otherwise be unnecessarily lost.

Mutual exclusion seems the obvious way to ensure that updates are safe. However, it has some unfriendly properties that can mean that it is worth considering speculation. In particular, exclusion forces serialization that may not really be required. Consider, for instance, using locks to make a serial hash-table implementation a thread-safe one. The obvious way to do this using the OpenMP* API is shown in Listing 6.8.

```cpp
class lockedHash {
  std::unordered_map<uint32_t, uint32_t> theMap;
  omp_lock_t theLock;

public:
  lockedHash(omp_sync_hint_t hint) {
    omp_init_lock_with_hint(&theLock, hint);
  }
  void insert(uint32_t key, uint32_t value) {
    omp_set_lock(&theLock); // Claim the lock
    theMap.insert({key, value});
    omp_unset_lock(&theLock); // Release the lock
  }
  uint32_t lookup(uint32_t key) {
    omp_set_lock(&theLock); // Claim the lock
    auto result = theMap.find(key);
    omp_unset_lock(&theLock); // Release the lock
    return result == theMap.end() ? 0 : result->second;
  }
};
```
Listing 6.8: Thread-safe implementation of `std::map`.

Since the OpenMP API does not have a standard reader/writer lock we have to use a simple lock, which forces complete serialization of the accesses if the lock is implemented normally. However, if the lock is implemented as a speculative lock using the hardware transactional memory support, then we can get the effect of fine-grained reader/writer locking without having to have access to the internals of the `std::map` implementation.

6.8.1 Transactional semantics

The transactional semantics we require are:
- **Isolation:** No stores that occur inside the transaction become visible until the transaction commits.
- **Atomicity:** When the transaction commits, all of its stores become visible at the same time.
- **Conflict detection:** Conflicting memory operations cause the transaction to abort, at which point the effects of any operations inside the transaction are discarded and never become visible.

The potential conflicts are (similar to what we have discussed in Section 3.1.3 already):

- **Write after Read:** A write-after-read conflict occurs when the transactional thread has read a location and another thread writes to it.
- **Read after Write:** A read-after-write conflict occurs when the transactional thread has written a location and another thread reads from it.
- **Write after Write:** A write-after-write conflict occurs when the transactional thread has written to a location and another writes to the same location.

In the abstract, we can implement these requirements by using a read set that holds the set of locations that have been read inside a transaction, and a write set, which does the same for modified locations. Race conditions can be detected by determining if a write set overlaps with any other read or write set and by ensuring that none of the writes that are performed become visible to other threads until the transaction commits.

6.8.2 Implementation in the MESI protocol

While we cannot know in detail how the different architectures implement transactional memory, we can only show one possible way in which transactional synchronization can be implemented and that reflects the general approach. It leverages the existing cache-coherence protocol and implements conflict detection at the cache-line level. This means that we may get false conflicts if two unrelated variables are allocated next to each other in memory and end up in the same cache line, but that is a safe failure; it does not allow for incorrect behavior.

The read sets and write sets can be maintained by adding an additional bit to the cache tags for each line to hold the information indicating that the line was accessed inside the transaction. The read set is then all of the unmodified lines in the cache that have the "transactionally accessed" bit set, while the write set is all of the modified lines with that bit set.

Isolation can be enforced by ensuring that no lines in the write set can be snooped out of the cache. Atomicity can be enforced by clearing all of the transactionally accessed bits simultaneously and by not allowing any external cache operations to occur while that is happening. Aborting can be implemented by marking all transactionally-modified lines as invalid and then clearing the transactionally-accessed bits of all lines. Conflicts can then be detected by checking the transactionally-accessed bit associated with a cache line as well as its other state when an external request occurs. If we see a write request for a line that has the transactionally-accessed bit set, then this is either a write-after-read or write-after-write conflict, either of which cause an abort. If we see a read request for a line in the "transactionally accessed, modified" state, then that represents the read-after-write conflict, and again the transaction should be aborted.

One important point here is that the transactional state is all local to the cache of the thread executing transactionally. There are no changes to the external cache protocol required, merely additions to the behavior of the local cache. That is good, because changing a cache protocol is a hard thing to do as even managers know that cache protocols are complicated and therefore changing them is dangerous. However, it means that there is no global awareness of transactions, and thus no easy way to ensure that any transaction completes and forward progress is made. As an example, the Intel architecture specification makes this clear: "Hardware provides no guarantees as to whether a transactional execution will ever successfully commit" [68]. The implication of this is that any transactional lock must also have a backoff implementation which uses a "real" lock, and ensure that transitions between the state in which the real lock is being used, and that in which transactions are being used are correctly handled.

Of course, other extensions to the hardware are possible. Since the implementation above stored the read set as bits associated with cache lines, it requires that none of these lines be evicted from the cache while the transaction is executing. This limit on the read set size can be overcome if an additional hardware resource is used instead of, or in addition to, using the cache to hold the read set. For instance, one could provide a hardware implementation of a Bloom filter [14], which would allow larger read sets to be handled. Although a Bloom filter is less precise than maintaining a bit per cache line, its failures are all in a safe direction, since it will never say that an address is not in the set when it is, though it may say a line is in the set when it isn't. That could cause unnecessary transactional aborts, but cannot cause the required semantics to be broken.

6.8.3 Transactional instructions

The instructions used by Intel and those specified by Arm to enable speculative execution are quite similar. Each provides:

- **A way to enter speculative execution:** xbegin in the x86_64 architecture, tstart in the Arm architecture.
- **A way to abort speculative execution:** xabort in the x86_64 architecture, tcancel in the Arm architecture.
- **A way to finish speculative execution and make changes visible:** xend in the x86_64 architecture, tcommit in the Arm architecture.
- **A way to determine whether the processor is executing speculatively:** xtest in the x86_64 architecture, ttest in the Arm architecture.

The instruction that starts speculation is rather unusual, since, in the case of a transactional abort, it will complete execution twice. The first time it completes, it will have caused the processor to enter speculative execution; the second is when the transaction has aborted, and the machine state has been restored. To tell us what is going on, these instructions set a value in a register that includes status bits to allow the code to detect whether execution is now speculative, or, whether a transactional abort has occurred, and, if so, to give more information about why that happened.

This information can then be used by the retry code to attempt to work out whether it is worth retrying the speculative execution, or whether the code should backoff to the underlying real lock and force serialization. (For instance, if the reason for the failure is an instruction which is not supported in the transactional execution has been executed, then there is little point in retrying the transaction, whereas, if the reason is a conflict or that an interrupt that happened to arrive, then retrying may be appropriate.)

6.8.4 Transactional locks

We can implement simple speculative versions of some of our locks by adding speculation around the existing lock. To do this, we need one more method in the underlying lock: `isLocked()`, which just tells us whether the lock is currently locked. We can then create a template like the one shown in Listing 6.9. To avoid the listing being too long, critical comments have been removed, so we have covered them here:

- **Issue 1:** When acquiring the lock speculatively, we must check that the lock was still unclaimed after we have entered speculation, since there is a window of time between us checking (on line 8) and this thread entering speculation (on line 13) during which another thread could claim the lock. Testing the lock inside the speculative region also ensures that the cache line holding the lock is in the read set of the transaction, and, therefore, that if any other thread claims the lock the transaction will abort.
- **Issue 2:** At this point (line 21) the thread is executing speculatively inside the critical section, so the library can return to user code to allow it to execute the code it wanted to protect with the lock.
- **Issue 3:** If we reach the `unlock()` function, we must determine whether we were executing the lock speculatively. We can do this by checking whether we have acquired the lock "for real", in which case the underlying, non-speculative lock will be acquired. If it is, we must release it; if it is not, we must be executing speculatively, so we can commit all of the changes made during the speculative execution. The hardware could still abort at this point, in which case the commit will fail and execution will continue at the return from `architecturalSpeculationStart()` with the speculative changes forgotten and the machine state (aside from the result of that function) restored.

```cpp
1  template <class baseLockType>
2  class speculativeLock : public abstractLock {
3    baseLockType baseLock;
4
5  public:
6    // Empty constructor/destructors elided
7    void lock() {
8      while (baseLock.isLocked()) {
9        // Wait for the base lock to be unlocked.
10       architecturalYield();
11     }
12     // Try to speculate
13     if (architecturalSpeculationStart() == -1) {
14       // Executing speculatively.
15       // Issue 1
16       if (baseLock.isLocked()) {
17         // Executing this causes
18         // architecturalSpeculationStart()
19         // to return again!
20         architecturalAbortSpeculation(0);
21       }
22       // Issue 2
23       return;
24     } else {
25       // Aborted. Claim the real lock.
26       baseLock.lock();
27     }
28   }
29
30   void unlock() {
31     // Issue 3
32     if (baseLock.isLocked())
33       baseLock.unlock();
34     else
35       architecturalCommitSpeculation();
36   }
37 };
```

Listing 6.9: Speculative lock template.

A subtle point in this code is that we have to ensure that we can recover from executing using the real lock and resume speculation. If we use a queuing lock as our backoff, and allow multiple threads to join the queue, then the lock will never be free under heavy contention. Therefore, no thread can ever start to speculate. To overcome this here, we don't let threads attempt to take the lock until they have seen it free, tried speculation, and aborted. This works, but means that we rarely get any benefit from the queuing lock. The problem is that a queuing lock is designed to ensure that only the thread at the head of the queue can take the lock when it is released, but the whole idea of speculative, transactional execution is that many threads are executing inside the critical section at the same time. Therefore, many threads need to be woken and allowed to enter the critical region speculatively. If we went deeper inside the lock implementation (rather than wrapping an existing implementation with speculation), we could perform additional optimizations here.

As with all of our examples to date, we are not performing sensible backoff, or, in this case, making decisions based on the reason why speculation failed. Both of these are obvious improvements that would be required before using this code in a real application.

To demonstrate the potential performance advantage of speculation, we can use these locks to provide thread-safe access to a `std::unordered_map`. In the test code, we populate a `std::unordered_map<uint32_t,uint32_t>` with 10,000 random entries, and then perform concurrent lookups and updates in multiple threads. In Figure 6.14, we show results with 0 %, 1 %, 2 %, or 5 % updates for our MCS lock using the speculative wrapper, and `std::mutex`. As you should expect, the non-speculative `std::mutex` lock has constant machine throughput; adding threads merely causes them to wait. However, the speculative lock shows significant performance improvements within a single socket even with 5 % update rate. When we move to the second socket, its performance when there are any updates is worse than the `std::mutex`, so, once again, one needs to tread carefully.

One could argue that this is unfair on the standard locks, since it seems a case where a `std::shared_mutex` reader/writer lock should be used. However, it is not clear that it is safe to wrap something like `std::unordered_map` with a reader-writer lock unless one has investigated the details of its implementation, since it could be updating its internal data-structures even on a lookup. Just consider an implementation that uses a small, cache-like hash-table to hold recently accessed entries in the expectation that they will be accessed again soon. It will be performing internal updates even though the external interface shows the lookup methods as `const` qualified. Indeed, the C++ standard supports the `mutable` type qualifier precisely to make it easier for code in `const` methods to update data in a class instance.

It is also worth noting that thread-safe implementations of data structures that are equivalent to `std::unordered_map` (such as Threading Building Blocks' `tbb::concurrent_unordered_map`) exist, and are very likely a better solution to the specific problem used for the benchmark example.

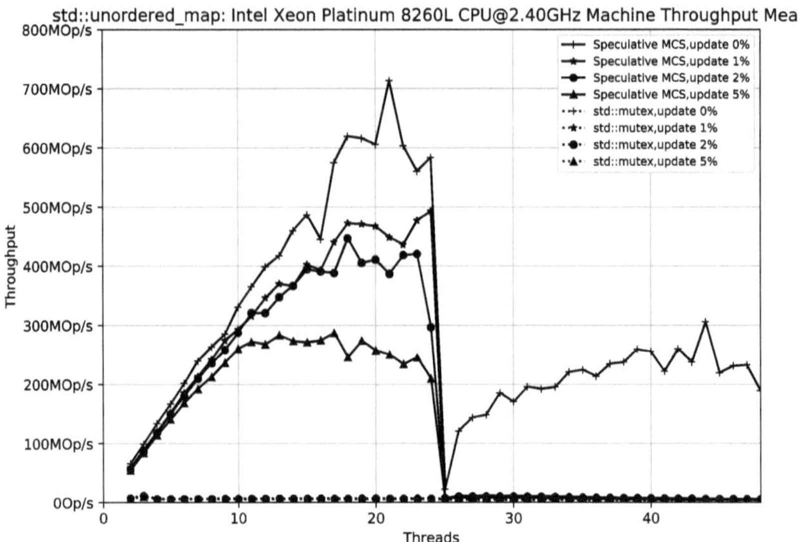

Figure 6.14: Speculative lock total machine throughput.

6.8.5 Comparison of mutual exclusion and speculation

Mutual exclusion seems to be the obvious way to ensure that updates are safe. However, it has some unfriendly properties which mean that it is worth considering speculation. In particular, mutual exclusion always forces serialization which may not be required.

On the other hand, speculation can be wasted work if a conflict occurs after a thread has nearly completed executing through the critical section. Whether that is real wasted work depends on whether the resources that were used could have been used more productively elsewhere. If the non-speculative lock we are using has no sensible backoff then the thread should have been spin-waiting for the lock anyway.

One other advantage which a speculative lock has for small critical sections that only touch only a few cache lines is that the speculative lock does not write to the cache line which holds the backup lock. Therefore that line does not have to be moved into an exclusive state and copied around the machine. Since data movement is often a performance-limiting factor, reducing the number of lines which have to be moved can be beneficial even if there is no contention, so the headline value of speculation (avoiding serialization) does not apply.

6.9 Other serializing operations

As well as the locking operations that we have discussed up to now that ensure that only a single thread can execute inside a particular region of code at a time, we can also serialize by ensuring that a particular dynamic region of code is only executed once, either by a specified thread (e. g., the main thread) or by the first thread that arrives at the given code block.

The OpenMP API provides the master and single constructs to enable this. (Note that the master construct does not have an implicit barrier after it, while the single construct does, therefore only the single construct can take the nowait modifier.) As we mentioned in Section 2.1.4, the master construct is likely soon to be subsumed by more general masked construct which is expected to be in the OpenMP 5.1 standard. The masked construct also does not have any implicit barriers.

6.9.1 The master and masked constructs

The implementation of the master and masked constructs is rather easy; simply check whether the executing thread is one of the ones that are allowed into the guarded region. For the master construct, that is always the thread for which omp_get_thread_num() returns the value 0. The GCC compiler simply inlines the test, so it converts

```
#pragma omp master
{
    ... body ...
}
```

into

```
if (omp_get_thread_num() == 0) {
    ... body ...
}
```

and does not generate any other runtime calls (as we showed in Section 4.4.4). The LLVM compiler does call into the runtime and invokes the function __kmpc_master(), whose minimal implementation can be something like this:

```
int32_t __kmpc_master(ident_t *, // where
                      int32_t)   // gtid
{
    return omp_get_thread_num() == 0;
}
```

The LLVM compiler also calls the function `__kmpc_end_master()`, whose minimal implementation can be completely empty. While the GCC implementation should run faster, the LLVM implementation has the advantage that it can generate the tool callbacks that OpenMP performance-tracing tools would like to see. This is not to say that the GCC implementation is incorrect; most tracing callbacks are optional as per the OpenMP specification.

Since the `masked` construct is (at the time of writing) only in a draft of the OpenMP specification, compilers do not yet support it, and there are no runtime interfaces to use. However, the runtime implementation (as a simple generalization of that for the `master` construct) should be obvious, though the compiler may need to generate code to evaluate an extra expression in the `filter` clause of the construct (see Section 4.4.4).

6.9.2 The `single` construct

The implementation of the `single` construct is slightly more complicated, since all of the threads are competing to try to execute each dynamic instance of a single region. The OpenMP specification is clear that all threads within a team must encounter every `single` region, so code like that below in which only half of the threads attempt to execute the `single` region is non-standard, and need not execute correctly:

```
if (omp_get_thread_num() & 1 == 0) {
#pragma omp single
  {
    ... body ...
  }
}
```

As a result, the implementation can identify each dynamic `single` region by keeping count of both the number of `single` regions that each thread in the team has seen, and the number which have been started. This allows a thread to determine whether it is the first one to arrive at a `single` region, and, therefore, whether it should execute it. The code for such an implementation is shown in Listing 6.10.

Here, you can see the atomic operation (which you should expect by now) that is required to resolve the potential race as many threads attempt to enter the `single` construct. The per-thread single counter need not be atomic since it is only visible to the thread that owns it. Note that we maintain a 64-bit count here, since we don't want the counter to overflow (which, in this context, means that one thread waits a long time while the global counter is increasing, and then the global counter wraps around, past the waiting thread's value). With a 32-bit counter, if it is incremented once per nanosecond, then the overflow happens after 4.3 s (not the New York Seconds of the

```cpp
int32_t __kmpc_single(ident_t *,    // where
                      int32_t *) { // gtid
  auto myThread = lomp::Thread::getCurrentThread();
  auto myTeam = myThread->getTeam();
  auto mySingleCount = myThread->fetchAndIncrSingleCount();

  return  myTeam->tryIncrementNextSingle(mySingleCount);
}


// In the team, we have a single std::atomic<uint64_t>
auto tryIncrementNextSingle(uint64_t oldVal) {
  // test and test-and-set
  if (oldVal != NetxSingle.load(std::memory_order_acquire)) {
    return false;
  }
  return NextSingle.compare_exchange_strong(oldVal,
                                            oldVal+1);
}
```

Listing 6.10: Implementation of the single construct (__kmpc_single).

multiverse!), which is certainly feasible. Whereas, with a 64-bit counter the overflow would take over 584 years, which does not seem like something we need to worry about (but we might be wrong with that assumption).

No specific operation is required in the __kmpc_end_single() function, since the LLVM compiler introduces separate explicit calls to the barrier functions if a barrier is required.

6.10 Atomic operations

We have seen above that implementing locks that can work well in the general case is extremely complex, to the level where Linus Torvalds explicitly told us not to do it. However, there are smaller operations that are either implemented directly by single hardware instructions or that can be implemented with small speculative sequences that do not require locking that are worth providing direct support for in our interfaces and implementations. These are operations such as simple arithmetic or logical operations (e. g., +, -, ,&, |), and other small operations, such as maximum and minimum. Of course, any of these can be implemented using locks, but much higher performance and ease of use can be achieved if we use hardware functionality directly.

The OpenMP API provides these interfaces through the `atomic` directive; C++ through methods defined as part of the `std::atomic` types.

6.10.1 Mapping of atomic instructions

In many cases, the mapping from an atomic operation in the source code to the required instruction sequence is obvious. For instance, an atomic add operation on an 8-, 16-, 32-, or 64-bit integer on x86 can be mapped directly to an `add` instruction with the `lock` prefix.

In other cases, where there is no single machine instruction, or where the machine architecture uses LL-SC, these operations are small enough that they can be implemented in a small CAS sequence.

You can see the general, simple CAS implementation in Listing 6.11. When compiled with `clang` 9.0.0 for x86_64, this generates the taut assembly code shown in Listing 6.12.

We do, however, need to worry about backoff here too. In a CAS implementation, we are guaranteed overall forward progress, since to cause a thread to fail to complete its speculative section, the CAS must fail at the end of it, which can only happen if some other thread has successfully modified the value. Therefore, this other thread is making progress. However, each unsuccessful attempt that a thread makes may cause the relevant cache line to be moved around the machine, causing lots of coherence fabric traffic, which we have already seen (in our discussion of the TAS lock) can cause poor performance. Therefore, we will also experiment with a TTAS analogue in which we perform a simple test before the CAS operation to see whether the CAS operation can succeed. This has the effect of reducing the number of atomic test-and-set operations performed, and, therefore, the number of times the cache line migrates around the machine.

The backoff here is somewhat different from what we ideally need for a lock, since here we are not waiting for a value to change, but rather for an interval during which it is not changing. Therefore the deeper backoff schemes we discussed for locks cannot be used, since here we cannot use any of the hardware instructions that monitor cache lines, or wait on a kernel event. Our simple exponential backoff class can be reused here, though, since all it does is provide random, exponentially increasing, delays.

To demonstrate some of these effects, consider the data shown in Figure 6.15. Here, we show the overall machine throughput on our Intel Xeon Platinum 8260L Processor and our Marvell TX2 Arm machine for the simple CAS atomic code that we showed in Listing 6.12 ("float (no backoff)"); for the same code, but with an additional load and compare after the addition, but before the CAS operation (adding an `if (*t == current.uintValue`), "float (TTAS)"); and, finally, a version with exponential backoff ("float (random e**x backoff)").

```cpp
static void atomicPlus(float *target, float operand) {
  std::atomic<uint32_t> *t = (std::atomic<uint32_t> *)target;
  typedef union {
    uint32_t uintValue;
    float typeValue;
  } sharedBits;
  sharedBits current;

  current.uintValue = *t;
  for (;;) {
    sharedBits next;
    next.typeValue = current.typeValue + operand;
    if (t->compare_exchange_weak(current.uintValue,
                                  next.uintValue))
      return;
  }
}
```

Listing 6.11: Floating-point addition with compare-and-swap.

```asm
atomicPlus(float*, float):    # @atomicPlus(float*, float)
        mov       eax, dword ptr [rdi]
.LBB0_1:                      # =>This Inner Loop Header: Depth=1
        movd      xmm1, eax
        addss     xmm1, xmm0
        movd      ecx, xmm1
        lock      cmpxchg dword ptr [rdi], ecx
        jne       .LBB0_1
        ret
```

Listing 6.12: Assembly code for Listing 6.11.

We can see that on the Intel machine, moving from the "no backoff" code to TTAS gives us a ~3× improvement in performance, while the exponential backoff adds a little more performance. On the Arm machine, there is little difference between no backoff and TTAS, while the exponential backoff has a huge impact. The exponential backoff provides the highest throughput on both machines, despite their different architectures and memory models.

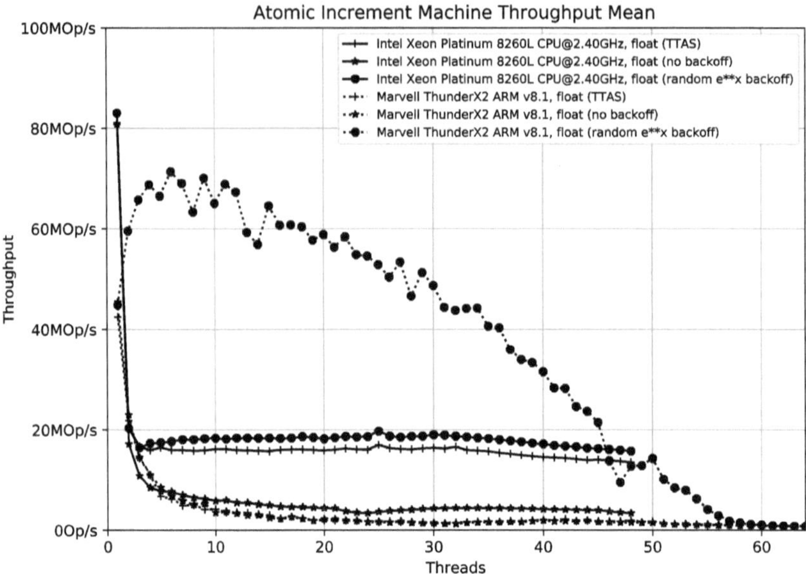

Figure 6.15: Atomic `float32` addition machine throughput.

6.10.2 Atomic implementation of minimum and maximum

There are some atomic operations such as minimum and maximum which clearly can be implemented using CAS in the obvious way, just as we would do for other integer or floating-point operations. However, there is an important additional optimization which should be applied to them, based on the observation that in many cases there is no need for an update at all.

Consider, for example, performing a max operation on the sequence $(2, 1, 4, 3)$. As we scan this, our initial running value will be 2; when we compare with 1, there's no need to update; when we compare this with 4, we need to update, but comparing 4 with the final 3 again needs no update. So, we only required one atomic update, rather than the three that we would naively have executed if we used the simple implementation above, replacing the floating-point addition with a maximum operation.

For these operations, we can work out the expected mean number of changes to the running value (and, thus, the number of atomic operations required) like this if we assume that any permutation of values is equally likely. Obviously the worst case would be if the values were sorted, but, if they were, and we knew that we wouldn't need to be executing a max or min operation at all because we could just pick the appropriate extreme element.

Thinking about max (the generalization to min should be obvious), consider the final comparison; it can only introduce a change if it is the maximum value in the whole list, and the probability of that is $1/N$ where N is the number of elements we're

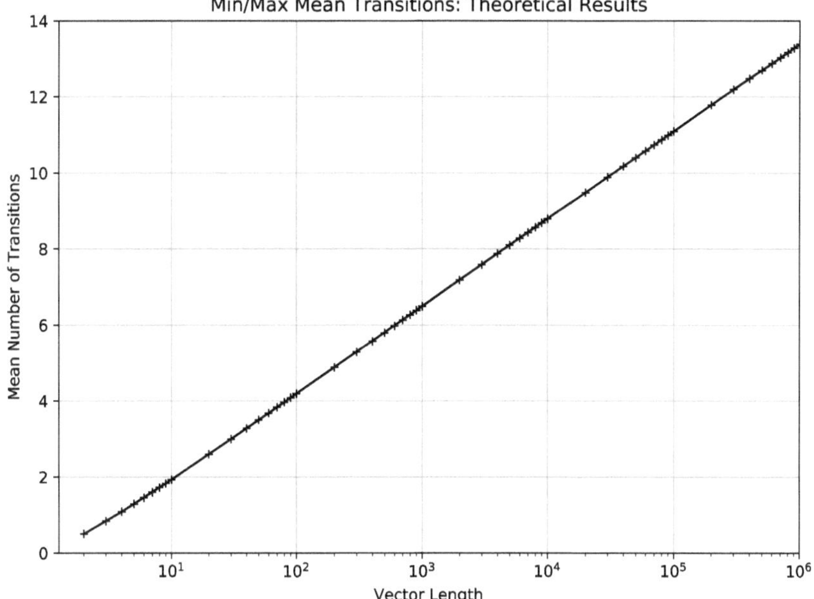

Figure 6.16: Mean required number of atomic operations.

operating on. This final element therefore contributes $\frac{1}{N}$ transitions. Now that we've handled the final element we can apply the same logic to the shorter vector of length $N - 1$. Putting those two cases together gives us the recursive solution:

$$T(N) = T(N - 1) + \frac{1}{N}.$$

We can clearly expand that into:

$$T(N) = \sum_{i=2}^{N} \frac{1}{i},$$

where we ignore the initial $i = 1$ case, since we're assuming that we initialize the running value with the first element of the sequence. If, instead, we initialize the accumulator to the lowest value (for max), or highest (for min), then we should consider that case that clearly and reassuringly adds 1 to the number of transitions. If the list contains repeated values, then this can only decrease the number of updates required, since any replicated value after the first can never require an update. Our formula is therefore pessimistic for those cases.

This formula is known in mathematics as the Harmonic Series [143]. Even though there is no simple closed form for the sum, we can easily compute it. This gives us the expected number of atomic operations shown in Figure 6.16. Note that this graph has a logarithmic x-axis, and so it shows results out to a large scale. We can see that even if we are performing a max or min operation on 1,000,000 elements, on average, we statistically expect less than 14 atomic operations that will be needed compared with the 999,999 that the naive implementation would have. This optimization is clearly important!

6.11 Conclusions

Let's conclude this chapter by summarizing a few of our observations. We will split them into two main topics: conclusions about locks and conclusions about atomic operations.

6.11.1 Conclusions: locks

First of all, Linus is right! We are probably wasting our time here and will normally be better off using an existing lock implementation and spending our precious time elsewhere. While we can sometimes beat the existing implementation on a specific benchmark, it is much harder to do so on a suite of real, and maybe even unknown, codes. You will need to demonstrate that you have a genuine problem there before thinking you can fix it with your own lock implementation.

Beware of naive profiling, because it can be evil and misleading! A profiler can reasonably show a lot of time spent in a lock function, but that almost certainly means that the application code is subject to contention on a lock, not that the lock implementation needs to be fixed. The point of a lock is to serialize and that naturally requires someone else to wait.

Consider speculative or transactional locking. Hardware support for transactional memory is still relatively new, and many people remain unaware of it. It is therefore an area worth considering, since, on hardware that has it, it is supported in (at least) the OpenMP API and Intel Threading Building Blocks, so can be used easily and may provide large performance gains.

6.11.2 Conclusions: atomic operations

Use compiler primitives where you can, as they provide portability and direct access to appropriate machine instructions while maintaining portable code. The compiler usually is smart to select the right machine instruction to implement the atomic operation or has a runtime system that implements a reasonably performing entry point for the atomic operation.

Use backoff and/or Test and Test-and-Set lock implementations. If you have to write your own atomics using compare-and-swap instructions, then you should provide sane backoff strategies. And, perhaps, hassle compiler folks to do that in their implementations if they do not.

Reduce the number of atomic instructions whenever you can. This applies both in the Test and Test-and-Set case, and, more dramatically in the one when computing a minimum or maximum value atomically. The principle is the same, though: these instructions are powerful but costly and can affect other computation, so should be used only where absolutely needed. Overall, it's about parallelism and anything that serializes gets in your way when scaling up the machine size and the number of threads or cores.

7 Barriers and reductions

Barriers are frequently used in parallel programming models to simplify the reasoning that is required to understand the behavior of the parallel code. They are one of the fundamental schemes used in the OpenMP[*] programming model, and many OpenMP constructs have implicit barriers that the user has to explicitly remove by using the `nowait` clause if they do not want them to be introduced by, e. g., the OpenMP `for`, `single`, and other worksharing constructs. Barriers can also be explicitly inserted by the programmer to ensure synchronization.

However, barriers are costly, and have unpleasant properties. As we will see, the cost of executing a barrier can be reduced if we are smart, but the barrier semantics *require* that threads wait, since the whole point of the barrier is to ensure that no thread can execute code after it until all threads have arrived at the barrier, and have, therefore, completed execution of the preceding code. This means that barriers necessarily amplify load imbalance. We can see this effect in Figure 7.1.

Since thread one arrives after the other three, each of them has to wait, thus the CPU time that is wasted is three times the delay induced by thread two arriving last. Of course, as threads are added, this multiplication factor increases with the number of threads. This demonstrates that the barrier is inherently not a scalable programming model, no matter how much we are able to reduce the barrier overhead by going to logarithmically scaling implementations that decrease the time spent inside the barrier code and provide better scalability.

As we saw when discussing locks, the fundamental problem for scalable performance of an application is not that our implementation of locks or barriers is poor, but rather that the semantics of these primitives is not scalable. However, since we can't eliminate them from existing programming models, we have to do our best to implement them well, and hope that competent 1970s technology like CSP's [57] channel communication will return. This need not be a completely idle hope, since CSP-inspired channels have reappeared in the Go [49] and Rust [134] programming languages. But, back to barriers now.

7.1 Barrier fundamentals

A barrier is necessarily a global operation, since information about the state of every thread must reach every other thread. Different barriers have different ways of passing that information around. For instance:

- **All-to-All:** Each thread sends a message to every other thread and receives a message from every other thread.
- **Centralized, or Two Phase:** Here, information is collected by one thread (the "root" thread), which discovers that all threads have checked in, and then broadcasts that information back to them. This can be implemented in many ways. For

https://doi.org/10.1515/9783110632729-007

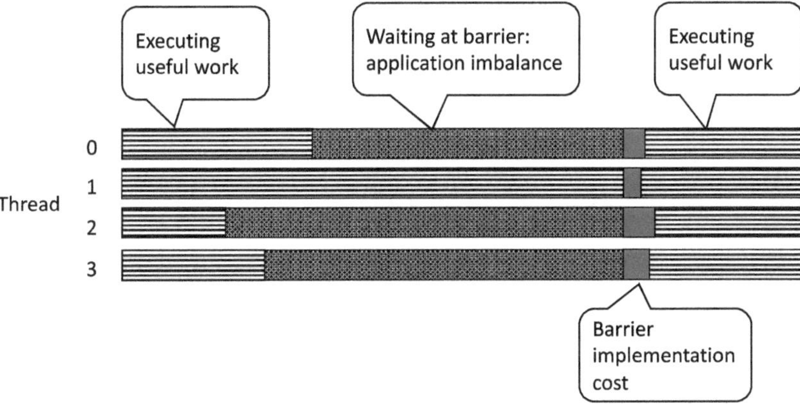

Figure 7.1: Barrier imbalance.

instance, a tree barrier is necessarily like this, but a counter in which one thread detects that it was the last to arrive and then broadcasts that knowledge is also a two-phase barrier, while a counter in which all threads poll the count is not.

– **Tree:** Threads pass information up a tree (either one that is pre-determined by a "fixed tree", or one that is determined based on arrival times that create a "dynamic tree"); a single thread ultimately knows that all of the others have arrived and then has to broadcast the "We're all here" information back to everyone. Note that there is no requirement that the broadcast operation uses the same tree (or, indeed, any tree) as the check-in phase.

– **More complicated logarithmic patterns:** In these, there is no central point, and therefore no separation of a check-in phase from a check-out phase. Instead, the communication topology ensures that, after some fixed number of communication rounds, information has passed from all threads to all others. (Barriers in this class include hypercubes and dissemination barriers.)

As well as the complexity of the barrier, i. e., does the number of operations on the critical path scale as $\mathcal{O}(N^2)$, as $\mathcal{O}(N)$, or as $\mathcal{O}(\log N)$ for N threads (in Landau "Big $\mathcal{O}$" notation [76]), optimizations based on hardware properties are important. For instance, requiring inter-socket, cross-NUMA domain communication is likely to be more expensive than keeping communication more local; similarly, an algorithm that only performs a single store operation before waiting on another thread is likely to perform worse than one that can get many store operations in flight at the same time, or an algorithm that does not force contention on a single cache line may beat one that does.

To be useful, barriers must also enforce full memory ordering to ensure that all writes in any thread are globally visible before the barrier completes, and no reads in any thread float up above the barrier. This is required because the objective of the barrier is to ensure that each thread sees the same, consistent memory state that reflects all computations completed before the barrier. In a barrier that is implemented in two phases with a separate check-in and check-out, the check-in should be a release operation, while the check-out should have the corresponding acquire operation.

7.2 Barrier performance measurement

To measure the true overhead of a barrier implementation, we cannot simply look at the time it takes for multiple back-to-back barriers to execute, since
1. it does not reflect the use in real codes,
2. it measures the performance when there is only a small amount of jitter, and
3. it also includes any jitter that may exist as part of the barrier implementation cost. However, this time should not be accounted to the barrier implementation as the implementation cannot, in general, do anything about it.

There are two apparently reasonable ways to measure the barrier implementation performance:
1. **Last In, Last Out (LILO):** This reflects the worst-case delay introduced by the barrier on any thread.
2. **Last In, Mean Out (LIMO):** This reflects the CPU time resources that are used by the barrier code itself. (The total CPU time consumed by the barrier is the LIMO time multiplied by the number of threads participating in the barrier.)

While we have said that jitter is not the barrier's problem, if the barrier itself introduces jitter by releasing one thread a long time after the others, then that is clearly untrue, since if the work that follows the barrier ends at another barrier, such jitter shows up as wasted resources when the threads enter the next barrier.

Consider Figure 7.2 that shows two possible barrier timings involving four threads. In the first, the LILO time is 4, whereas the LIMO time is 2, so this barrier itself consumes 8 units of CPU time. In the second, both the LILO and LIMO times are 3 (so the barrier consumes 12 units of CPU time). However, although the second barrier has higher LIMO time and therefore consumes more CPU time, it introduces no jitter.

As an example, we can think of this barrier being followed by 4 units of work in each thread, and then another barrier, and look at the overall execution time with the two different sets of barrier properties. With the first (LIMO = 2, LILO = 4) barrier type, the second barrier in the execution will be entered after $4+4 = 8$ units of time, whereas with the second (LIMO = 3, LILO = 3) type, we'll get there after $3 + 4 = 7$. Thus, the overall execution time is determined by the LILO time and the second barrier provides better performance. So although the LIMO time of the second barrier is worse, the achieved performance will be better. In effect, the total resources used by the barrier are $N_{\text{Threads}} \cdot T_{\text{LILO}}$, and that is what we want to reduce. We therefore use LILO times when comparing our barrier implementations.

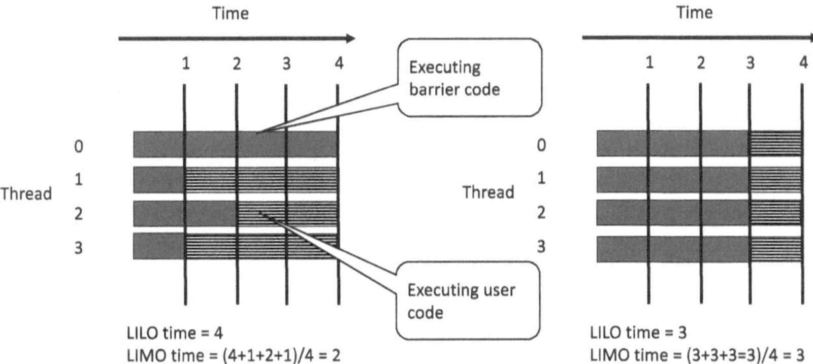

Figure 7.2: LILO and LIMO times.

As ever, there is no point optimizing any of our code for cases where its performance is effectively irrelevant anyway. If the LILO times of our barriers are < 10 μs, they become negligible when dealing with applications that run for > 1 ms between barriers, since then the barrier overhead is < 1 % of the execution time, and even halving the barrier LILO time would only improve overall application performance by 0.5 %. That's likely to be within the normal run-to-run variation, and, therefore, effectively unmeasurable. This implies that the times when barrier performance matters are also the times when their data is likely to remain in cache, and we don't need to optimize for cases where all of the barrier data has been evicted to memory. In effect, we're saying that barriers will be slow when they are rarely used, but, because they are rarely used, this won't have a big impact on overall performance, so it isn't worth optimizing for.

In a two-phase barrier, we can further split the LILO time into two components, the "Last In, Root Out" (LIRO) time and the "Root In, Last Out" (RILO) time. The LILO time is then the sum of the LIRO and RILO times. Since centralizing barriers can use many different components for check-in and check-out, this makes it easier to analyze the performance, as we can analyze each component separately.

7.2.1 Barrier micro-benchmark

The micro-benchmark we use explicitly measures the entry, root (where appropriate), and exit times of all threads participating in a barrier, and then computes the performance metrics. To do this, we obviously need a high-resolution clock, since (as we'll see below) we're measuring sub-μs times. Luckily for us, both the Arm[*] and x86_64 processors provide high-resolution timers that can be accessed directly from user space, are monotonic, and have constant tick time. One additional aspect we have to be careful about, however, is that although the Linux[*] kernel attempts to synchronize all of the clocks, it may not completely succeed.

```cpp
static void timeFullBarrier(int numThreads,
                            lomp::Barrier::barrierFactory factory,
                            lomp::statistic * LILO,
                            lomp::statistic * LIMO,
                            int64_t const * offsets) {
  auto B = factory(numThreads);

  // This has false sharing, but it's not in the timed code.
  uint64_t entryTime[MAX_THREADS];
  uint64_t exitTime[MAX_THREADS];

#pragma omp parallel num_threads(numThreads)
  {
    // Execute an empty parallel region to make sure that
    // OpenMP has initialized
  }

#pragma omp parallel num_threads(numThreads)
  {
    auto me = omp_get_thread_num();
    lomp::randomDelay delayer(1023); // 1023*100ns = ~100us
    auto myEntry = &entryTime[me];
    auto myExit = &exitTime[me];
    auto myOffset = offsets[me];

    for (int i = 0; i < NUM_REPEATS; i++) {
      // Jitter which thread arrives last
      delayer.sleep();

      // Measure the times
      auto myEntryTime = lomp::tsc_tick_count::now();
      B->fullBarrier(me);
      auto myExitTime = lomp::tsc_tick_count::now();

    // to be continued
```

Listing 7.1: Barrier timing code

```cpp
    // continued

      // Store them
      *myEntry = myEntryTime.getValue() + myOffset;
      *myExit = myExitTime.getValue() + myOffset;

      // Ensure that all of the exit times have been
      // filled in!
      B->fullBarrier(me);

      // Now compute the statistics in thread zero.
      if (me == 0) {
        uint64_t li = entryTime[0];
        uint64_t lo = exitTime[0];
        uint64_t sumO = 0;

        for (int i = 1; i < numThreads; i++) {
          li = std::max(li, entryTime[i]);
          lo = std::max(lo, exitTime[i]);
          sumO += exitTime[i];
        }

        LILO->addSample(lo - li);
        LIMO->addSample((sumO / numThreads) - li);
      }

      // Barrier again so that there's no bias towards
      // thread zero arriving last because it was doing the
      // computation
      B->fullBarrier(me);
    }
  }

  delete B;
}
```

Therefore, if we are trying to measure the time between events in different cores, we need to account for this, otherwise we may see negative times. In the cases where we do this, we run code to compute the offsets between the different (logical) cores. Of course, this implies that all of our measurements must be made with threads that are tightly affinitized to specific logical cores (since otherwise, if a software thread moves between logical CPUs, we will apply the wrong clock offset to its measurements). In Section 5.2.2, we have shown how to achieve the pinning of threads to cores. The code to measure LILO and LIMO barrier time is shown in Listing 7.1.

The `lomp::tsc_tick_count::now()` function of this book's example runtime reads the high-resolution clock and returns a 64-bit value. The variable `myOffset` holds the offset that we have measured between a thread and thread zero, so we're moving all of the times into thread zero's timeline. The `delayer` introduces a random delay of up to ~100 μs to ensure that different threads arrive last. `LILO` and `LIMO` are our statistics objects that compute the min, max, mean, and standard deviation of the samples they are fed.

7.2.2 Barrier performance modeling

As we have seen consistently, the most time-consuming operations that our code performs are data transfers between caches. Since a barrier is entirely about moving information between threads, it should be no surprise that we can model barrier performance by analyzing the data movements on the critical (LILO) path to estimate the fundamental performance we should expect from our barriers.

7.3 Barrier components

Since barriers communicate between threads, we need to consider how to implement such communication efficiently. In the abstract, all barriers are built from a few basic components, which can themselves be implemented in different ways, each of which will have a different performance. Different barriers communicate in different topologies, and it is the topology of the communication that determines the type of the barrier, not the details of the components used to implement it. Here, we consider possible implementations of some of these components; many of these will already be familiar, since, for instance, waiting politely is a general problem.

7.3.1 Counters and flags

One obvious way to implement a barrier is to have some centralized state in which we keep track of whether all of the threads have arrived. Then, a single thread takes the responsibility for noticing when all threads have arrived and waking them all up again. We call this state a counter as all it has to check is that all of the threads have arrived, though, as we will see, we can implement it at the scale we need without actually counting.

As well as being used in a completely centralized barrier, we need the same operation in a check-in tree; at each level, we have to check whether all of the children have arrived yet.

We can implement a counter using atomic operations, where each thread increments the number of threads that have arrived when it arrives, with either the final thread to arrive noticing that it is the last or a specific manager thread polling the value to notice when all threads have arrived. In the dynamic case, where the last to arrive will also move on, it does not need to perform an atomic operation, but rather merely check that the count is one fewer than the total expected, since it knows it is present and does not need to tell anyone else about that fact. This can allow us to avoid an expensive atomic operation on the critical LILO path.

However, at reasonably small scales—that is, small numbers of threads— a "counter" can also be implemented as a set of flags where each thread sets its flag, while the manager polls until all flags are set. By using a union, like the following one:

```
typedef union {
  std::atomic<uint64_t> allFlags;
  std::atomic<uint8_t> flags[8];
} byteWord;
```

the manager thread can check the status of eight workers at once with a single comparison.

On the machines we are working on with a 64-byte cache-line size, up to 64 threads can share a single cache line, at the cost of the manager thread having to make up to eight comparisons. If we have wide vector instructions available, the number of operations can be reduced even further. In any case, the cost of the comparisons is likely to be smaller than the cost of moving the cache line from the last thread to update it. If we're using a counter in a tree, where each non-leaf node in the tree has at most the tree fan-in many children, the flag counter may be a good choice, since a tree fan-in of eight or less may be a sane choice in most cases. Using this approach is still possible in the case where we don't need to broadcast the result, but is slightly more complicated. In this case, each thread needs to have a pre-computed mask that includes all other threads, but not itself. It can then compare against that to check that all other threads have arrived.

Since we are concerned with LIRO time, what interests us about these is the time between the last worker thread arriving, and the central manager thread knowing this. For the atomic counter, there are two cases to worry about: either the manager arrives last or some other thread arrives last. If the manager arrives last, the relevant cache line will be in a modified state in one other place (whichever other thread last incremented the counter), so we might expect that the time will be that required for the

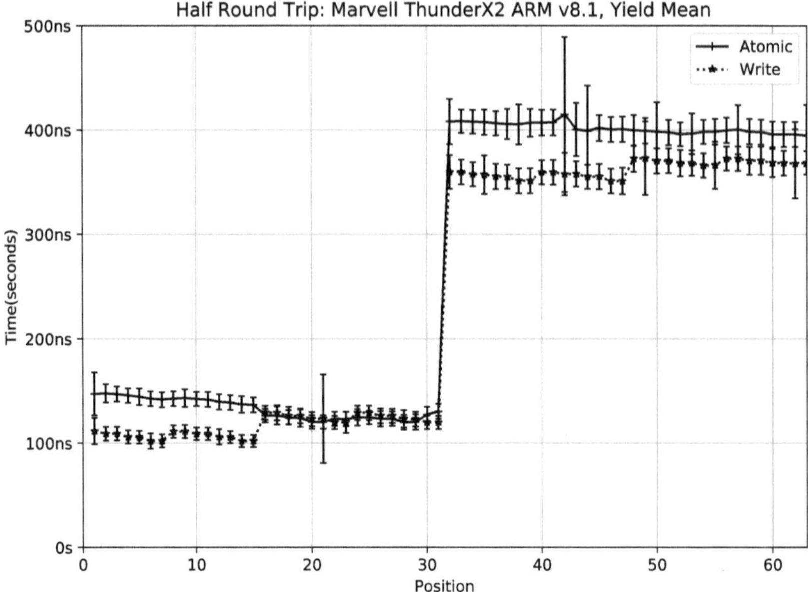

Figure 7.3: Atomic vs. store half round-trip time.

manager to perform a single atomic operation and look at the result. However, we can improve on this and avoid the atomic operation in the critical path by using a "test and test-and-set" approach. Instead of performing the atomic operation and then checking the counter value, a thread can check whether the current value of the counter is $N_{\text{Threads}} - 1$ before incrementing it, and can thus detect that it is the last to arrive without having to execute the atomic increment.

If a non-root thread arrives last, which is, of course, much more likely assuming that the thread that arrives last is random, then the time is the transfer time for the root to see the atomic operation complete.

For the flag counter, the two cases are similar: either the root thread arrives last or some other thread arrives last.

We can relatively easily measure the expected time for this as being the half round-trip time, measured using either a write operation or an atomic operation. Figure 7.3 shows the time for these two cases on our Arm machine as we consider the position of each possible last thread to arrive. Here, we can see, as we should expect, that the inter-socket communication is slower than communicating inside a socket, and that the write operation is slightly faster than the atomic (arithmetic mean over all 63 other places of 241 ns vs 263 ns). This $T_{\text{HalfRoundtrip}}$ time represents the fundamental time we can use for estimating barrier performance, and this data shows that it is ~100 ns inside a socket and ~380 ns between the two sockets, or, if we have traffic equally over both, we can use their mean of 240 ns.

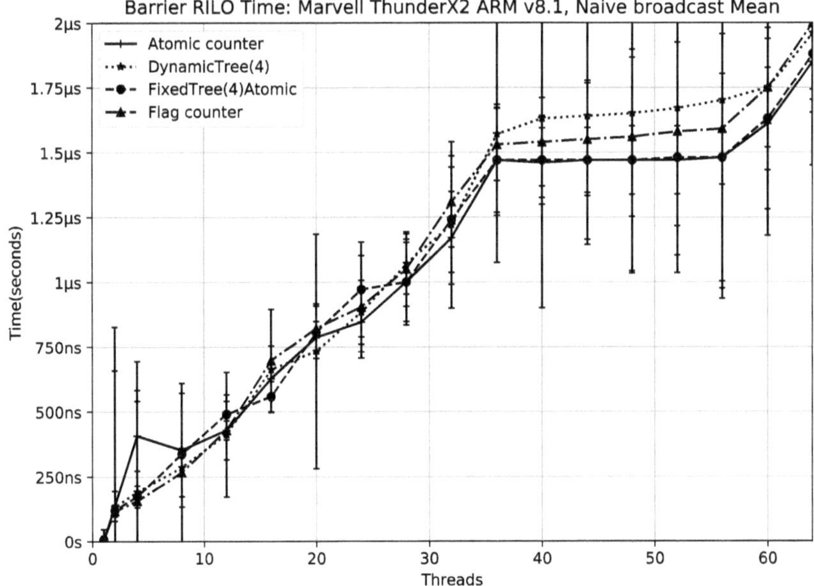

Figure 7.4: Naive broadcast Root In Last Out time.

7.3.2 Broadcast

In any centralized barrier, one thread ends up knowing that the barrier entry phase is complete and then has to pass that information to all of the other threads so that they can leave the barrier. This is a broadcast operation. One obvious way to implement a broadcast is to have the central thread store a value in a single variable and have all other threads poll that value, waiting for it to change.

We can measure the performance of the naive broadcast directly in the context of our barriers by looking at the RILO time of centralizing barriers that use the naive broadcast. This data is shown in Figure 7.4. We can see that the time is rather better than the simpler visibility measurement suggested (see Figure 3.18), and that the cache state caused by the check-in side of the barrier has an effect; however, we are still looking at between ~1.8 μs and ~2.0 μs for this operation at 64 threads.

At the other extreme, each thread can have its own flag, and the releasing thread writes individually to each of them.

In between these two extremes is an implementation in which there is more than one thread polling each cache line, but also more than one line. This reduces the number of writes by however many threads we think should look at a single line, a parameter we call "line broadcast width" (LBW). All of these implementations are flat; there is no tree involved, but the performance will be different for each implementation (and CPU), since it depends on the balance between the cost of updating a single line that is heavily contended, and the number of independent writes that a CPU can get in flight

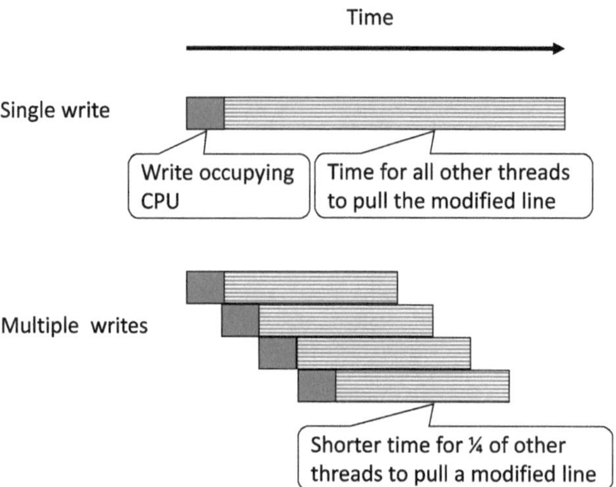

Figure 7.5: Improving broadcast time by overlapping writes.

simultaneously. If the LBW is larger than the number of threads being synchronized, this degenerates into the naive broadcast; if it is 1, then it is the extreme where each thread polls its own cache line and the root thread must perform one write per thread.

By using a smaller line broadcast width, we are hoping to achieve a faster broadcast because although we have to perform more writes, the latency between the write and the final thread seeing it is reduced. Figure 7.5 shows how this works.

We can see the effect here (including for the flat, write a line for each thread, case) if we plot the RILO time for different broadcast widths, as we do in Figure 7.6. As we expected, the LBW64 case gives similar results to the naive broadcast (the precise code is different, but the most important aspect of having a single line polled by everyone is the same), and we can see that the balance between the intermediate cases change at different scales. This should be unsurprising, since one of the factors here is the number of writes that the root core can have in flight at once, which is limited by the micro-architectural design. Since

$$\text{writes required} = \frac{N_{\text{Threads}}}{\text{LBW}}$$

as the number of threads increases, we will also have to increase LBW if the writes required are not to exceed the hardware limit and cause the hardware to pause the thread while it waits for space in the write buffer.

The most important conclusion here is that we have reduced the broadcast time when dealing with 64 cores in the two-socket Arm machine from the 1.6 μs (best case) that the single-line, naive broadcast required to ~500 ns.

As well as these flat implementations, in which a single thread performs all of the writes, one could also use a tree, in which we can potentially achieve more writes

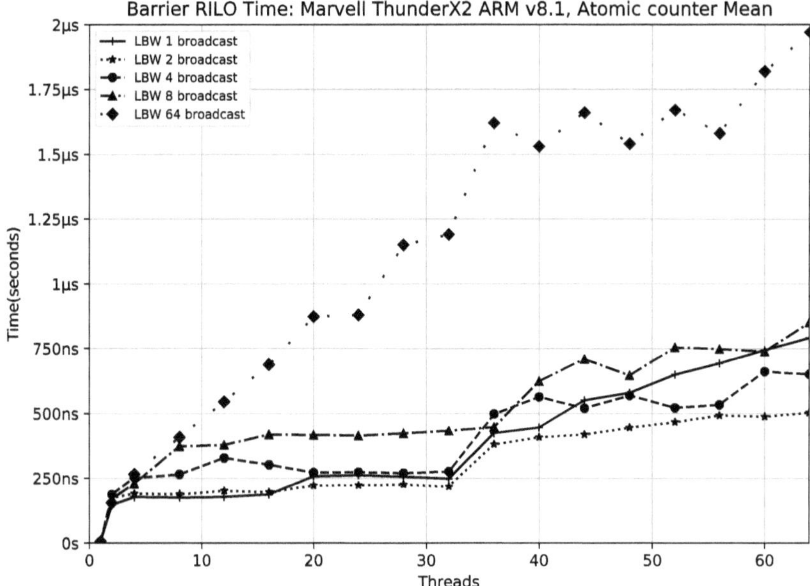

Figure 7.6: RILO time for different line broadcast widths.

in flight, since we can have more than one thread generating them; again, we have the option of having more than one thread polling the same line (reusing our LBW broadcast code), so we can have a tree where the fan-out is wider than the number of store operations required at each level.

7.4 Categorization of barrier algorithms

Barrier algorithms are best categorized by the topology of communication they embody, rather than the details of how the individual components of the communication are implemented, since, as we saw above, we can implement the components (such as counters or broadcast operations) in more than one way. Similarly, for centralizing barriers we can implement the check-in and check-out phases in different ways (for instance, using a tree check-in with a flat-broadcast check-out, or a counter check-in with a tree-broadcast check-out).

The most critical difference between barrier algorithms, then, is whether they are centralized or distributed. In other words, does the barrier have two phases, in the first of which a single thread discovers that all other threads are present, and in the second of which it informs all other threads that they can now continue; or does it operate in a distributed manner in which all threads behave identically and information is passed between threads without requiring centralization?

In general, OpenMP implementations prefer centralized barriers, since:

1. The fork/join operation implied by the `parallel` construct is effectively a centralized barrier where the fork operation is a check-out while the join at the end of the parallel region is a centralized-barrier check-in operation.
2. Reduction operations can be specified as associated with OpenMP barriers, and those map most naturally onto tree barriers.

However, in other environments, or even in OpenMP code in cases where a full barrier with no reduction operation is required, a distributed barrier may perform better.

7.5 Barrier algorithms

After the discussion of the fundamental aspects of barrier performance and the behavioral aspects of the basic mechanisms, let's now turn towards the actual implementations. Many options for implementation choices and algorithms exist and barriers are no different in that regard. Thus, we have picked a selection of barrier algorithms that we feel represent common categories of algorithms that are used in the wild.

7.5.1 Counting barrier

A counting barrier can be either distributed or not, depending on whether the final "everyone is here" state is tested by all threads or only by a single thread.

The decentralized, distributed version, in which all threads observe the state, is worth discussing because it shows a general problem that we have to overcome with distributed barriers, and motivates a general pattern that we can use for other barriers as well.

Our first and simple approach to this barrier might look something like the atomic counter shown in Listing 7.2. Our barrier would then consist of each thread calling `checkIn()` followed by `wait()`. If we only used the barrier once, this will work perfectly well. However, we normally want to be able to use the barrier more than once, so we have to reset it or come up with some other way of reusing it.

One approach might be to think that we can count up in the first barrier, then down in the second, up in the third, and so on. We'd therefore add a per-thread count of how many barriers each thread has executed and choose the direction of the operation based on the low bit of the count. (Obviously, one could use a single value that is toggled instead of a counter; for reasons that will soon be apparent, we describe this as a counter.) However, this doesn't fix the problem, since we still have a race condition, as we can have an execution like that shown in Table 7.1, where we end up with both of our threads waiting for the other and, therefore, there is no forward progress.

```cpp
class AtomicCounterBarrier {
  CACHE_ALIGNED std::atomic<uint32_t> present;
  uint32_t num;

public:
  AtomicCounterBarrier() {
    init(0);
  }

  AtomicCounterBarrier(int count) {
    init(count);
  }

  void init(int count) {
    num = count;
    present = 0;
  }

  void reset() {
    present.store(0, std::memory_order_release);
  }

  void checkIn(int) {
    ++present;
  }
  void wait() {
    while (present.load(std::memory_order_acquire) != num)
      Target::Yield();
  }
};
```

Listing 7.2: Code for a barrier with atomic counter.

Table 7.1: Racy barrier execution.

Thread 0	Thread 1
Enter barrier increment counter (now 1)	
	Enter Barrier increment counter (now 2)
See value 2; leave barrier	
Enter next barrier decrement counter (now 1)	
	Wait for value 2
Wait for value 0	

```cpp
class AtomicUpDownBarrier : public Barrier,
    public alignedAllocators<CACHELINE_SIZE> {
  enum { MAX_THREADS = 64 };
  AtomicUpDownCounter counters[2];
  AlignedUint32 barrierCounts[MAX_THREADS];
public:
  AtomicUpDownBarrier(int NumThreads) {
    for (int i=0; i<NumThreads; i++)
      barrierCounts[i].Value = 0;
    counters[0].init(NumThreads);
    counters[1].init(NumThreads);
  }
  void fullBarrier(int me) {
    auto myCount = barrierCounts[me].Value++;
    auto countIdx= myCount & 1;
    auto activeCounter = &counters[countIdx];
    auto countUp = (myCount & 2) == 0;
    if (countUp) {
      activeCounter->increment();
      activeCounter->waitAll();
    }
    else {
      activeCounter->decrement();
      activeCounter->waitNone();
    }
  }
  // ... uninteresting code elided ...
};
```

Listing 7.3: Atomic Up-Down barrier code.

One way we *can* overcome this is to have two separate counters and alternate between
them, so that we can avoid the race between one thread leaving a barrier and then
entering the next before all threads have left the first. We can then either reset the
counter that is not in use and always count in one direction or track the sense in which
we should use each counter so that we count up on a counter, and then the next time
we use it, we count down. This approach is frequently called *sense reversing*. Since we
have already notionally added a per-thread count, this is only adding one extra word
to the barrier, which now looks something more like Listing 7.3.

Since every thread is polling the same location, the time we expect for this barrier
is that of an atomic instruction on a remote line in a heavily shared state (~190 ns
cross-socket on our Arm machine) plus the naive broadcast time (around 2 µs at 64

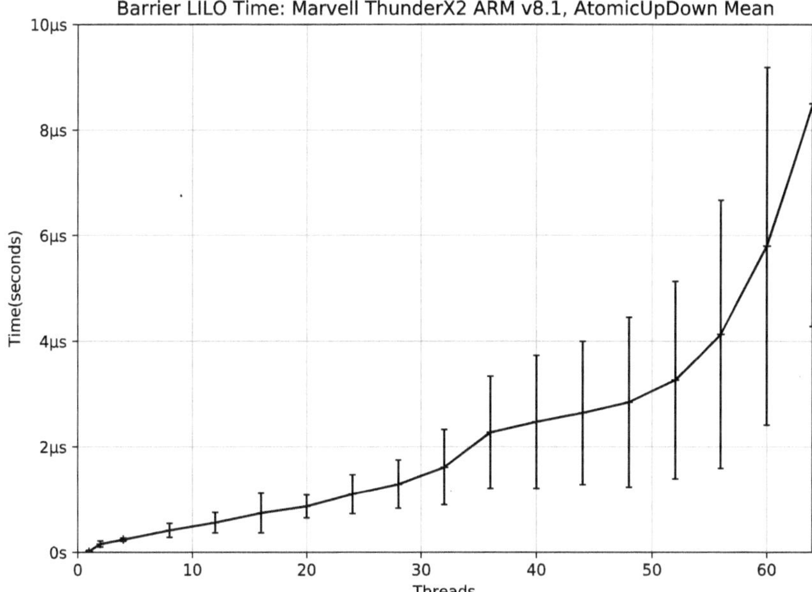

Figure 7.7: Atomic Up-Down barrier performance.

cores). The actual performance is shown in Figure 7.7, which shows that our estimate is overly optimistic for the mean, since we measure that as 8.5 μs at 64 cores. However, the best measured time at that scale is 2.9 μs, which is within 15 % of our estimate.

This general principle of having two sets of state (the two counters here), and alternating between them is a pattern that is used in all of the decentralized barriers. The second idea of avoiding having to reset the state by toggling the direction in which it is used (embodied in the countUp variable here) can also be useful in general. Of course, instead of doing this, one could explicitly reset the inactive state; however, this causes some centralization, since only a single thread needs to do this, so there will be a specific test for that thread, making it behave differently from all of the others, and, potentially, introducing more jitter into the LILO time.

In centralized barriers, provided that the broadcast uses different data from the check-in, there is no need to do this to avoid race conditions. One of the threads knows that everyone has checked in and thus it can reset the check-in state before it releases anyone. Since it knows that all of the other threads are waiting to be released, none of them can get to the next barrier, and all have finished with the check-in state of this one, so resetting the check-in state is safe. However, it may still be worth using the alternating sets of data approach, since in barriers where there is more than one cache line of a shared state, this allows more than one thread to do some of the resetting, thereby reducing the critical-path time. Similarly, the up/down trick can also be useful since it avoids the need for a complete state reset, at the cost of a (probably badly predicted) branch.

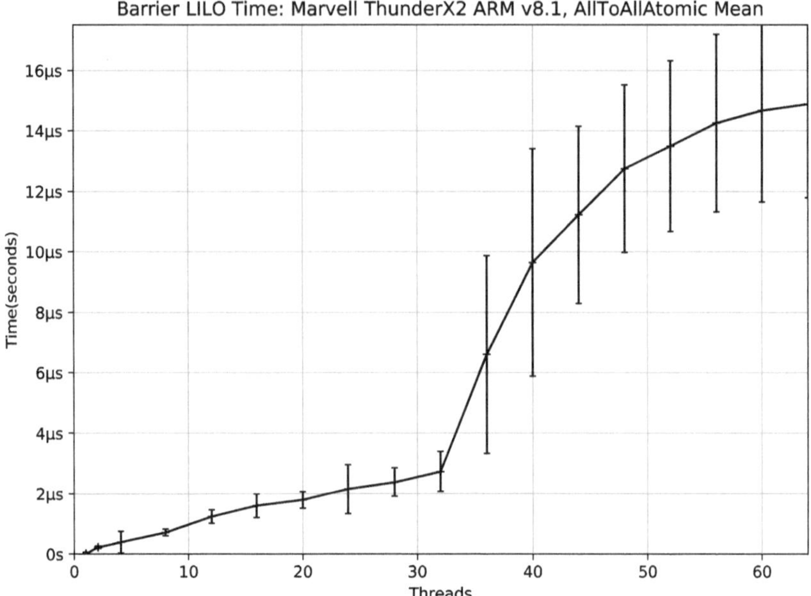

Figure 7.8: All-to-All atomic barrier performance.

7.5.2 All-to-All barrier

The simplest distributed barrier is a flat All-to-All barrier, in which each thread sends a message to every other thread and then waits until it has received one from every other thread.

Such an implementation is relatively simple and is shown in Listing 7.4. Although this is simple, the critical path (LILO) time is that for a single thread to perform N_{Threads} stores and have them all become visible, which is $N_{\text{Threads}} \cdot T_{\text{HalfRoundtrip}}$.

As we saw in Section 7.3.1, our Arm machine has a $T_{\text{HalfRountrip}}$ of ~100 ns inside a socket, and ~380 ns between the two sockets. When we have a thread on every core, half of the traffic will be in-socket and half between sockets, so it's reasonable to use the average time of 240 ns. We therefore expect this barrier to have a LILO time of 63 times 240 ns, i. e., ~15.4 µs at 64 threads. Figure 7.8 shows the measured performance on our Marvell[*] Arm machine achieving a LILO time at 64 threads of 14.9 µs, so our estimate is very close to the measured performance.

Since we have already seen that the atomic up-down barrier is much faster than this, we won't consider it further.

```cpp
class AllToAllAtomicBarrier : public Barrier,
    public alignedAllocators<CACHELINE_SIZE> {
  enum { MAX_THREADS = 64 };
  uint32_t NumThreads;
  AlignedAtomicUint32 flags[2][MAX_THREADS];
  AlignedUint32 sequence[MAX_THREADS];
public:
  AllToAllAtomicBarrier(int NThr) : NumThreads(NThr) {
    for (uint32_t i = 0; i < NumThreads; i++) {
      flags[0][i].value.store(0, std::memory_order_relaxed);
      // No need to clean flags[1] here; they'll be cleared
      // by each thread when it checks in.
      sequence[i].value = 0;
    }
  }
  void fullBarrier(int me) {
    auto odd = sequence[me].value & 1;
    // If I am checking in to barrier n, no one can be
    // checking in to barrier n+1 yet, so I can clean it
    // here before telling everyone I have arrived.
    flags[!odd][me].value.store(0,
                                 std::memory_order_relaxed);
    // Tell everyone we're here.
    for (uint32_t i = 0; i < NumThreads; i++)
      flags[odd][i].value++;
    // Use the other set of flags next time.
    sequence[me].value++;
    // and wait until everyone else is here
    while (flags[odd][me].value != NumThreads) {
      Target::Yield();
    }
    return 0;
  }
};
```

Listing 7.4: All-to-All barrier code.

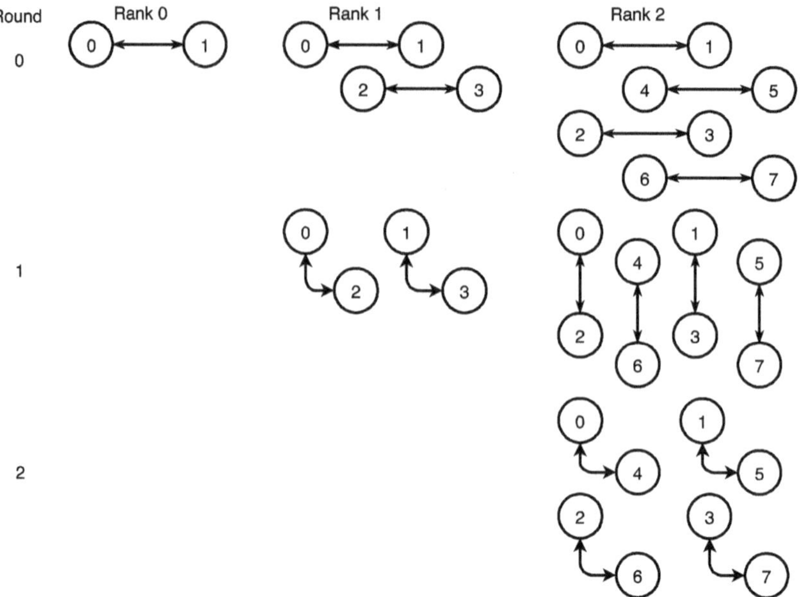

Figure 7.9: Communication in a hypercube barrier.

Table 7.2: Incomplete hypercube barrier communication example.

Phase	Communication
1	0<->1, [2<->3]
2	0<->2, [1<->3]

7.5.3 Butterfly/hypercube barrier

The hypercube barrier is a distributed, logarithmic barrier. In each phase, a thread sends a message to and receives a message from its neighbor in that dimension of the hypercube. You can see this in Figure 7.9. However, the naive hypercube does not work for numbers of threads that are not a power of two. Consider even the simplest such case with three threads. The communication pattern is shown in Table 7.2 (if we ignore the non-existent thread 3, whose communications are shown in "[ ]", since they don't occur). You can see that thread 1 can leave the barrier after the first round without having any knowledge of thread 2.

At each phase in the hypercube barrier, the thread's neighbor is given by

```
int neighbor(int me, int round) const {
  return me ^ (1 << round);
}
```

For the power-of-2 cases, the LILO path will be the same whichever thread arrives last, since the barrier is completely symmetrical. In this case, the LILO time will be

$$T_{\text{LILO}} = (\log_2 N_{\text{Threads}}) \cdot T_{\text{HalfRoundtrip}}.$$

For other cases (where one thread has to stand in for another non-existent one and perform more communication), the performance will be worse than this (as with the fixed tree, we should average over all possible last-in threads in that case, but we can easily see that it can only be worse than the formula above).

The butterfly barrier [17] generalizes the hypercube so that it will work for any number of threads; however, since a dissemination barrier should have identical performance, and can more easily handle arbitrary numbers of threads, we will not investigate this barrier further here, though, for implementing MPI [90] barriers on machines with a hypercube-network topology this might be the ideal barrier.

7.5.4 Dissemination barrier

The dissemination barrier is very similar to the hypercube barrier. In our implementation, they both derive from the same base class and the only difference between them is in the `neighbor()` function, which determines the communication pattern of the participating threads. Unlike the hypercube barrier, the communication in the dissemination barrier at each phase is not reflexive (where if A sends to B, then B sends to A), but rather a thread sends to a thread that is not (in general) the same as the one it receives from.

The `neighbor()` function for the dissemination barrier looks like this:

```
int neighbor(int me, int round) const {
    return (me + (1 << round)) % NumThreads;
}
```

The resulting communication pattern can be seen in Figure 7.10.

The expected performance here is the same as that of the hypercube barrier, so we predict a LILO time of six times the half-round trip time $T_{\text{HalfRoundtrip}}$ for the 64-core case. Since there is no attempt to optimize to reduce cross-socket communication (it is hard to see how one can do that with this communication pattern), we can use our estimate of 240 ns for $T_{\text{HalfRoundtrip}}$, suggesting a LILO time of ~1.4 µs. This is, indeed, the minimum time that we measure, with the mean time (which is what we have shown for all the other barriers) of ~1.9 µs, as we show in Figure 7.11. We can also see that when the communication is limited to being inside a single socket, the performance is much better, managing ~600 ns for the 32 cores in a single socket.

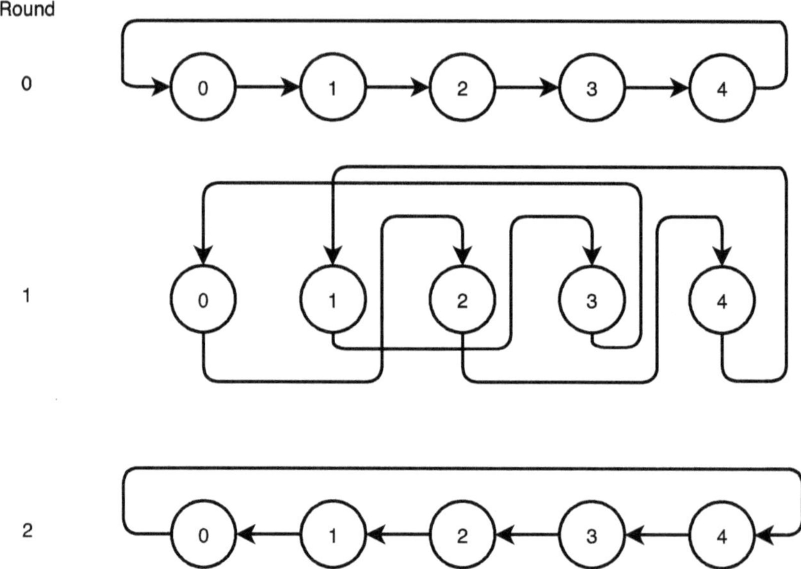

Figure 7.10: Communication in a Dissemination barrier.

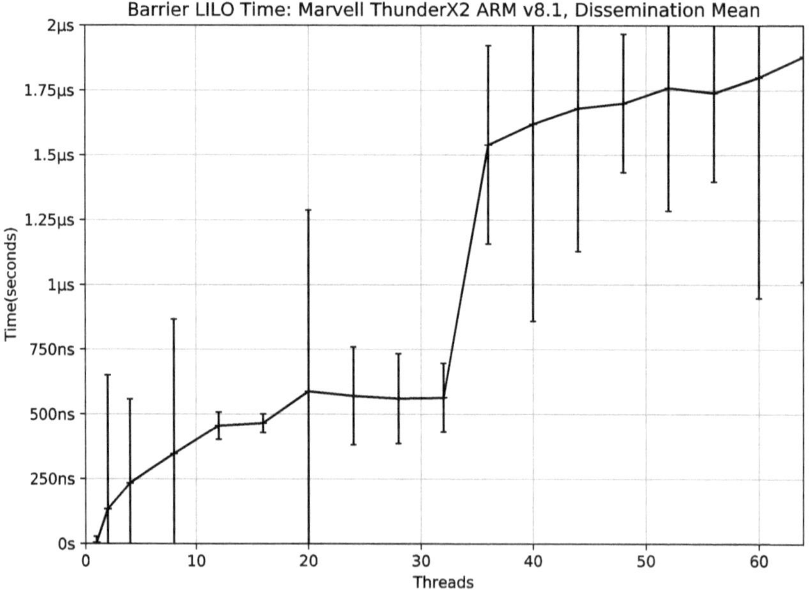

Figure 7.11: Dissemination barrier performance.

Table 7.3: Dissemination barrier knowledge propagation.

Phase	To	From	Know about
		Thread n	
0	n+1	n-1	n-1
1	n+2	n-2	n-3,n-2,n-1
2	n+4	n-4	n-7..n-1

7.5.4.1 Improving the dissemination barrier

One of the problems with the dissemination barrier is that each core can only have one store in flight at a time, so we see the full write latency. If we could pipeline more writes (as we did in our LBW broadcast by writing to more than one target cache line), we could potentially improve the performance. To do this, we must consider the propagation of information in the dissemination barrier so that we can see how this can work.

Table 7.3 shows this, and you can see that after n phases, a thread knows about all of the threads within 2^n before it (in a cyclical modular fashion). Or, from an alternative point of view, information about thread T has propagated to all threads in $[T, T + 2^n$ mod $N_{\text{Threads}}]$.

We can therefore condense multiple dissemination barrier phases if a thread communicates with more than one other thread in each phase; so, for instance, in phase 1, instead of sending only to thread $(T + 1)$ mod N_{Threads}, it also sends to $(T + 2)$ mod N_{Threads}. Now the number of phases required can be halved. Similarly, we could send to four other threads, reducing the depth even more.

This is known as an "n-way dissemination barrier" [58], and can increase the performance of the barrier by reducing the number of phases.

7.5.5 Tree check-in barriers

There are two slightly different approaches to implementing a check-in tree:

1. **Fixed Tree:** In a fixed tree, each thread has a pre-determined location in the tree, and is responsible for waiting for its children and then passing on the fact that all of them have arrived to its parent.
2. **Dynamic Tree:** In a dynamic tree, all of the threads enter the barrier at the leaf nodes in the tree, and as they arrive, the last to arrive at each node proceeds upwards to contribute to the next level.

We'll now discuss their properties in more detail.

7.5.5.1 Fixed-tree check-in

Since the threads occupy the nodes in the tree, a tree with branching ratio, or *radix R*, and a depth of *depth D*, can handle up to

$$N_{\text{Threads}} = \sum_{i=0}^{D} R^i$$

entries.

As each thread has a fixed position in the barrier, this barrier is convenient when the implementation wants to ensure that a specific thread is the root of the tree. This is particularly useful when implementing the OpenMP API, because we want to use the barrier check-in as the join operation at the end of a parallel region and have the serial thread that forked the parallel region leave to execute the next serial region. OpenMP semantics require that this is always the same thread, since the thread identity is user visible and since the user can access TLS variables (as we saw in Section 5.3.5).

While these OpenMP restrictions may seem unnecessary they make sense from a performance point of view, since when threads are tightly affinitized, they ensure that the serial code always executes in the same place, so its NUMA domain and cache locality are preserved.

Considering the LIRO path through the fixed tree, its length depends on the position of the "last in" thread in the tree. If the last entrant is the root, then the data transfer required is a single $T_{\text{ReadModified}}$. For a non-leaf, non-root thread, the time will be $T_{\text{ReadModified}} + \text{Depth(Thread)} \cdot T_{\text{HalfRoundtrip}}$, while for a leaf, it will simply be $\text{Depth(Thread)} \cdot T_{\text{HalfRoundtrip}}$, since the leaves do not need to check the state of any other threads before it stores its own state.

7.5.5.2 Dynamic tree check-in

The dynamic tree has all of the threads as leaves. An arriving thread checks whether all of its siblings at the leaf have arrived; if not, then it notes that it has arrived so that a later arriving thread can see that, then leaves the check-in to wait to be woken. If it is the last to arrive, then it moves up one level and checks whether it is the last to arrive there, and so on until the final thread to arrive will see that the counter at every level up its path is just awaiting its arrival, and it will leave at the top of the tree.

Since it has no threads bound to internal nodes in the tree, a dynamic tree with a branching ratio R of depth D can handle

$$N_{\text{Threads}} = R^D$$

threads at maximum. This means that the dynamic tree is frequently one level deeper than the fixed tree, as we can see in Table 7.4.

However, the dynamic tree has the advantage that the LIRO path does not require store operations; at each level, the last thread to arrive only has to read the counter

Table 7.4: Fixed and dynamic binary trees: maximum entries vs. depth.

Tree Depth	Capacity	
	Fixed Binary Tree	**Dynamic Binary Tree**
1	3	2
2	7	4
3	15	8
4	31	16

and check whether it reflects the arrival of all of the other threads at that level. Since it is the last, it knows that it is present itself and doesn't need to tell anyone else about that fact. This means that although more operations are required, they are cheaper operations, and the dynamic tree can potentially outperform the fixed tree. When the tree is full, or when no threads get a pass and skip the first round, the time will be $T = D \cdot T_{\text{ReadModified}}$; when there are threads that get a pass in the first round, then the mean number of operations will be fewer, since such threads need only perform $D - 1$ read operations to reach the root of the tree.

Figure 7.12 shows the expected mean number of communications required by the fixed and dynamic trees if there is an equal probability of each thread arriving last. You can see that the dynamic trees of a given radix always require more communication—for instance, at 32 threads the radix-2 fixed tree requires ~3.2, whereas the dynamic tree requires 5. We can also see that the higher-radix trees require fewer communications.

In Figure 7.13, we show possible fixed and dynamic tree mappings of ten threads into a radix-4 tree. Here, the threads are distributed cyclically, just as iterations in an OpenMP `schedule(static,1)` loop would be. If threads are tightly bound and enumerated consecutively over cores before sockets, then a binding more like that for `schedule(static)` is likely to perform better, since it would require fewer cross-socket communications. (The dynamic trees we are measuring do *not* have that optimization.)

The achieved LIRO performance of these alternative tree implementations is shown in Figure 7.14, where we measure it when used in conjunction with a naive broadcast. You can see that whatever the tree-branching factor, the dynamic tree is faster than the fixed tree, and that as we increase the radix, the trees generally perform better. The geometric mean performance for all of these measurements (at 1, 2, 4, 8, 12,..., 64) is ~160 ns for the radix-16 dynamic tree and ~320 ns for the fixed tree. At 64 cores, the RILO time is 270 ns for the dynamic tree and 520 ns for the fixed tree. Remember, though, that this is only the check-in phase of the barrier, and our best (LBW4) broadcast takes 510 ns at this scale.

Finally, we can compare the performance of our two trees using the same, LBW4, broadcast scheme, with the results shown in Figure 7.15.

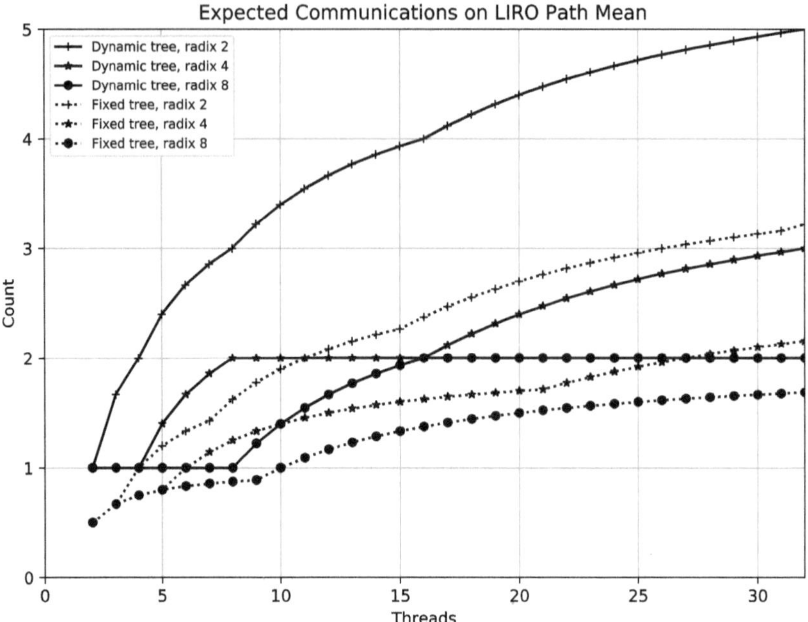

Figure 7.12: Fixed and Dynamic tree mean communications on LIRO path.

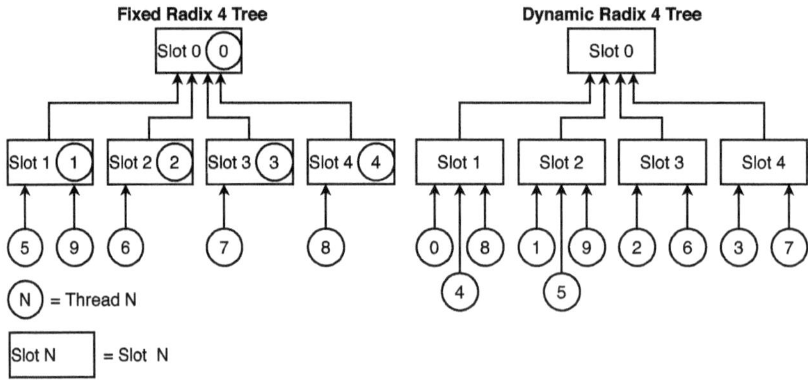

Figure 7.13: Possible radix-4 fixed and dynamic binary trees for ten threads.

We can see that at scale, the high-radix tree barriers all outperform the other barriers we have considered so far, achieving a <1 μs LILO time with 64 threads in the two-socket, 64 CPU machine.

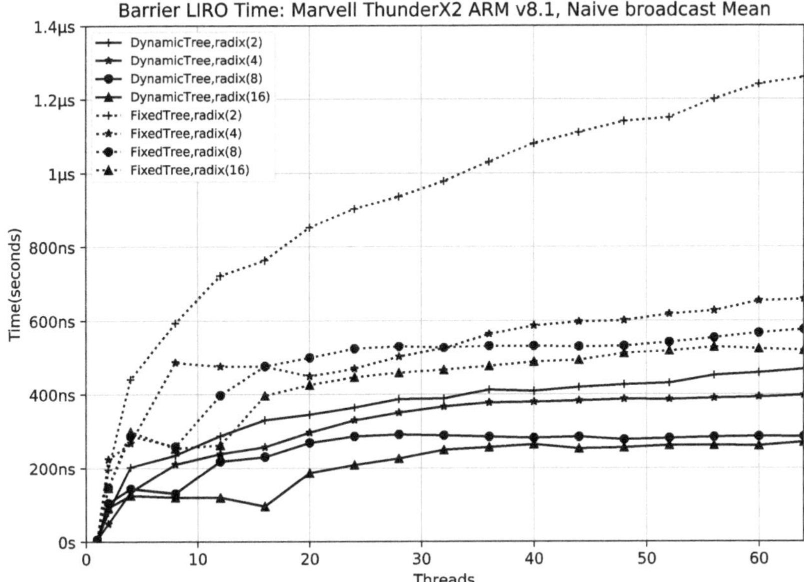

Figure 7.14: Fixed and dynamic tree LIRO times.

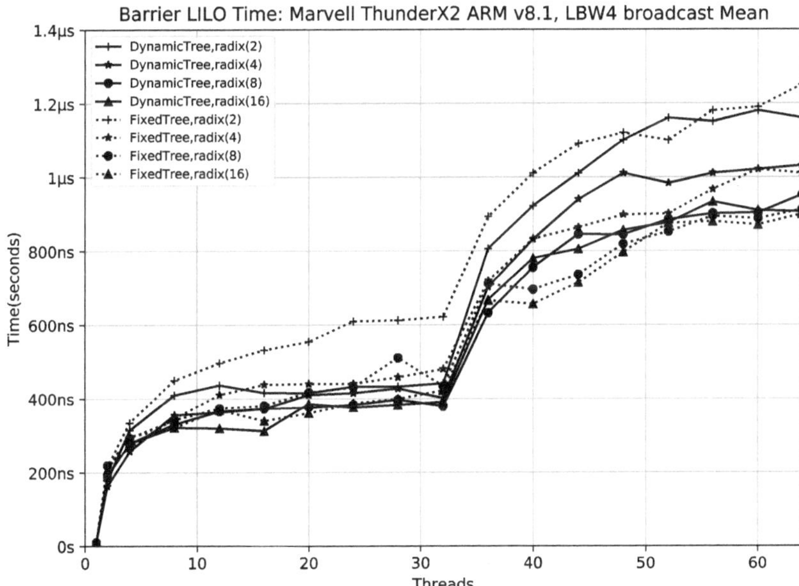

Figure 7.15: Fixed and dynamic tree LBW4 LILO times.

7.5.5.3 Tree check-in conclusions

Ultimately, which of these trees should we use?

We have shown that the dynamic tree is often faster than the fixed tree; however, in an OpenMP context it has the disadvantage that the root thread from the check-in can be different each time that the barrier runs. Since OpenMP semantics require that the root thread is always the same, if we use the dynamic tree for the OpenMP join at the end of a parallel region, we will have to remove the OpenMP root thread (thread zero) from the barrier, and then have the winner of the check-in tell the OpenMP root thread that the barrier is complete. This will add one half of the round-trip time to the LIRO. Since that is at least 100 ns on this machine, that would remove much of the dynamic tree's performance advantage.

We can also improve the effective performance by choosing appropriate parameters for the tree at the point when the barrier is constructed, since all of our barriers need to know the number of threads that will be participating. Thus, one might choose the radix of the tree based on the number of threads, or even choose another barrier completely if you were only dealing with two or four threads.

7.6 Reductions

One of the classic patterns in scientific computation is a *reduction*, in which values from many computations are combined into a single value. For instance, in a particle-in-cell (PIC) code, when examining the gravitational motion of the planets at each time step, one must compute the gravitational force on each planet, which is the sum of the forces generated by each of the other planets (and the sun). Similarly, in a molecular-dynamics (MD) code, the total force on an atom is the sum of all of its interactions with other atoms, or in a Monte-Carlo optimization problem, one may want the minimum or maximum value of a set of different sample runs. In all of these cases, the code is reducing many values to a single value. While what is being computed is often a single scalar value, we have already seen cases where it is a vector (the force on a planet or atom is the sum of a set of other individual force vectors), but no matter what the dimensionality of the item being combined is, in all of these cases, we are taking many values and reducing them to a single value.

In scalar code, such a reduction is simple, since one can write code like this (for a simple scalar sum over a vector of elements with type `double`):

```
double sum(int count, double const * vec) {
  double total = 0.0;
  for (int i=0; i<count; i++)
    total += vec[i];
  return total;
}
```

Table 7.5: Accumulation from left or right.

	Accumulate from Left		Accumulate from Right	
Step	**Addend**	**Running Total**	**Addend**	**Running Total**
0	0.49	0.49	100	100
1	0.49	0.98	0.49	100
2	100	101	0.49	100

Before we introduce a parallel solution, we must also consider what the accuracy requirements of the code are, since any parallel solution will clearly accumulate the values in a different order from the one that was implemented in the serial code, and, in many cases, that order may differ from run to run.

In case this seems a marginal difference (and you have not read "What Every Computer Scientist Should Know about Floating-Point Arithmetic" [47]), consider a simple example in a floating-point implementation that is base 10 with three significant digits. Now, think about the result of a sum reduction over a vector containing the numbers (0.49, 0.49, 100). Table 7.5 shows that, depending on the order of the additions, we can get the result 100 or 101. The problem is that since we only have three significant digits, 100 + 0.49 rounds down to 100. Clearly, in this system we could have as many 0.49 values on the left as we like and the total will still seem to be 100 if we add from the right.

Of course, the scale of the problem is not as dramatic as this in normal IEEE-754 arithmetic [60], but the same effect is still there, so if one is trying to meet a bit-identical acceptance test comparison with a serial code, then any parallel reduction may prove impossible to use. Overcoming this problem is an active area of research (e. g., "Numerical reproducibility for the parallel reduction on multi-and many-core architectures" [23] and "Fast, good, and repeatable: Summations, vectorization, and reproducibility" [96]).

Assuming that the change in result is acceptable, we can attempt a parallel implementation. However, this will necessarily be more complicated than simply adding `#pragma omp parallel for` on the loop, since that obviously introduces a horrible race condition on the variable `total` that is being updated in every thread.

A variety of solutions are possible:

- **Use an atomic:** We could protect the update of `total` by using an atomic. However, we know that highly contended atomics instructions are, in general, slow. This also assumes that the reduction operation is simple enough that it can be implemented as an atomic, which is not true in general (consider the vector sum of forces as an example).
- **Use a critical section:** While this allows any code to be used for the reduction, it still serializes, and so is unlikely to achieve good performance, and means that,

```
double sum(int count, double * vec) {
  int nThreads = omp_get_max_threads();
  double totals[nThreads];
  for (int i=0; i<nThreads; i++) {
    totals[i] = 0;
  }
#pragma omp parallel
  {
    int me = omp_get_thread_num();
#pragma omp for
    for (int i=0; i<count; i++)
      totals[me] += vec[i];
  }
  double total = 0.0;
  for (int i=0; i<nThreads; i++)
    total += totals[i];
  return total;
}
```

Listing 7.5: Reduction idiom causing false sharing and serialization.

as well as the accumulator variable, we are also bouncing a cache line containing
a lock around the machine.
- **Use a per-thread accumulator:** The idea here is that each thread can accumulate
a partial result and then, at the end, we combine the results into a single, final
value.

Since the per-thread accumulation with a final cross-thread aggregation can maintain
at least some parallelism, we'll look at that in more detail.

One naive way to implement this (if one hasn't bothered to read the OpenMP specification, and so didn't realize that the OpenMP API has built-in support for reductions)
might look like the code in Listing 7.5. This code should work, but has significant performance issues:

1. **False sharing:** The totals array has not been padded in any way. Therefore, on
   a machine with 64-byte cache lines, eight threads will be contending to update
   each cache line, since eight of the values of the array will be in each cache line.
   That additional cache-to-cache movement has a high cost.
2. **Serial final addition:** The final addition is still serial, and so might become a
   performance bottleneck.

Although we just said the serial reduction of the per-thread values is a potential bottleneck, it is also a potential advantage, since it means that at least that part of the

```
double sum(int count, double * vec) {
  double total = 0.0;
#pragma omp parallel
  {
    double myTotal = 0.0;
#pragma omp for nowait
    for (int i = 0; i < count; i++)
      myTotal += vec[i];
#pragma omp atomic
    total += myTotal;
  }
  return total;
}
```

Listing 7.6: Using OpenMP reductions to avoid false sharing.

reduction has a fixed order of operations, and so is deterministic and reproducible. (Of course, if the loop were run with a dynamic schedule, the overall reduction would still be non-deterministic because there'd be no inter-run reproducibility of the operations executed by each thread, and this only gives reproducibility when the same number of threads are used, but it may still be an advantage.)

We can (relatively easily) eliminate the false sharing by writing cleaner and more idiomatic OpenMP code, like that shown in Listing 7.6.

Here, we have created the per-thread accumulators in their natural scope on the stack of each thread, so they will not suffer from false sharing. By using `nowait` on the `for` directive, we allow each thread to leave the loop as soon as it has finished work on its chunk of the array, and, finally, we use an atomic operation to ensure that the updates to the true accumulator are thread safe. (If the update was more complicated then we'd have to use a critical section instead of an atomic update, of course.) The cost of this final accumulation is then going to be N_{Threads} atomic instructions; each, almost certainly, with an associated cache miss.

The code also looks simpler, since it avoids the OpenMP runtime library calls to `omp_get_num_threads()` and `omp_get_thread_num()`, something that is generally advantageous, as code that has to know about thread identities and which thread is executing is often harder to understand and debug than code that treats all threads the same.

Of course, although this code now has each thread perform its own part of the final reduction, that code is still somewhat serialized, since it has to use an atomic operation (or critical section) to avoid race conditions. If there is load imbalance in the (in this case, statically scheduled) loop, then this code does, at least, exploit the imbalance time to perform the reduction, rather than wasting it waiting at the barrier and then performing the final reduction serially.

On the other hand, the serial reduction here is only over the number of threads, and, as we saw, reducing serially can provide more determinism (even if it can't reproduce the serial result), which may be a valuable property. If you were to implement that approach, you should still ensure that there is no false sharing by allocating whole cache lines for each thread's partial result, or, more easily, having each thread accumulate into a thread-local variable (as we did in the later code), and then copy that result to the shared array only when the loop is finished.

One potential issue in all of these reductions is that we need a per-thread copy of the entity into which we are performing the reduction. If that is a scalar or a small vector, that's fine. However, if it is a large vector, that may require more memory than we have available in the stack of each thread (or, indeed, in our machine). In such cases, it may be necessary to revert to using atomic instructions or locks to guard critical sections that perform the reduction into the array. If locks are necessary (because the reduction operation is not supported as an atomic primitive), then using an array of locks and a hash from an array index to a lock (or, on machines where there is hardware support this, a speculative lock) may be required to maintain parallelism. Since, by definition, these cases have large arrays, we have to hope that the accesses are well scattered across the array and that therefore true conflicts are rare.

Although we have shown how one could implement a reduction operation at the user level in an OpenMP threading environment, please remember that you don't need to do that, because the OpenMP language already has support for reductions, including those that require a vector as the reduction target, and those where the reduction operation is user defined. Therefore, if you need reduction in your OpenMP application code, use those interfaces, and don't write your own code. As ever, remember that "The best code is the code I don't have to write".

Since code will not be performing a reduction unless it needs the result, there will normally be a barrier associated with the reduction to ensure that the result has been computed before any thread attempts to use it. This leads us to consider whether we can exploit the structure of our barriers to implement an efficient, parallel reduction of the values of the per-thread accumulators.

7.6.1 Piggy-backing reductions

One obvious way to move reductions into barriers is to move code very similar to our previous example into the barrier implementation. Then, when a thread is about to check-in to the barrier, it will also perform the accumulation of its partial result into the global accumulator before really entering the barrier. As with our code, it will need to do this with an atomic operation or by taking and releasing a lock to create a critical section.

Another possibility is to exploit the topology of the barrier check-in process to remove the need for atomic operations. This can most easily be done when using a tree check-in barrier. Then, the reduction can be performed at each slot in the tree into the local accumulator, by either the thread that owns that slot in a fixed tree or the thread that is the current winner at that level in a dynamic tree. This reduces the number of reduction operations by one (since we don't start with a zeroed accumulator, but rather one that already contains a single thread's accumulated value), and, more importantly, removes the need for any atomic operations, since the logic of the barrier ensures that no other threads can be concurrently updating the relevant accumulator, and allows multiple reduction operations to execute simultaneously.

It is much harder to piggy-back reductions on decentralized barriers, since the more general communication patterns mean that there is no obvious way to choose which thread should perform a particular reduction operation, or to ensure that whichever thread computes and stores the final result can do that before other threads have left the barrier. This is another reason why tree barriers are popular in OpenMP implementations. Whichever type of tree (fixed or dynamic) is used, the reduction up the tree can be written to ensure that it is deterministic.

7.7 Additional optimizations

There are a number of potential barrier optimizations that we have not implemented (such as the n-way dissemination barrier or hierarchical barriers). Some of these are discussed in "Effective Barrier Synchronization on Intel* Xeon Phi* Coprocessor" [111].

7.7.1 Hierarchical barriers

As we have continually emphasized, to achieve high performance in these communication-intensive operations we must minimize inter-thread communication or overlap it, and, where possible, try to avoid more expensive communications. This leads to the obvious conclusion that we can improve barrier performance by using a hierarchical barrier in which we use different barrier implementations for threads at different levels in the hardware hierarchy. Thus, we could implement a simple counting barrier for threads that share the same physical core (and, therefore, all levels of the cache), then have one thread (either the last to arrive or a fixed-root thread) participate in a higher-level in-socket (or in-tile) barrier, before yet another barrier is used in which only one thread per socket participates. On the AMD* machine, where groups of four cores share an L3 cache, it may make sense to lift the leaf barrier to the four-core level (8 hardware threads).

This approach allows us to choose barriers at each level of the hardware hierarchy that work well for the number of entities at that level and with the communication

Table 7.6: Barrier summary.

Barrier	Style	LILO Path Scaling	$T_{\text{LILO}}(64)$
All-to-All	Distributed	Linear	14.9 µs
Atomic Counter	Distributed	Linear	8.5 µs
Dissemination	Distributed	$\log_2 N_{\text{Threads}}$	1.9 µs
Fixed-Tree	Centralized	$(\log_R N_{\text{Threads}})$ + Broadcast	900 ns
Dynamic Tree	Centralized	$(\log_R N_{\text{Threads}})$ + Broadcast	905 ns

costs of that level. Since fixed-tree barriers are already hierarchical it is possible to get some of the benefits of a hierarchical barrier by choosing how threads are mapped into the tree based on how they are bound to hardware resources. Thus one might minimize cross-socket communication by having immediate descendants of the root node of the tree be in different sockets, each with their own sub-tree of threads inside that socket. That may require dealing with different fan-in and fan-out values for different levels of the tree.

Of course, if we are working in an environment in which threads are not tightly bound to hardware locations, then these optimizations are unlikely to work, since the OS may choose to reschedule a thread far away in the machine from where the runtime thought it was running. They are thus more suitable for single-user HPC applications than operations in a shared and, therefore, noisy machine, or in a virtual-machine environment where even tight binding to the virtual hardware need not ensure binding to the underlying physical cores.

7.8 Conclusions

Table 7.6 summarizes the properties of the barriers, while Figure 7.16 shows the performance of our tree barriers against that of the LLVM OpenMP runtime. You can see that our barriers are faster (both take ~900 ns at 64 threads, whereas the LLVM implementation takes ~2.7 µs at the same scale). However, this is, perhaps, an unfair and invalid comparison, since the LLVM implementation handles tasking and reductions, while the barriers measured here do not.

Let's re-iterate that barriers are evil and can cause a lot of trouble. Despite all of our work here to implement barriers well, they remain problematic from a performance point of view, and ideally, people should use a programming model that does not require them.

Consider communication patterns, because the communication pattern along the LILO path gives the upper bound on the performance of the barrier. So thinking about that before implementing your new barrier is a sensible thing to do.

Also, take into account what happens when you scale up the number of threads. Which barrier performs best will depend on the machine's properties and the number

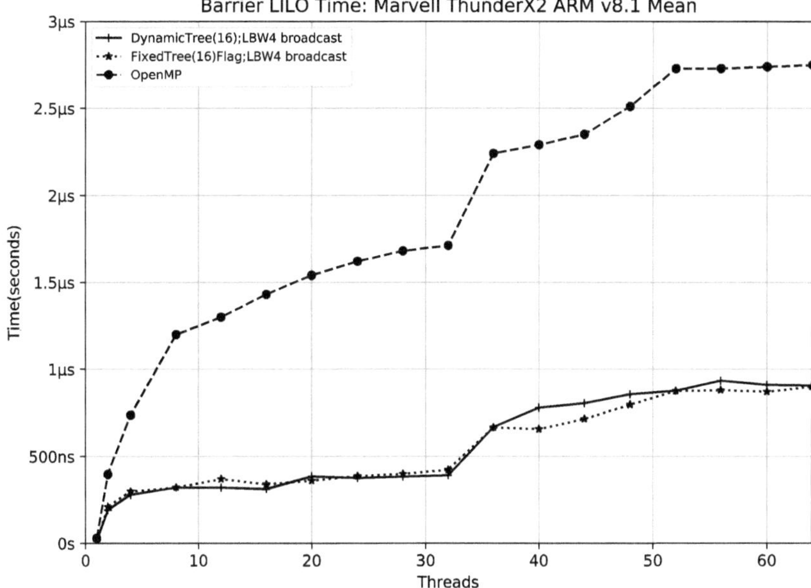

Figure 7.16: Barrier LILO time (including LLVM's OpenMP barrier).

of threads involved. Therefore, it is very likely that there is no single barrier that is best across all thread counts. Since a barrier has to know the number of threads on which it will operate when it is created, the code at that point can choose an appropriate implementation.

8 Scheduling parallel loops

The OpenMP[*] API explicitly supports parallel loop constructs and describes in some detail how loop iterations must be allocated to the available threads. Other parallel systems take a different approach; for instance, task-based systems (such as TBB [108]) do not treat loop-iteration scheduling as a special operation, but rather apply some subtlety to creating tasks that cover the iterations and then allow those tasks to be scheduled just like any other.

In this chapter, we will consider the issues with scheduling loops that have a well-determined iteration count that can be calculated when the loop is entered. Such loops normally represent iteration over a fixed space (such as the indices in an array), rather than iteration down a list of indeterminate length or a search in which a loop can terminate after any iteration. To ensure a fixed iteration count, early exit from the loop body (whether by `break`, `return`, `goto`, or any other trickery, such as `longjmp()` or `throw`) is forbidden.

8.1 Aims of scheduling

The aim of scheduling work onto threads is to execute all of the work in the minimum time. At first glance, this seems a simple thing to do (though the general mathematical problem is NP-hard [142]). However, there are a number of issues that make it more complicated:

1. **Machine state:** We have seen previously that the location of data in processor caches can have a significant effect on code performance. It should therefore be obvious that if a set of iterations operate on a common set of data, they will run faster if they are scheduled onto threads that are bound to hardware threads that share the same caches, since then data in the cache can be reused and will not have to move around the machine. However, we normally have no information about the data being accessed by a given loop iteration, so that information is not available as an input to our scheduling choices; instead, it becomes another cause of load imbalance as iterations that the user may expect to take the same time to execute instead perform differently.

2. **Interference:** Aside from self-induced cache effects where there is data affinity between iterations, the elapsed time it takes to execute an iteration can be affected by external interference. For instance, an interrupt could occur that causes a device driver to execute for some time, stealing the hardware thread away from the thread executing the iteration, or code executing in another hardware thread could be a "noisy neighbor" that creates jitter and churn the shared cache, or execute instructions that lower the processor's clock. These are unpredictable effects, at least for the runtime library doing the scheduling.

https://doi.org/10.1515/9783110632729-008

3. **Incomplete information:** In most cases, at the time that we are performing the scheduling, we do not know how long the tasks we are scheduling will take to execute. Even if the user expects that each chunk of work will take the same amount of time to execute, because of interference or machine-state affinity, this may not actually turn out to be the case at runtime. We are therefore dealing with an "online" version of the scheduling problem where we have incomplete information [3].
4. **Scheduling overhead:** There is always a balance between the time we spend in the scheduler attempting to find a schedule that would minimize runtime and the time spent in the useful user code. The extreme case here would be to construct a 100 % accurate scheduler that never ran any user tasks at all, but could predict their completion time perfectly! It is often the case that using simple schemes with low overhead can provide better overall throughput, even if the schedule they produce is mathematically sub-optimal.

8.2 Theoretical limits on scheduling efficiency

Before we consider the complexities of different loop schedules, it is worth thinking a little in the abstract about the maximum theoretical efficiency that we can expect. If we consider an OpenMP loop without the nowait clause, we can see that there is a barrier at the end of the loop. This makes it much simpler for the programmer to reason about their code, since no threads can leave the loop until all threads have completed their work, and therefore all code after the loop can rely on all iterations of the loop having completed execution. However, like all barriers, it will amplify the resources lost to load imbalance.

We therefore consider the highest efficiency that can be achieved if we are attempting to execute n chunks of work on N_{Threads} threads, where each chunk of work takes the same amount of time to execute, and there is a barrier at the end.

This is clearly closely related to bin-packing problems; we are trying to pack the n chunks of work onto the N_{Threads} threads in such a way that we minimize the amount of time the threads spend waiting at the barrier. The optimal solution is clearly to give each thread as close to the same amount of work as we can. However, if n is not divisible by N_{Threads}, we will have $n \bmod N_{\text{Threads}}$ chunks of work left.

Figure 8.1 shows how we can most efficiently pack seven pieces of work, each of which takes the same amount of time to execute, onto three threads. The useful work is represented by the seven gray squares and the time wasted where there is no work for the threads to execute (but they must wait until the other threads finish) by the two white ones. This therefore uses only seven of the available nine units of work, giving a parallel efficiency of $\frac{7}{9} = 77.8\,\%$.

Clearly, any packing that has more than one chunk difference in the number of chunks executed by the threads (e. g., a (1,2,4) distribution) will be less efficient. This would have a parallel efficiency of $\frac{7}{12} = 58.3\,\%$.

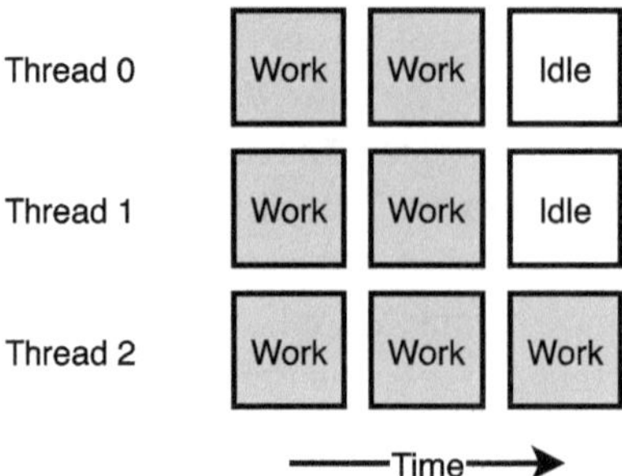

Figure 8.1: Packing seven chunks of work onto three threads.

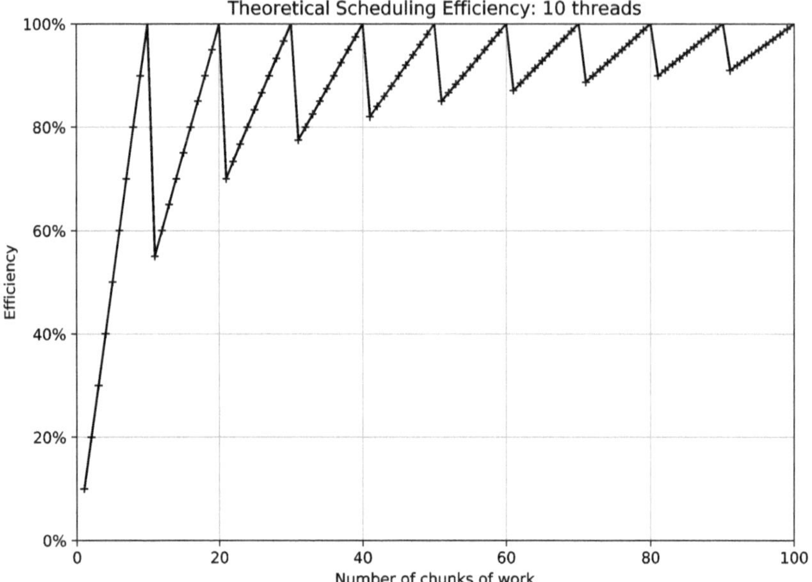

Figure 8.2: Maximum efficiency when executing n equal chunks on ten threads.

In Figure 8.2, we plot the efficiency when executing work on ten threads. As you should expect, it shows a sawtooth effect, reaching 100 % efficiency whenever we have a number of pieces of work that is a multiple of 10. You can also see, though, that if we cannot ensure that we have a number of chunks that is divisible by the number of threads (and that imbalance is normally the case, since we can't choose our problem size to match our machine), then to guarantee an efficiency of more than 90 % we need to ensure that $n > 10 \cdot N_{\text{Threads}}$.

More generally, we can express the efficiency as

$$E = 1 - \frac{N_{\text{Threads}} - 1}{N_{\text{Threads}} \cdot \lceil n/N_{\text{Threads}} \rceil}.$$

At large numbers of threads, we can simplify this to the rule of thumb that to guarantee efficiency E, we need

$$n > \frac{N_{\text{Threads}}}{1 - E}$$

pieces of work. (Clearly, this is also telling us that we can never guarantee complete efficiency, since as $E \to 1$, it follows that $n \to \infty$.)

It is important to note that this is a fundamental mathematical constraint, which is independent of the particular scheduling approach we take. It doesn't matter how the specific pieces of work are allocated to the threads (which will be affected by the choice of scheduler), but merely how many are allocated to each. With these simplified assumptions, no scheduler can perform better than this limit.

While this is really a constraint on the application code that it must create enough parallel work, it is useful to be aware of it, since application programmers are always happy to find inefficiencies in a (parallel) runtime system, rather than their code, and may not understand that the runtime can only exploit the parallelism that has been described by the programmer or found by the compiler. Thus, running parallel loops with an iteration count of 30 may work fine on a 2-core laptop, but cannot possibly perform well on a 64-core server.

8.3 Fundamental scheduling approaches

There are two fundamental approaches to scheduling:
1. **Static:** The distribution of iterations to threads is determined at the start of the loop with no consideration of the dynamic behavior of the code.
2. **Dynamic:** The distribution of the iterations is determined as the execution of the loop takes place so that it can react to the progress of execution so far.

8.3.1 Static loop scheduling

Static loop scheduling is the most straightforward type of scheduling, and has the lowest scheduling overhead, since each thread can completely determine which iterations of the loop it should execute with only a single calculation and without needing to share any information with other threads.

The OpenMP API provides two different static schedules:
- **Blocked:** The blocked schedule is invoked by simply requesting a static schedule (`schedule(static)`), or, in most OpenMP implementations, not even bothering with a `schedule` attribute on a parallel loop. This schedule requests that the implementation allocates a single, large contiguous chunk of iterations to each

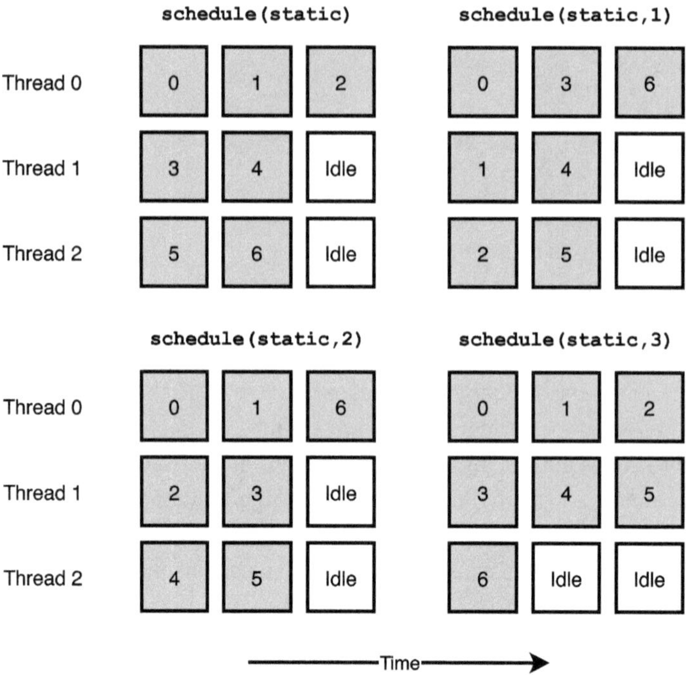

Figure 8.3: Different static allocations of seven equal chunks to three threads.

thread. This schedule has the advantage that iterations that are nearby in the iteration space are very likely to be executed by the same thread, which may improve cache performance. Since this schedule allocates a single consecutive block of iterations to each thread, there are only $N_{\text{Threads}} - 1$ places where iteration n and iteration $n + 1$ execute on different threads. It is impossible to have fewer switches between threads.

– **Block cyclic:** The block-cyclic schedule is invoked by giving the static schedule a chunk size (even if that is 1, which seems rather perverse, but syntactically distinguishes the two possible approaches here)—for instance, schedule(static,16). It allocates chunks of iterations cyclically to threads. If we take the same example that we used in Section 8.2, then we can illustrate how the block size affects the allocation of iterations to threads. Figure 8.3 shows the distributions for the same test case as before with chunk sizes of 1, 2, and 3. You can see that the cyclic distributions have less locality than the blocked distribution. Since there is a switch between each chunk of iterations, a (static,1) distribution has a switch between threads after each iteration, which is the most switches there can possibly be.

The cyclic schedules are potentially useful when the time of each iteration is known to increase or decrease consistently, as happens in loops that operate over a triangular subset of an array, like this:

```
#pragma omp for
for (int i = 0; i < LIMIT; i++)
  for (int j = 0; j <= i; j++)
    /* ... do something taking constant time ... */
```

Here, the simple schedule(static) schedule will allocate all of the shortest iterations to the first thread, and all of the longest to the final thread, so the imbalance will be large. As a practical example, if we were to make LIMIT equal to 1,000 in the above example and distribute the work over 32 threads, then we can easily compute that the theoretically expected efficiency of the schedule(static) loop is only 51 %, whereas that of the schedule(static,1) loop is 97 %. Of course, other factors (in particular, false sharing that causes cache-line motion) can impact this and make the cyclic distribution perform much worse; however this does suggest that where we know the distribution of iteration times, static schedules can still be useful if there is predictable imbalance and neighboring iterations do not share data.

8.3.2 Dynamic loop scheduling

Dynamic loop scheduling requires that information about the current state of the loop is maintained and shared between threads so that they can take decisions based on that information. As a result, the threads have to perform a scheduling operation (normally, a call into the runtime library) to discover which iteration in the canonical space to execute before every such iteration. This inherently means that there is more scheduling overhead than in the static schedules. However, the dynamic schedules have the advantage that they can react to imbalance, whether it is algorithmic or the result of running on a noisy system with a high amount of jitter. In the real world, the benefits of being able to adjust to such induced imbalance can often overcome the additional cost of dynamic scheduling.

However, the schedule code must be careful how it manipulates the shared state, both to ensure correctness and to achieve performance, since the cache line(s) that contain this shared state will necessarily be moving around the machine. The various dynamic schedules are therefore attempting to minimize the number of atomic operations required while maintaining the ability to perform dynamically.

The idea of exploiting the runtime state when scheduling goes back to at least the 1980s (see [50, 85, 120, 127, 128]).

8.4 Mapping to canonical form

To simplify the discussion, we will describe all of our loop-scheduling decisions as if the loops were in canonical form and follow the pattern:

```
for (i = 0; i < loopCount; i++)
```

Any loop can be reduced to this form, and it will then be possible to map back from an iteration in this canonical space to the iteration values the user code expects to see. Indeed, many compilers reduce loops to this form internally to simplify their loop optimization passes. The LLVM OpenMP implementation always passes descriptions of loops to the runtime in this dense canonical space, generating code to map from this back into the iteration space expected by the user's code. This has the advantage that the compiler knows more about the user's loop than the runtime does; so, for instance, it may know that the increment really is the constant value 1, which will then allow it to remove unnecessary multiplication or division-by-one operations from the scheduling computations, whereas the runtime has to assume that the increment could have any value, and so has to either treat the case where the increment is one specifically or perform the unnecessary multiplications and divisions.

As we have already mentioned in Section 8.3.1, there is one additional complexity to handle when implementing the OpenMP API, which is that the `schedule` clause allows a chunk size to be specified. This means that when allocating iterations to threads, rather than allocating a single iteration, a chunk of iterations is allocated. When mapping into and out of the canonical form, this reduces the "real" iterations that we are concerned about. For instance:

```
#pragma omp for schedule(static,4)
for (i = 0; i < 10; i++)
```

This allocates iterations in chunks of four, so the chunks that can be scheduled are `[0,1,2,3]`, `[4,5,6,7]`, `[8,9]`, and we only have three chunks to deal with.

The OpenMP API specification [100] mandates that if the number of loop iterations is not a multiple of the requested chunk size, then the remainder iterations should be scheduled as a single small, final chunk and therefore a balanced approach is not permitted. This makes sense when considering that the user (or compiler) may be using the chunk size to ensure alignment, or to ensure that iterations are handed out in SIMD-width chunks.

In the book's example runtime, we use the class `CanonicalLoop` to represent a loop, as shown in Listing 8.1. This data structure is initialized using the `init()` member function shown in Listing 8.2, which maps the more general loop description into the canonical space. We will see in Section 8.8 why the class uses an `init()` function instead of a constructor for initialization.

```cpp
// Store information about iteration space in canonical loop
// form for (<loopVarType> j = 0; j < iterationCount; j++)
template <typename loopVarType>
class CanonicalLoop {
  typedef typename
    typeTraits_t<loopVarType>::unsigned_t unsignedType;
  loopVarType base;    /* Initial value */
  loopVarType incr;    /* Value of the loop increment */
  loopVarType end;     /* Initial value */
  loopVarType scale;   /* Scale factor */
  unsignedType count; /* Number of iterations */
  ...
}
```

Listing 8.1: The CanonicalLoop class.

```cpp
// Initial form is
//    when i > 0 : for (j = b; j <= e; j += i)
//    when i < 0 : for (j = b; j >= e; j += i)
// Where
//    b is the non-canonical base,
//    e is the non-canonical end,
//    i is the non-canonical increment.
void init (loopVarType b, loopVarType e,
           loopVarType i, uint32_t chunk) {
  base = b;
  end = e;
  incr = i;
  if (incr > 0)
    count = 1 + (end - base) / incr;
  else
    count = 1 + (base - end) / -incr;

  // Scale down by the chunk size.
  count = (count + chunk - 1) / chunk;
  scale = chunk * incr;
}
```

Listing 8.2: Initialize the CanonicalLoop.

8.5 Compiler loop transformations

In Chapter 4, we saw how the compiler handles high-level transformations to enable code to be executed in parallel and share some variables. Here, we'll talk a little about how loops are transformed so that they can make runtime calls to ensure that they execute worksharing loops correctly.

The fundamental problem here is that the compiled code does not know about the environment in which it will be running. In particular, it cannot know how many threads will be sharing the loop iterations, which thread is executing, or sometimes even which schedule the user has requested; therefore, it needs to make calculations at runtime. When the schedule is known, and is static, the compiler could make a call into the runtime library to discover the information it needs and then compute the set of iterations in the compiled code before executing them.

We can see GCC 8.2 doing exactly this for this simple example:

```
extern void f(int i);
void applyF(int *p) {
  #pragma omp for schedule(static,1) nowait
  for (int i=0; i<100; i++)
    f(p[i]);
}
```

which generates code like this for the Arm[*] processor for the chunked loop schedule:

```
applyF():
            stp       x29, x30, [sp, -32]!
            mov       x29, sp
            stp       x19, x20, [sp, 16]
            bl        omp_get_num_threads
            mov       w20, w0
            bl        omp_get_thread_num
            cmp       w0, 99
            bgt       .L1
            mov       w19, w0
    .L3:
            mov       w0, w19
            add       w19, w19, w20
            bl        f(int)
            cmp       w19, 99
            ble       .L3
    .L1:
            ldp       x19, x20, [sp, 16]
            ldp       x29, x30, [sp], 32
            ret
```

We can see that the compiler calls `omp_get_num_threads()` to determine how many threads will execute the loop, then invokes `omp_get_thread_num()` to determine the index of the thread, performs the computation to decide the iterations this thread should execute, and finally executes the loop body (at label `.L3`) for the thread's assigned iterations.

If we change the compiler to `clang`, then the compiler instead generates code that calls the runtime library function `__kmpc_for_static_init_4()` to perform the calculations (we don't show the code here, since it is uninteresting).

Note that if you want to investigate (and compare) the code generated by the different compilers for small functions such as these, then the "Compiler Explorer" website [45] is an excellent resource, since it allows you to see the assembly code generated by different compilers displayed next to each other, with the code generated from individual source lines shown with a different background color, and also allows you to choose compilation flags or compiler versions. All of this can be achieved from your browser, without you having to install all of the compilers on your machine, and also allows you to compare different target architectures.

In the case of dynamic schedules, the execution requires the use of shared data, which is hard for the compiler to handle on its own, since it cannot use static allocation as the same function may be called recursively. It is therefore more natural to implement these schedules as calls into the runtime library. We can see this happening if we change the loop in our example above to use `schedule(dynamic)`. The corresponding code that is generated by GCC is in Listing 8.3.

Listing 8.3 shows calls to initialize the loop (`GOMP_loop_dynamic_start()`), claim each iteration (`GOMP_loop_dynamic_next()`), and finally, allow the runtime to clean up any internal state (`GOMP_loop_end_nowait()`). The LLVM compiler makes similar calls to the functions `__kmpc_dispatch_init_4()` and `__kmpc_dispatch_next_4()`, but does not make a finalization call since the runtime must know when each thread has finished, as in the LLVM interface, it has to return `false` from the next-chunk function when there are no remaining iterations to be executed. It is therefore easy for the runtime to execute per-thread cleanup code.

Chapter 4 discussed the generation of SIMD instructions that allow multiple iterations of a loop over a vector to be executed by a single instruction operating on multiple elements of the vector simultaneously. When this is applied, the number of times the loop body code has to be executed is reduced by the number of lanes in the SIMD instructions, since each execution of the body is operating on that many elements. This clearly affects the number of iterations that can be scheduled by the runtime in a way that is similar to the chunking, which we have already described. However, in general, it remains up to the compiler to handle those complexities by describing the coarser-grained iteration space to the runtime library.

```
applyF ():
        stp     x29, x30, [sp, -48]!
        mov     x3, 1
        mov     x1, 100
        mov     x29, sp
        add     x5, sp, 40
        add     x4, sp, 32
        mov     x2, x3
        mov     x0, 0
        bl      GOMP_loop_dynamic_start
        tst     w0, 255
        beq     .L2
        stp     x19, x20, [sp, 16]
.L4:
        ldr     w19, [sp, 32]
        ldr     w20, [sp, 40]
.L3:
        mov     w0, w19
        add     w19, w19, 1
        bl      f
        cmp     w20, w19
        bgt     .L3
        add     x1, sp, 40
        add     x0, sp, 32
        bl      GOMP_loop_dynamic_next
        tst     w0, 255
        bne     .L4
        ldp     x19, x20, [sp, 16]
.L2:
        bl      GOMP_loop_end_nowait
        ldp     x29, x30, [sp], 48
        ret
```

Listing 8.3: GCC code for a dynamic loop schedule.

When generating SIMD loops, the compiler often also has to handle alignment issues to make sure that accesses to data are aligned with the requirements of the instruction set architecture. It is also likely that the width of the SIMD register does not evenly divide the chunk size that is assigned to a thread. In this case, the compiler will have to emit remainder loops to process the excess loop iterations. It also may have to emit peel loops to process iterations in the beginning of the loop iteration space to ensure alignment properties or to move the loop index to the beginning of the first full SIMD register.

8.6 Monotonicity of loop scheduling

In this section, we will describe one aspect of the user-visible semantics of a schedule, which is *monotonicity* and its absence, *non-monotonicity*. From a user point of view, this may seem a small and irrelevant issue, since we often don't worry about monotonicity when working on user code, as either the code requires stricter semantics, which can be forced by using the ordered clause on the loop, or the order of execution of iterations can be completely arbitrary. The more subtle distinction between monotonic and non-monotonic schedules (which is only about a thread-local ordering, not a global one) is harder to exploit. However, since the difference is detectable, the OpenMP standard allows users to ensure monotonicity in case they need it, and as implementers, we must ensure that the code we write meets the semantic requirements of the language.

We do not discuss the implementation of ordered loops, since they are rare, and their semantic constraints almost guarantee that limited parallelism will be available.

The original OpenMP specification describes the semantics of dynamic loop scheduling by referring to an example implementation in which each thread atomically claims a chunk of loop iterations (in our canonical loop representation, a single value) from a shared loop counter. When that counter reaches the loop limit, no more iterations are available. This implies the property of monotonicity, which means that each thread only ever executes iterations that are beyond the ones it has already seen. The code in Listing 8.4 shows how this property can be detected by the user, and therefore, that they might be depending upon it.

While at first it was not entirely clear whether this property was required by the OpenMP specification, in the 5.0 and later versions, the semantics have been clarified to state that a simple dynamic schedule can be implemented with non-monotonic semantics. (So the OpenMP implementation is allowed to cause the example code in Listing 8.4 to abort.) The standard has also introduced the schedule modifiers monotonic and nonmonotonic so that if monotonicity is required, a user can request it by using a schedule(monotonic:dynamic) schedule. Similarly, they can explicitly state that they are happy with a non-monotonic schedule by using schedule(nonmonotonic:dynamic).

Although non-monotonicity can now be the default, it is possible that OpenMP implementations will not actually make it the default, since implementing schedule(dynamic) as schedule(nonmonotonic:dynamic) can break existing code (such as our example), and compiler customers do not like their code that used to work to be

```
#pragma omp parallel
{
  // Each thread remembers the last iteration it has
  // seen, ...
  int myLast = -1;
#pragma omp for schedule(dynamic)
  for (int i = 0; i < LIMIT; i++) {
    // checks whether the current iteration is below it, ...
    if (i < myLast)
      abort(-1);
    // and then remembers the new, highest, value.
    myLast = i;
  }
}
```

Listing 8.4: Code to detect nonmonotonicity in a parallel loop schedule.

broken by a compiler upgrade, even if the code is not conforming to the specification. It is therefore worth being explicit about the monotonicity you require if you are writing OpenMP code with dynamic schedules, since, as we will see, the implementation of schedule(nonmonotonic:dynamic) often has lower scheduling overhead than that of schedule(monotonic:dynamic). Note that the runtime that accompanies this book implements schedule(dynamic) as schedule(nonmonotonic:dynamic). In our performance discussions, we will always refer to these as "monotonic" and "nonmonotonic".

8.7 Static-schedule implementation

Now that we are equipped with the specific code patterns and performance aspects of loop scheduling, we can start looking at actual implementations. We will start with static loop scheduling, as it's a much simpler case and well-suited to set the stage for dynamic loop scheduling.

8.7.1 Blocked loop schedules

The implementation of the blocked schedule that minimizes the imbalance in the number of iterations per thread looks something like what is shown in Listing 8.5, where we are returning results to the compiled code through the pointers plower, pupper, and pstride. Here, we are operating in the canonical iteration space that we

```cpp
auto myThread = Thread::getCurrentThread();
auto me = myThread->getLocalId();
auto numThreads = myThread->getTeam()->getCount();
auto wholeIters = count / numThreads;
auto leftover = count % numThreads;
// One contiguous chunk per thread balanced as well
// as possible.
loopVarType myBase;
loopVarType extras;

if (me < leftover) {
  // Hand out remainder iterations one to each thread to
  // maintain balance as best as we can.
  myBase = me * (wholeIters + 1);
  extras = 1;
}
else {
  myBase = me * wholeIters + leftover;
  extras = 0;
}
*plower = getChunkLower(myBase);
*pupper = getChunkUpper(myBase + wholeIters - 1)
          + extras * incr;
*pstride = count;
```

Listing 8.5: Code to implement the blocked static schedule.

have discussed above, and so have to translate back to the space expected by the compiled code using the getChunkLower() and getChunkUpper() methods (not shown).

As with all of these code snippets, this code is extracted from a member function of the CanonicalLoop class, whose definition and initialization we saw in Section 8.4; therefore, count represents the number of iterations to be executed.

While the OpenMP specification does not require that the "leftover" iterations are allocated so as to minimize imbalance, most compilers (at least GCC, LLVM, and the Intel[*] compilers) do use this implementation, as shown by using a small test code to count the iterations executed by each thread with an imbalanced static schedule.

```
//   Allocate work block cyclically over the threads; this
//   expression works when the schedule is visible to the
//   compiler, but not for the same schedule when invoked
//   (via schedule(runtime)) from the dynamic runtime
//   scheduling interface. There, we have to return one chunk
//   per call to the dispatch-next function...

auto myThread = Thread::getCurrentThread();
auto me = myThread->getLocalId();
*pstride = numThreads * scale;
*plower = base + me * scale;
*pupper = base + (me + 1) * scale - incr;
```

Listing 8.6: Code to implement the block-cyclic static schedule.

8.7.2 Block-cyclic loop scheduling

Once we have the transformed the loop into the canonical space, which we described above in Section 8.4, the block-cyclic schedule that can be enabled by using `schedule(static, chunk_size)` is easier to implement than the balanced schedule we described in the previous section. Our implementation is shown in Listing 8.6.

8.8 Dynamic loop schedule implementation

Since dynamic loops require a shared state, the runtime has to allocate and release the state object and ensure that each thread can find the state associated with the loop that it is executing. As the OpenMP API allows the `nowait` clause at the `for` construct to remove the implicit barrier at the end of the loop, it is possible for different threads to be executing different loop instances at the same time. At first sight, this seems as though it shouldn't be a problem, since, as this is a dynamically scheduled loop, no thread should leave it until after all of the iterations in it have been started. However, if there are some very long-running iterations or if a thread is not scheduled for a long time, we can still have many different loops executing simultaneously.

The following code shows such an extreme corner case. Even though nobody with a sane mind should write such a code, it still shows that threads can be executing many different loop instances, if the load imbalance becomes to large and the execution thus is skewed:

```
#pragma omp parallel
{
  int me = omp_get_thread_num();

  for (int outer = 0; outer < 100; outer++) {
#pragma omp for schedule(dynamic) nowait
    for (int i = 0; i < 100; i++) {
      if (me == i) {
        sleep(1);
      }
      else {
        // Do something...
      }
    }
  }
}
```

The simple solution, which is used in this book's runtime, is to pre-allocate a fixed number of loop descriptors, each of which maintains a reference count and a sequence number, into which the thread that initializes the structure stores its count of how many dynamically scheduled loops it has executed. That's why we didn't use a constructor to initialize the `CanonicalLoop` object, but rather rely on an `init()` function to (re-)initialize the loop state, when the code enters a new parallel loop.

As each thread arrives at a dynamically scheduled loop, it looks up the loop descriptor that will be used by the loop that matches its own count of the number of dynamically scheduled loops it has executed, and then checks the state of that descriptor to see whether:

1. **The descriptor is initialized for this loop:** In which case it can proceed into the loop using the descriptor, since some other thread arrived earlier and has already finished the initialization.
2. **The descriptor is free:** In which case this thread must try to initialize it.
3. **The descriptor is being initialized:** In which case it must wait until the thread performing the initialization has completed it.
4. **The descriptor is still in use for another loop:** In which case it must wait until the descriptor is freed and then initialize it.

Since we expect that there will normally not be much difference in the number of dynamically scheduled loops that each thread has executed (i. e., the threads remain in reasonable synchrony), we allocate 16 descriptors and map from the loop sequence number to the descriptor by using a simple `Sequence % Max_Concurrent_Loops` computation. Of course, one could use a more complicated data structure here, but this lookup is on the critical path that each thread has to take when entering the loop, so simplicity and speed (or lowest possible latency) are important.

Both the guided and monotonic:dynamic schedules described below are examples of "self-scheduling" [128], which denotes a scheduling approach that is based on local decisions according to a set of rules and constraints; each worker computes its work assignment from the loop iteration space using the same rule.

8.8.1 Guided scheduling

Guided scheduling is invoked in an OpenMP code by requesting `schedule(guided)` or `schedule(guided,chunk_size)` on a worksharing loop. As with all of the OpenMP schedule definitions, the chunk size has the effect of reducing the number of chunks of work that the scheduler can allocate (as we showed in the mapping to canonical form in Section 8.4).

The idea behind guided scheduling is to hand out work in large chunks when there is a lot of work to be done, but smaller ones when the work is nearly exhausted [102]. This reduces the number of scheduling calls that have to be made, while, we hope, achieving some reasonable load balance and immunity to jitter. The OpenMP specification specifies that at each point, the number of loop iterations assigned to a chunk is determined as

$$
n_{\text{assigned}} = \left\lceil F \cdot \frac{n_{\text{remainining}}}{N_{\text{Threads}}} \right\rceil,
$$

where N_{Threads} is the number of threads, executing the loop, $n_{\text{remaining}}$ is the number of unprocessed loop iterations, and F (for fraction) is a value between 0 and 1 that determines what proportion of the thread's share of the available iterations should be allocated immediately. Many implementations set $F = \frac{1}{2}$. Making it large reduces the number of atomic instructions required, but at the cost of reducing the ability to adapt well at runtime.

This schedule can perform better than a simple `monotonic:dynamic` schedule because of its reduction in the number of scheduling calls, and, therefore, contention on the single, centralized iteration counter. We can easily calculate the number of atomic operations required as we change the number of schedulable chunks of work. That number is shown in Figure 8.4 for the case where $F = \frac{1}{2}$. You can see that the required number of atomic operations increases only logarithmically with the number of schedulable chunks of work once there are a sufficient number of them, and so is significantly smaller than the number required for a simple monotonic schedule that requires one atomic operation for each chunk.

However, although the guided schedule can lower the scheduling overhead, like any schedule with monotonicity, it cannot balance a work distribution where the final

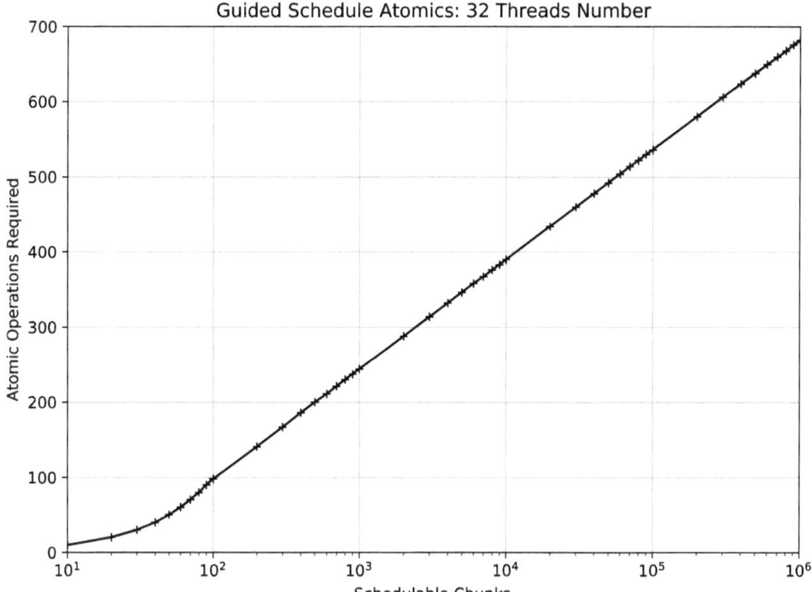

Figure 8.4: Atomics required for a guided schedule on 32 threads.

iterations take the most time, since the monotonicity requires that their execution start late in the sequence, after all the other work has already been started.

The guided-schedule implementation is shown in Listing 8.7. On line 2 you can see the code read the current value of the shared `std::atomic nextIteration`, which is then subtracted from the canonical loop cl's count to compute the number of chunks remaining. On lines 10 through 12, the number that this thread should claim is computed, and then on line 16, the `compare_exchange` atomic operation is executed to make that claim visible to all other threads. If another thread updates the shared counter (`nextIteration`) while this thread is computing its share, then the `compare_exchange` operation will fail, so this thread has to restart. If the `compare_exchange` succeeds, then this thread maps the results back into the iteration space expected by the compiled code and returns those results through the pointer arguments it was passed.

8.8.2 `monotonic:dynamic`

The monotonic dynamic schedule must issue iterations to threads in an increasing manner. The simplest implementation is to maintain a single iteration counter that is atomically incremented by each thread when it requires an iteration on which to work. The problem with this (as with any scheme that relies on a single, centralized counter)

```
 1  for (;;) {
 2    auto localNextIteration = nextIteration.load();
 3    auto remaining = cl->getCount() - localNextIteration;
 4    if (remaining == 0) {
 5      return false;
 6    }
 7    // What is my share of the remaining iterations?
 8    // Round up to ensure that the last iterations are
 9    // consumed.
10    auto myShare = (remaining + threadCount - 1)
11                     / threadCount;
12    // Arbitrarily choose to take (rounded up) 1/2 of it.
13    auto delta = (myShare + 1) / 2;
14    // Execute the atomic CAS to update the shared state
15    // (or fail if another thread has updated while we
16    // were computing this thread's share).
17    if (nextIteration.compare_exchange_strong(
18                          localNextIteration,
19                          localNextIteration + delta)) {
20      auto lastIteration = localNextIteration + delta - 1;
21      *p_lb = cl->getChunkLower(localNextIteration);
22      *p_ub = cl->getChunkUpper(lastIteration);
23      *p_st = cl->getStride(localNextIteration,
24                            lastIteration);
25      if (p_last) {
26        *p_last = cl->isLastChunk(lastIteration);
27      }
28      return true;
29    }
30    // Compare exchange failed, should maybe perform
31    // better backoff.
32    Target::Yield();
33  }
```

Listing 8.7: Code to implement the guided dynamic schedule.

is that moving the cache line containing the counter around the machine between the caches local to each core is expensive, and with this schedule, it is likely to happen for almost every iteration.

Our measurements show that on the Arm machine, an atomic increment on a line that was in a modified state in another core's cache on the same die takes ~58 ns (com-

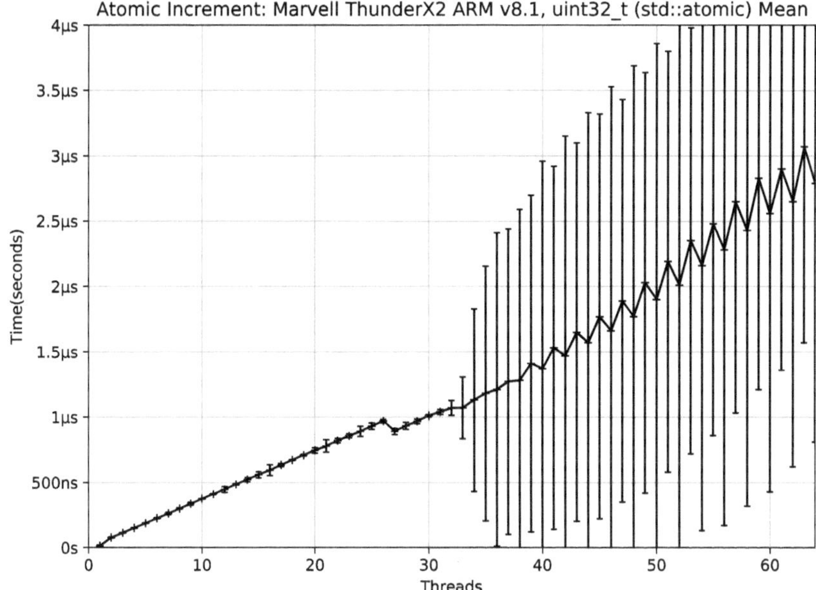

Figure 8.5: Contended atomic increment scaling on the Arm processor.

pared with a simple load that takes ~2.7 ns), whereas across sockets the atomic operation takes ~177 ns while the regular load is only ~9.6 ns. Thus, the lowest overhead (even with no contention) is significant if we are attempting to schedule loop iterations that take <1 µs. Under contention, the atomic operation time increases rapidly, as shown in Figure 8.5, where we can see that it exceeds 2 µs under complete contention when using more than 50 cores in this machine. This implies that this schedule cannot work well here with less than that amount of work/iteration.

The `monotonic:dynamic` schedule is easy to implement, as each thread is merely incrementing a single global counter in the canonical iteration space (see Listing 8.8). You can see that it is very similar to that for the guided schedule, the only difference being in the choice of how much to change the shared counter. The rest of the code is the standard code to map the result back from the canonical iteration space (handling the base of the loop and any chunking), which is identical to that for the guided schedule.

8.8.3 `nonmonotonic:dynamic`

The non-monotonic dynamic schedule has the fewest semantic requirements of the schedules described by the OpenMP standard, and, therefore, the largest number of possible potential implementations. If one is interested in adding a new scheduling implementation to an OpenMP runtime implementation, it is here that this may be a

```
for (;;) {
  auto localNI = nextIteration.load();
  if (UNLIKELY(localNI == cl->getCount())) {
    // No iterations remain.
    return false;
  }
  // Using compare_exchange rather than add avoids going
  // past the end, which could, potentially, be a problem
  // if iterating over the full range of the type.
  if (nextIteration.compare_exchange_strong(localNI,
                                            localNI + 1)) {
    *p_lb = cl->getChunkLower(localNI);
    *p_ub = cl->getChunkUpper(localNI);
    *p_st = cl->getStride(localNI, localNI);
    if (p_last)
      *p_last = cl->isLastChunk(localNI);
    return true;
  }
  // Compare exchange failed, should maybe perform
  // better backoff.
  Target::Yield();
}
```

Listing 8.8: Code to implement the `monotonic:dynamic` schedule.

good place to implement it, where there are few constraints beyond those of sanity (i. e., each iteration must be executed precisely once, and only iterations in the loop should be executed).

One, relatively obvious, way to implement this schedule is to use *static stealing*. The idea here is based on a technique that is used in tasking systems such as Intel TBB [66, 151] and [16], where each thread maintains its own task pool (see Chapter 9). It then updates one end of a double-ended queue while other threads update the other if they steal tasks. In the loop scheduling case, each thread is initially allocated the set of iterations it would receive from the blocked schedule, i. e., those it would execute in a `schedule(static)` loop. However, when a thread exhausts those initially allocated iterations, it searches for iterations that have not yet started and are currently owned by another thread. It then steals some to execute itself. Since the iterations are contiguous, they can be represented by a base and an end, rather than the double-ended queue that is required when dealing with arbitrary tasks. We do want to maintain the separation of updates, though, by having the owning thread increment the base, while other threads that are trying to steal work pull the end downwards.

In such an implementation, it should be possible to remove all atomics from the common path on which a thread executes work it already owned. The details of such an implementation are left as an exercise; here, we only consider the simplest implementation in which both the owner and thief of the work execute atomic operations. Although this is sub-optimal, it can still perform much better than the implementation of the monotonic schedule, since there are now N_{Threads} different cache lines involved in atomic instructions, rather than only one, and, many iteration executions only require an atomic operation on the local work descriptor, which is likely to remain hot in the owner's cache.

This implementation has the advantage that if the work is well balanced, then it is very similar to a `schedule(static)` well-blocked schedule, and therefore has the same locality of reference and affinity benefits, while, when there is load imbalance, it has the advantage over any static schedule in being able to adapt to the current state of execution. However, it differs from the guided and monotonic schedules in that for each of these, we can compute the number of atomic operations that will be required, whereas with this work-stealing implementation, we cannot. It might be that no atomic operations at all are required (if the balanced static schedule works perfectly), or that many are needed if there is significant imbalance. This "pay-as-you-go" aspect of the implementation is one of its advantages.

The whole code of this implementation is too large to show, so we will only show the key elements. Listing 8.9 shows the code for the case of consuming local work. You can see that in this implementation, the local consumption of work is still using a compare-exchange operation, so it can be delayed by other threads that are stealing. The code that tries to steal remote work is shown in Listing 8.10. As the comment in the code says, this is also overly cautious!

To complete the picture, we need to show how a thread with no work finds a victim thread from which to steal work. There are many possibilities here, and which one will work best is likely to depend on the workload. In the book's runtime, we currently choose a random victim and then search forward from there, skipping threads that are also searching. Other possibilities would be to simply choose random victims without searching forward, or to try to steal within a NUMA domain initially. Experience from task scheduling can clearly be applied here (see also Section 9.4.3).

We must also determine when there is no work left to schedule (i. e., all canonical iterations have started execution). One approach would be to wait until all threads are searching for work, at which point we certainly know that no threads have any local work. However, this determines whether all iterations have finished execution, whereas the test we actually want is that all iterations have started. We need the more subtle test so that threads can move on out of the loop and potentially execute tasks at the barrier following the loop, or, if the loop has the `nowait` clause, move on to execute the code after the loop.

To allow us to detect when all canonical iterations have started, each thread maintains a count of the number of iterations it has started. This is a local count that is only

```
// For simplicity use compare_exchange here as well as in
// the steal operation.  This should not be *too* bad,
// since most of the time, stealing is not happening, and
// a thread is operating on local data.
template <typename UT>
    for (;;) {
      contiguousWork oldValues(this);
      auto oldBase = oldValues.base;
      auto oldEnd = oldValues.end;
      // Have we run out of local iterations?
      if (oldBase >= oldEnd) {
        return false;
      }
      auto newBase = oldBase + 1;
      contiguousWork newValues(newBase, oldEnd);
      // Did anything change while we were calculating our
      // parameters?
      if (atomicPair.compare_exchange_strong(oldValues.pair,
                                     newValues.pair)) {
        // Nothing changed; we succeeded in incrementing the
        // base without anything else changing around us.
        *basep = oldBase;
        return true;
      }
    }
}
```

Listing 8.9: Consuming local iterations for `nonmonotonic:dynamic`.

updated by the owning thread, so these updates do not need to be atomic add instructions. Then, when iterating over potential victims, a thread that is looking for work can accumulate the total number of iterations that have been started, and, when it has looked at all threads, compare this with the total number of iterations required. If there is no work left, it can set a flag (so that other threads can detect this immediately) and then leave the loop.

While it might seem that scanning all other threads and finding no available iterations should be sufficient here, there is a potential race condition in which a stealing thread has removed iterations from a victim, but has not yet made them visible again, which means that this may not work.

```cpp
template <typename UT>
bool contiguousWork<UT>::trySteal(UT * basep, UT * endp) {
  for (;;) {
    contiguousWork oldValues(this);
    auto oldEnd = oldValues.end;
    auto oldBase = oldValues.base;

    // We need this >= to handle the race resolution case
    // mentioned above, which can lead to (1,0)
    // (or equivalent) appearing...
    if (oldBase >= oldEnd)
      return false;

    // Try to steal half of the available work. By doing
    // that, we distribute stealable work across the
    // machine, reducing contention and meaning that
    // stealing threads need to look less far to find it.
    auto available = oldEnd - oldBase;

    // Round up so that we will steal the last available
    // iteration.
    auto newEnd = oldEnd - (available + 1) / 2;
    contiguousWork newValues(oldBase, newEnd);

    // Did anything change while we were calculating our
    // parameters? This is slightly over cautious. If we are
    // stealing iterations 1000:2000, it doesn't matter if
    // the owner claims iteration zero, so we should be able
    // to be smarter about this.
    if (atomicPair.compare_exchange_strong(oldValues.pair,
                                           newValues.pair)) {
      // Nothing changed, so we succeeded in stealing.
      *basep = newEnd;
      *endp = oldEnd;
      return true;
    }
  }
}
```

Listing 8.10: Stealing iterations in the `nonmonotonic:dynamic` schedule.

8.9 Evaluation of loop schedules

To evaluate the different loop schedules, we look at three different benchmarks.

In each case, the inner loop calls the same function, which performs an unoptimizable operation on a value to consume some apparently constant amount of time:

```
static uint64_t loadFunction (uint64_t v) {
  for (int i = 0; i < 15; i++) {
    v = ((v + 4) * (v + 1)) / ((v + 2) * (v + 3));
  }
  return v;
}
```

The benchmarks then use this function in the following contexts to reproduce three different algorithmic behaviors:

- **Square:** Here, we iterate over a square array parallelizing only on the outer loop, and performing a small amount of work at each cell. We expect the inner iterations here to have approximately constant time:

```
#pragma omp for schedule(...)
for (int i = 0; i < DIM; i++)
  for (int j = 0; j < DIM; j++)
    array[i][j] = loadFunction(array[i][j]);
```

- **Increasing:** Here, we iterate over an increasing triangular section of a square array, again parallelizing over the outer loop:

```
#pragma omp for schedule(...)
for (int i = 0; i < DIM; i++)
  for (int j = 0; j <= i; j++)
    array[i][j] = loadFunction(array[i][j]);
```

This gives outer iterations, each of which takes longer than the one before.
- **Random:** Here, we perform the same inner computation, but choose a random number (between 1 and 15) of inner iterations, and execute each set of inner iterations in parallel by collapsing the two outer loops:

```
#pragma omp for schedule(...) collapse(2)
for (int i = 0; i < (DIM / 2); i++)}
  for (int j = 0; j < (DIM / 2); j++) {
    int loops=int(array[i][j]) & 15;
    for (int k = 0; k <= loops; k++)
      array[i][j] = loadFunction(array[i][j]);
  }
```

Table 8.1: Scheduling experiment iteration times.

Experiment	Iteration Time		
	Min	Mean	Max
Square	230 µs	230 µs	230 µs
Increasing	95 ns	115 µs	230 µs
Random	95 ns	700 ns	1.4 µs

In all of the benchmarks, we re-initialize the contents of the array before each timed execution of the loop.

On the Arm processor, the `loadFunction()` calculation takes ~93 ns, so the iterations we are scheduling have the properties shown in Table 8.1, with `DIM=2500`, which is the size that we are using:

We show the performance of the loop schedules when running on our Arm machine using code compiled with `clang` 9.0.0 and the associated LLVM OpenMP runtime library. Performance with the book's runtime library is similar. In all cases, thread affinity is forced so that we use only one of the four SMT threads implemented in each Marvell[*] core, and strongly bind the thread to that single logical core.

In each case, we plot the parallel efficiency of the different schedules compared with the best single thread (but still OpenMP code) case as we increase the number of cores. Note that the bottom of the y-axis is not zero, and the top is not 100 % so you need to be careful when comparing the different results.

Figure 8.6 shows the results for the square case, in which we might expect each chunk of work to take the same amount of time. However, we can see that our dynamic schedules (monotonic, non-monotonic, and guided) all perform better than the static ones. They show parallel efficiency of ~105 % within a single socket, which means that the code is achieving superlinear speedup, so the mean time to execute an individual chunk of work when running in parallel is less than it was when running on a single core. This strongly suggests that the code is cache sensitive, and that having more total cache available is beneficial. The drop off in scaling when we enter the second socket (i. e., use >32 cores) also suggests that NUMA effects may be in play. Overall, the dynamic schedules are performing best here, though the static schedule performs reasonably inside a single socket, while `static,1` is worst (again implying that cache locality is important).

Figure 8.7 shows the results for the increasing work case. The dynamic schedules' performance is similar to that of the previous case, as is that of the `static,1` schedule, while the simple static schedule now performs horribly, as we expected, since all of the largest pieces of work will be assigned to the same core.

However, if we look at the performance of the random test shown in Figure 8.8, we see rather different results. Here, the `monotonic:dynamic` schedule performs awfully. This is because it requires an atomic operation on a single, contended cache line for

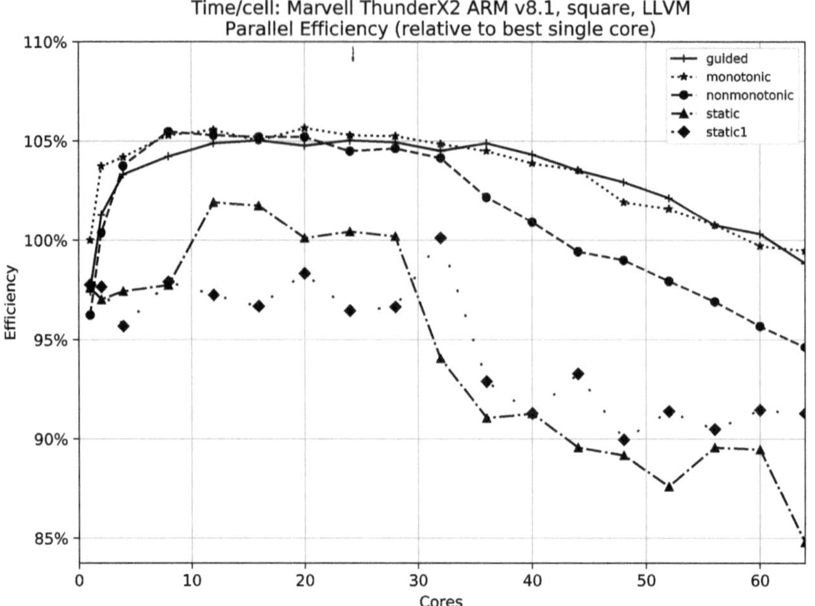

Figure 8.6: Parallel efficiency of loop scheduling (Square).

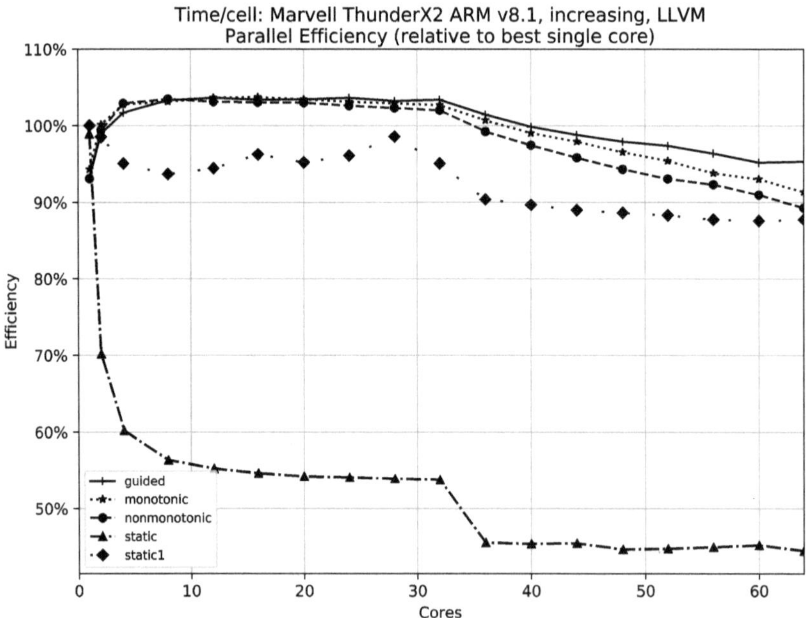

Figure 8.7: Parallel efficiency of loop scheduling (Increasing).

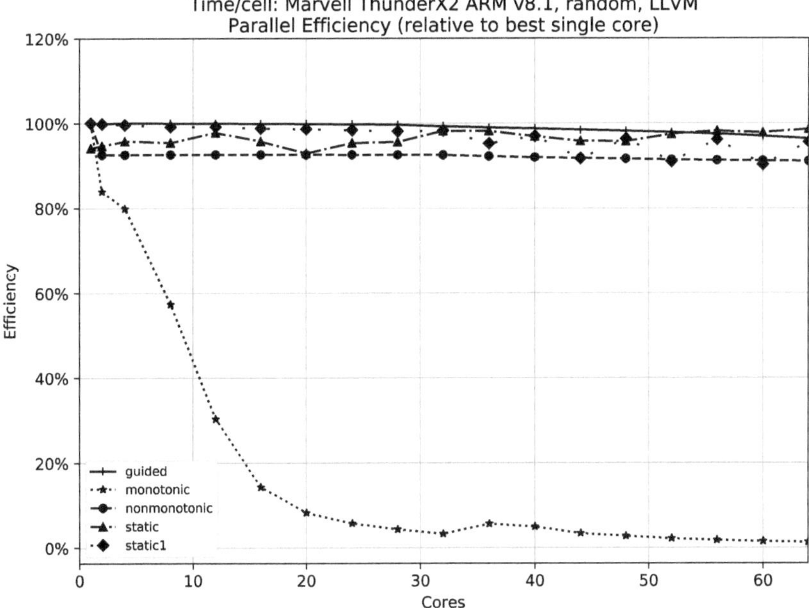

Figure 8.8: Parallel efficiency of loop scheduling (Random).

each scheduled chunk of work. In this case, where we have smaller chunks of work, we overload the throughput of the machine for these operations, and as a result, have extremely poor performance. If we look at the time it takes to execute a completely contended `std::atomic<uint32_t>::operator++(int)` (the postfix increment operator) on this machine (Figure 8.5), we can see that the cost of this operation becomes larger than the mean iteration time at around 16 threads. It is therefore unsurprising that this schedule cannot work well with such small pieces of work.

As we saw above, the `guided` schedule avoids this problem by reducing the number of atomic operations required while leaving a single point of contention, while the `nonmonotonic:dynamic` schedule removes the single point of contention and can also reduce the number of atomics.

Of course, we are not the first people to compare OpenMP scheduling implementations, and will no doubt not be the last. Many of the papers we have already cited include benchmarks, and there are also micro-benchmark suites such as "The EPCC micro-benchmark suite" [18], which are publicly available.

8.10 Other loop-scheduling approaches

As well as the schedules we have already discussed, the OpenMP API has a `schedule(auto)` that allows the compiler to choose any schedule that it prefers. This potentially allows the addition of other schedules; however, some compilers currently treat it the same as `schedule(static)`.

The OpenMP API also supports `schedule(runtime)`, which causes the compiler to generate code similar to dynamic schedule (with a call into the runtime for each chunk of iterations), but which schedule is actually used is determined by setting the environment variable `OMP_SCHEDULE` or calling the `omp_set_schedule()` function. While this looks like a simple way to experiment with different schedules in your code, it is slightly unfair to the static schedules, since if the static schedule had been known at compile time, the compiler can generate a single call into the runtime in each thread to handle the whole loop, whereas when invoked via `schedule(runtime)` and `OMP_SCHEDULE="static"`, there has to be one call into the runtime for each chunk of iterations. The static schedules will therefore perform slightly worse than they would have done if they had been inlined by the compiler. (In our benchmark, we have separate functions for each schedule to avoid this.)

There is continuing work on evaluating and adding new schedules to the OpenMP API—see, for instance, the proposals described in [22] and [73].

8.10.1 Using history

None of the schedules that we have looked at make use of information about how this loop ended up being scheduled the last time it was executed. Instead, they all treat each loop execution as if it is unique and has never been executed before. However, since we often expect that the worksharing loops that we are executing will occur inside other outer loops, it may be possible to store measured iteration times and make use of that information to improve the scheduling the next time the loop executes.

There are potential problems with this, though:

- **Overhead:** Collecting timing information itself takes time, and remembering it takes memory, which will have to be allocated and increase the cache pressure when accessed by the runtime system. These extra costs have to be overcome before you win, and if these costs apply to all loops, then improving the performance of some may not be enough.
- **Non-repetition:** Although the same code may be executing, it is entirely possible that a given loop in the source code will have different load imbalance each time it is executed, since the time to execute an iteration may very well be data dependent. To try to overcome this, one might remember the performance based on a backtrace through the call tree, so that the scheduler can better distinguish different invocations of the same function.

8.10.2 Exposing scheduling to the user

Another potentially interesting idea is to expose control of loop scheduling to the application programmer to allow them to provide their own functions to allocate iterations to threads, so that rather than merely allowing them to choose from a menu of prepared schedules, they are allowed into the kitchen to implement their own recipes.

This is unlikely to be useful to most users; however, it may be interesting for people writing heavily-used libraries where the significant work this represents can provide enough overall gain, and there is already expertise in this low-level control. It also provides a good research environment where schedules can be experimented with and then potentially later incorporated into the standard (or used via `schedule(auto)`, `schedule(runtime)`, or `schedule(nonmonotonic:dynamic)`) if they can demonstrate performance when used by real applications.

The OpenMP API specification seems likely to add features to enable this in a future version of the specification [72].

8.11 Conclusions

Do not ignore the mathematical limits that create an upper bound for what a loop schedule can achieve. No loop schedule can overcome the limitations caused by the underlying imbalance and bounded by the mathematics when there is insufficient parallelism available. This is then turns out to be an issue that is best solved by the application programmer, not the runtime implementation.

Dynamic schedules can perform well, even when work appears to be perfectly balanced. In the real world, there is imbalance caused by things like interrupts or the different cooling rates of different chips leading to clock-speed differences. Static schedules cannot react to this, whereas dynamic ones can, so they may perform better, although that seems unlikely in an abstract, theoretical world.

Static schedules provide affinity, as they pre-determine the mapping of iterations to threads. Thus, static schedules have the advantage that each time a loop is executed, the distribution of iterations to threads remains the same. If the iterations represent array indices, then that can mean that data can remain in the processor's caches, leading to faster execution even if there is more load imbalance than a dynamic schedule would have generated. The work-stealing implementation of the `nonmonotonic:dynamic` schedule attempts to maintain this property, since until work is stolen, the allocation of work to threads is that of `schedule(static)`.

The best parallel loop schedule may depend on other parameters. The number of cache misses and amount of false sharing when threads execute parallel loops that iterate over arrays are clearly affected by the loop schedule. However, it is also affected by the affinity mapping that is in place. In a `static,1` schedule, iterations are mapped to successive threads as enumerated by the OpenMP implementation, but whether those threads are physically near each other (and so likely to be in the same socket) or far away (potentially in another socket) will depend on what the mapping from the OpenMP thread enumeration to the underlying hardware is, and that is controlled by the affinity settings.

Loop scheduling is a complex and complicated topic. There are many possible implementations, and which will work best for a given code is not at all obvious, as we can see from the previous points and the copious literature on the subject.

Finally, know the history before you try re-inventing it. As has been stated in another context, "A couple of months in the laboratory can frequently save a couple of hours in the library." [26] Here, spending some time reading relevant papers on scheduling can save you from a lot of coding time to re-invent a scheduling scheme whose properties are already well known. This is obviously not to say that you shouldn't be measuring the performance of different schedules on your machine and with your application, merely that this is a reasonably well-plowed field, so understanding the history is essential.

9 Runtime support for task-parallel models

In this chapter, we describe the implementation of tasking models. As we learned in Chapter 4, a task-based parallel programming model creates a closure (i. e., data plus a pointer to code) for each of the generated tasks. These closures are then stored in a task pool for execution by the threads in the system.

The first thing we look at is the implementation options for task pools. This is the foundation that we need for even the most basic task-based execution models. We will then discuss how we can extend this basic (and maybe naive) implementation to improve performance and scalability. The next topic will be how an implementation can handle scheduling constraints—that is, task synchronization between the executing tasks.

One additional concern that we should worry over is that by creating tasks and executing them in other threads we may increase the amount of memory required to solve our initial problem. It is unsurprising that we may need more memory for the parallel code than the serial one, since each thread requires its own stack, and we have seen earlier that threads (and tasks) often require their own copies of variables. Our concern is therefore that the amount of space required when executing a code with N_{Threads} should not be more than N_{Threads} times greater than the memory required in the serial case. "Space-Efficient Scheduling of Multithreaded Computations" [15] demonstrates that the task-stealing implementation used in Cilk can be formally proved to meet this requirement.

9.1 Task descriptors

As we briefly mentioned in Section 4.2 and Section 4.4.6, there are runtime entry points that the compiled code uses to hand a created task over to the runtime system for later execution. Since the tasks need to be stored, we have to design a data structure that we can use to hold them so that threads that have no work can then search it to find a task to execute.

Before we do this, let's briefly recap what a *task* is. As we have seen in Section 4.2.1, tasks consist of a pointer to a piece of executable code and a pointer to data that the task needs for execution. The *task descriptor* is the data structure that is supposed to store these two pointers. See Listing 9.1 for the most basic version of this structure. It simply contains a single object called `Closure`. In that object, the member `routine` of type `ThunkPointer` points to the executable code (the thunk function) that the compiler outlined. The `data` member in the `Closure` class is the pointer to the data that we need for task execution.

You may think that this trivial task descriptor is too simplistic, and you are absolutely right. During the course of this chapter, we will add further elements to the task descriptor as we go through the implementation of the different tasking features that

https://doi.org/10.1515/9783110632729-009

```
typedef int32_t (*ThunkPointer)(int32_t, void *);

struct TaskDescriptor {
  struct Closure {
    void * data;
    ThunkPointer routine;
  };

  Closure closure;
};
```
Listing 9.1: Most basic task descriptor.

we need. In the end, the task descriptor will be the central place to record all information about the current state of a task throughout its lifetime in the execution of the application code.

9.2 Task pool implementation

The *task pool* is the central container used to store all created tasks so that they can later be executed. There is one thing to note, though: some people call the task pool the *task queue*. However, the term queue implies a specific access behavior, usually first in, first out (FIFO), while a task pool may have different access behavior depending on the runtime implementation and the task scheduling approach chosen for the runtime system. While it seems that *task queue* is the more widely used term, we will be using *task pool* instead, as we believe it is the more accurate term.

The basic operations on a task pool are put() to add a task into the container and get() to retrieve a task for execution. The put() is performed by the parent task that creates the respective new child task, while the get() is performed by a thread that will execute it. The put() method returns a bool value to indicate whether the task was stored in the pool (true) or whether the pool was full and the task descriptor was not stored (false). If the pool is empty, then the get() method returns a nullptr. Table 9.1 summarizes the signature of the operations and their semantics.

Listing 9.2 shows the basis of our implementation of the interface for the operations in Table 9.1. Here, we use the C++ pattern of importing the implementation through private inheritance and the using statement. The actual implementation of the task pool is injected through the template parameter TaskPoolImpl and will be discussed below.

Table 9.1: Task poll fundamental operations.

Operation	Semantics
`bool put(TaskDescriptor * td)`	Called by the parent task to store the pointer to the task descriptor of the created task in the task pool. Return `true` if the task descriptor was stored; `false` otherwise.
`TaskDescriptor * get()`	Called by an executor thread to retrieve a pointer to a task descriptor from the task pool for execution. The return value may be a `nullptr` if the pool is empty.

```cpp
struct TaskPool : private TaskPoolImpl {
  using TaskPoolImpl::put;
  using TaskPoolImpl::get;
};
```

Listing 9.2: Interface to the task pool class.

9.2.1 Single task pool

The first, and simplest, implementation of a task pool is a single, centralized data structure that is shared between all the threads. When a task is created by any of the threads, it is put into this pool. When a thread needs a task to execute, it will look in the pool and remove a task from it. As you can imagine, there are multiple ways in which a centralized pool could be implemented. We will present two implementation strategies and discuss their merits and drawbacks.

9.2.1.1 LIFO linked list

The first data structure we will consider is a non-intrusive linked-list implementation of a pool based on a last in, first out (LIFO) linked list [24, 117]. Listing 9.3 shows the implementation of the `put()` and `get()` operations that we need for the task pool interface. The implementation is very straightforward.

The `put()` method receives the pointer to the task descriptor as an argument and allocates a new list node to store it. The `ListNode` class consists of a member to store that pointer as the payload and a pointer to the next `ListNode` object in the linked list. The new node is allocated outside the critical section, since allocation may be expensive and should not be on the serialized critical path. Then, the lock is claimed, the current `head` is stored in the new node, and, finally, the head pointer in the task pool is updated to point to the newly inserted node. Finally, the method returns `true` to indicate that the insertion was successful.

The `get()` operation is also simple. If the head node is set, there's at least one task descriptor in the pool. Thus, the code dequeues the corresponding task descrip-

```cpp
template<class Lock>
struct TaskPoolLinkedListLIFO {
  TaskPoolLinkedListLIFO() : head(nullptr) {}

  bool put(TaskDescriptor * task) {
    auto node = new ListNode{nullptr,task};
    {
      const std::lock_guard<Lock> guard(lock);
      node->next = head;
      head = node;
    }
    return true;
  }

  TaskDescriptor * get() {
    ListNode * node = nullptr;
    TaskDescriptor * task = nullptr;
    const std::lock_guard<Lock> guard(lock);
    if (head) {
      node = head; head = head->next;
    }
    if (node) {
      task = node->task;
      delete node;
    }
    return task;
  }

private:
  struct ListNode {
    ListNode * next;
    TaskDescriptor * task;
  };
  ListNode * head;
  Lock lock;
};
```

Listing 9.3: Linked-list task-pool implementation with LIFO semantics.

tor, removes the node from the linked list, and deallocates the ListNode object. It then returns the pointer to the task descriptor (or a nullptr, if the task pool was empty).

You may have noticed the use of the `std::lock_guard` template. The task pool is necessarily a shared data structure, since its whole objective is to be a place from which multiple threads can pull work. Therefore, there may be several threads creating tasks, and attempting to store a new task descriptor in the pool simultaneously while several other threads may be requesting tasks for execution. To avoid these race conditions, the implementation uses a Lock object to prevent the `put()` and `get()` operations from executing concurrently. To be flexible about the lock implementation, we pass the lock type as a template parameter so that we can easily change it when the template is instantiated. When relying on a `lock_guard` object, the lock is automatically acquired when the `lock_guard` is instantiated and it is unlocked when the variable guard goes out of scope. Chapter 6 goes into the problem of mutual exclusion and the details of the implementation of locks; you should consult that chapter if you want to refresh your memory of those details, though that is not necessary to understand the issues here.

We can now instantiate this implementation of a task-pool interface from Listing 9.2 using a file-scope variable in one of the compilation units of the parallel runtime, like this:

```
typedef TaskPoolLinkedListLIFO<std::mutex> TaskPoolImpl;
struct TaskPool : private TaskPoolImpl {...};
TaskPool taskPool;
```

For simplicity, we are using a `std::mutex` lock to instantiate the linked-list task-pool template. Any lock which meets the C++ `BasicLockable` requirement can be used with the `std::lock_guard` class; all of the locks described in Chapter 6 meet that requirement, so any of them could be used. We will revisit the locking topic for the task pool a few more times throughout the remainder of this chapter, but the next step is to implement the code to store a task that was passed into the runtime library in the task descriptor.

9.2.1.2 Storing tasks

The `put()` and `get()` interface operations make task creation and code storing relatively simple (see Listing 9.4). The generated application code calls the corresponding runtime function—that is, `task_base::spawn()` in TBB and `__kmpc_omp_task()` (and its variants) in LLVM's OpenMP[*] runtime. For TBB, the compiler passes a C++ function object directly, whereas when implementing the OpenMP API, the compiler typically passes the pointers to code and data separately. The C++ technique is cleaner for the runtime, but requires more compiler support, so early OpenMP implementers adopted the scheme that is simpler to implement, which became part of the *application binary interface* (ABI), and, therefore was hard to change. If one were implementing the OpenMP API from scratch now, one would almost certainly have the compiler gener-

```cpp
using namespace lomp::Tasking;

TaskDescriptor::Closure *
__kmpc_omp_task_alloc(void *, int32_t, void *,
                      size_t sizeOfTaskClosure,
                      size_t sizeOfShareds,
                      ThunkPointer thunk) {
  auto task = AllocateTask(sizeOfTaskClosure, sizeOfShareds);
  InitializeTaskDescriptor(task, sizeOfTaskClosure,
                           sizeOfShareds, thunk);
  return TaskToClosure(task);
}

int32_t __kmpc_omp_task(void *, int32_t,
                        TaskDescriptor::Closure * closure) {
  auto task = ClosureToTask(closure);
  StoreTask(task);
  return 0;
}

bool StoreTask(TaskDescriptor * task) {
  return taskQueue.put(task);
}
```

Listing 9.4: Creating and storing a task in the task pool.

ate code similar to that used by the C++ function-object implementation using lambda expressions.

Because of this dated interface, we are also forced to explicitly convert the task descriptor to a closure object via TaskToClosure() and back via ClosureToTask(). As we have mentioned before in Chapter 1, we wanted our implementation of the book's example codes to be compatible with the LLVM and Intel[*] compilers. Alas, these compilers do not easily allow us to extend the task descriptor to hold additional information. Instead, the implementers have used pointer-arithmetic trickery to place the additional state before the pointer to the descriptor. To be compatible, we have to repeat that same trick even though we consider it ugly and evil.

The AllocateTask() functions allocates the memory for a task descriptor. The InitializeTaskDescriptor() function then initializes all of the fields to their respective default values and makes sure that the task descriptor is in a consistent state before it is added to the task pool.

```cpp
TaskDescriptor * AllocateTask(size_t sizeOfTaskClosure,
                             size_t sizeOfShareds) {
  auto allocSize = ComputeAllocSize(sizeOfTaskClosure,
                                    sizeOfShareds);
  auto task =
  static_cast<TaskDescriptor *>(malloc(allocSize));
  return task;
}


void InitializeTaskDescriptor(TaskDescriptor * task,
                              size_t sizeOfTaskClosure,
                              size_t sizeOfShareds,
                              ThunkPointer task_entry) {
  if (sizeOfShareds > 0) {
    size_t offset = sizeof(task->metadata) +
                    sizeOfTaskClosure;
    task->closure.data = ((char *)task) + offset;
  }
  else
    task->closure.data = nullptr;
  task->closure.routine = task_entry;
}
```

Listing 9.5: Allocating and initializing task descriptors.

The `__kmpc_omp_task()` entry point converts the incoming task closure into a task descriptor and then calls `StoreTask()`, which calls the `put()` to store the task descriptor in the task pool. For now, we ignore the return value of all of these functions, as the implementation of the task pool is using a normal memory allocator that can allocate an arbitrary amount of memory (at least we assume that memory is not being exhausted).

9.2.1.3 Executing tasks

The code for the consumer threads that retrieve a task for execution is depicted in Listing 9.6. The `ScheduleTask()` function is called anytime a thread is idle, and thus requires a task to execute. It invokes the `get()` function from the task pool to get a task descriptor. Because this function may return a `nullptr` if the pool is empty, the `ScheduleTask()` function returns if no task is available. If a task is returned, it is invoked.

Task invocation at this point is rather simple. The `InvokeTask()` routine's only action is to perform an indirect call through the thunk pointer `routine` from the task

```cpp
bool ScheduleTask() {
  bool result = false;
  auto task = taskPool.get();

  if (task) {
    InvokeTask(task);
    result = true;
  }

  return result;
}

void InvokeTask(TaskDescriptor * task) {
  int32_t gtid = 0;

  task->closure.routine(gtid, &task->closure);

  FreeTask(task);
}
```

Listing 9.6: Finding and executing a task from the task pool.

descriptor's closure object and pass the data pointer as an argument to that indirect function call. The specific calling convention of the LLVM and Intel compilers mandates that the full closure object is passed, while GCC's implementation expects the thunk to be called with the data pointer directly:

```cpp
task->closure.routine(task->closure.data);
```

After the thunk call returns to InvokeTask(), the task has completed its execution and can thus be removed from the system. This is done by calling FreeTask().

Listing 9.7 shows how the allocation and deallocation of tasks is performed. Since we are dealing with some very low-level memory layout, the functions use the malloc() and free() API to deal with the underlying memory management. A more sophisticated implementation would use an internal memory allocator, as discussed in Section 5.3.

The AllocateTask() function exposes the hack that we mentioned, which is required to make the runtime implementation provided here compatible with the LLVM and Intel compilers. The sizeOfTaskClosure is passed by the generated code to tell the allocation function how much data the task closure requires. The sizeOfShared argument contains the number of bytes required to store additional data (see Section 4.4.6) for the created task. AllocateTask() then computes the amount of memory that must

```cpp
TaskDescriptor * AllocateTask(size_t sizeOfTaskClosure,
                              size_t sizeOfShareds) {
  size_t allocDiff =
      sizeof(TaskDescriptor) -
      sizeof(TaskDescriptor::Closure);
  size_t allocSize =
      std::max(sizeof(TaskDescriptor::Closure),
               sizeOfTaskClosure) +
      sizeOfShareds + allocDiff;
  TaskDescriptor * task =
      static_cast<TaskDescriptor *>(malloc(allocSize));
  return task;
}

void FreeTask(TaskDescriptor * task) {
  free(task);
}
```

Listing 9.7: Function for allocating and deallocating task descriptors.

be allocated to hold both the space the compiled code has requested and that our
runtime-library data requires.

One topic that we will have to leave open for a bit longer is where to actually put
the calls to `ScheduleTask()`. We'll first refine our implementation of the task pool to
make sure that we have a correct, high-performance implementation. We will then
attempt to demonstrate where to put `ScheduleTask()` invocations to ensure that the
created tasks are actually executed.

9.2.1.4 Discussion

With this first implementation of a runtime for a task-parallel programming module,
we can execute simple task programs. But is this simple implementation reasonable
and does it perform well? Unfortunately, as well as the fact that we have not yet implemented several key tasking features (e. g., synchronization mechanisms), our first
implementation also has some serious flaws.

First, the runtime library has to perform many memory allocations to keep track
of the tasks and their descriptors. If you recall Listing 9.3 and Listing 9.4, you will see
that for each task in the system, two memory blocks have to be allocated and deallocated: the node in the linked list and the task descriptor. While we cannot avoid the
allocation of the task descriptor, the allocation for storing the descriptor in the linked
list seems avoidable. We could, of course, use the techniques of Section 5.3 to reduce

the allocation/deallocation latency, as we know that the `ListNode` objects all have the same size and thus can easily be part of a special memory pool. We could also switch to an intrusive linked-list implementation that uses the task descriptor to store the links between the list items.

Another concern with the linked-list implementation is the poor efficiency with respect to memory usage. Looking at the `ListNode` class, one can see that only half of the memory is used for the actual payload—that is, the pointer to the task descriptor. The `next` field used to connect the nodes in the linked list consumes the other half. As discussed in Section 5.3.1, it's important to use memory as effectively and efficiently as possible to reduce the effect of the parallel runtime system on the memory subsystem, especially on the caches. Thus, we need to find a more efficient data structure.

This implementation has another critical issue that is associated with the `std::mutex` object that makes the task pool thread safe. As we will potentially have many concurrent producers, as well as many consumers of tasks, all threads that participate in task execution will compete for the `std::mutex` object. Section 6.6.2 showed that contended locks have significant overhead, so picking the right implementation will be crucial for this data structure. Even more problematic is the simple fact that the task pool serializes task management, as it locks the entire pool for both producers and consumers. This turns out to be problematic for large numbers of threads and/or short-running tasks.

Our next attempt to come up with a better implementation of a task pool will slightly improve the efficiency. Then we will investigate whether we can improve scalability and reduce the need for serialization.

9.2.2 Multi task pool

Next in the journey to an efficient implementation of a task pool, we will learn how to replace the linked list with an array-based pool. This data structure will store pointers to task descriptors in an array-like data structure, and thus will make much better use of the memory because there will be no additional data needed per task. It only needs the lock to protect the task pool and a counter to keep track of the stored task descriptors. Listing 9.8 shows the implementation.

We have opted to use `std::array` to store the pointers to the task descriptors. Another option would be to use `std::vector` or a regular C-style array. The implementation of the `put()` and `get()` methods is relatively straightforward. They again rely on a lock to make the task pool thread safe, and thus re-use the `std::lock_guard` to protect the methods.

When a task is pushed to the task pool, the code first checks that there's a free slot in the task pool. If that is not the case, then the `put()` method returns `false` to indicate that the task descriptor was not stored in the pool. This puts the responsibility for handling this case on the caller; we'll see how it handles this in Section 9.2.2.1.

```cpp
template <class Lock, size_t maxSize>
struct TaskQueueArrayLIFO {
  TaskQueueArrayLIFO() : taskCount(0) {
    queue.fill(nullptr);
  }

  bool put(TaskDescriptor * task) {
    const std::lock_guard<Lock> guard(lock);
    if (taskCount >= maxSize) {
      return false;
    }
    queue[taskCount] = task;
    taskCount++;
    return true;
  }

  TaskDescriptor * get() {
    TaskDescriptor * task = nullptr;
    const std::lock_guard<Lock> guard(lock);
    if (taskCount > 0) {
      task = queue[taskCount - 1];
      queue[taskCount - 1] = nullptr;
      taskCount--;
    }
    return task;
  }

private:
  size_t taskCount;
  std::array<TaskDescriptor *, maxSize> queue;
  Lock lock;
};
```

Listing 9.8: Array implementation of a task pool with LIFO semantics.

The remainder of the method stores the task descriptor, does the bookkeeping by incrementing the counter, and returns `true` to tell the caller that the task descriptor is now in the pool.

To retrieve a task descriptor from the task pool, `get()` first checks whether the pool is empty, and if so, returns `nullptr`. If there's at least one task descriptor in the pool, the pointer to it is retrieved, the counter is decremented, and the pointer to the descriptor is returned.

9.2.2.1 Handling full task pools

When the task pool is full, the code in Listing 9.8 requires the caller to handle the situation. If we were using `std::vector`, it seems that we could avoid the problem, as the storage automatically grows and makes room for more task descriptors. However, this is not an ideal solution either.

Even though the data structure extends automatically, we do not want it to grow too large. As you may recall, the memory consumption of the parallel runtime system should be kept low to reduce its effect on the memory footprint of the application. In addition, whenever the data structure grows, the previously allocated memory likely has to be copied to the new, larger memory area. This is an expensive operation, and thus increases the overhead of task management in the runtime system. You certainly don't want something like this happening on the critical path. It can also have dramatic effects on the contents of the caches and may cause the eviction of data that was useful to the application program, and therefore, may also cause additional cache misses. As a consequence, the size of the `std::vector` should be bounded to stop it from expanding above a well-defined limit. As soon as we enforce that requirement, we must again handle the case where there is no room to store more task descriptors.

One solution would be to use a series of `std::array` (or `std::vector`) instances that are then connected via a linked list to form a larger virtual structure. However, this makes the code a lot more complex, as it has to deal with the multi-level nature of the resulting data structure. At the same time, it does not solve the problem of the growing memory footprint of the parallel runtime system.

To answer the question of what the runtime system should do if the task pool overflows, we can go back to the semantics of most task-based programming models. Tasks *may* be executed concurrently, but they don't *have* to execute that way. This means that it is perfectly legal to not store a task descriptor for later execution, but rather to execute it immediately if the task pool is full. Listing 9.9 shows the code of the `StoreTask()` function with this overflow-handling mechanism added. The code attempts to store a task descriptor in the pool by calling the `put()` method. If it returns `false`, then the attempt to store the task descriptor failed and the `InvokeTask()` method is called to execute the task right away.

One of the tuning parameters of the runtime system will be to determine the number of task descriptors that can be stored in the task pool. Because a producer can execute tasks immediately, the number of task descriptors in the pool can be rather small without causing too much harm.

9.2.2.2 Going multi pool

Taking the array-based task pool as a starting point, we can extend our implementation from a single task pool to multiple task pools. For now, we will provide one task pool per thread. Instead of relying on a language feature for thread-local storage, e. g.,

```
bool StoreTask(TaskDescriptor * task) {
  bool result = true;

  if (!taskPool.put(task)) {
    InvokeTask(task);
    result = false;
  }

  return result;
}
```

Listing 9.9: Storing tasks and handling overflows in the task pool.

`thread_local` in C++, we use the `Thread` object that the runtime system already maintains as a place to hold its per-thread data. Listing 9.10 shows the additions made.

The first addition to make is a member variable to hold a pointer to the task pool object and a getter function to retrieve that pointer. In the `Thread` object constructor, the code calls the `TaskPoolFactory()` method to create a task pool object with the desired implementation. That way, the `Thread` object does not need to know about the actual implementation details of the task pool.

The main difference when using multiple task pools to store a task is that we need to determine the task pool instance into which the store operation should put it. Listing 9.11 shows one way to do this. When a new task is created, and its task descriptor is passed into the `StoreTask()` function, we first get a pointer to the thread that is trying to store the task (the currently executing thread). `StoreTask()` then retrieves the pointer to the thread's task pool and uses that pool to store the task descriptor. If this is not possible because the pool is already full, then the task is executed immediately as before. In this way, a newly created task is stored in the task pool of the thread that is executing its parent task at the time that the parent created the child.

When a thread wants to schedule a task for execution, it retrieves a pointer to the thread object and the task pool (see `ScheduleTask()` in Listing 9.11). It then tries to retrieve a task descriptor from the pool and invokes the task just as before.

You may have spotted something wrong with this implementation strategy. Since a thread only stores tasks in its own task pool and then only looks there for tasks, it will never see tasks created by any other thread. This is clearly suboptimal, since running out of locally created tasks says nothing about the global state of the computation; other threads may still have many tasks that need to be executed. To overcome this, we need to add the idea of *task stealing*. With task stealing an idle thread can access another thread's task pool and *steal* a set of tasks from that task pool to execute. Listing 9.10 silently introduced a function `steal()` for this purpose.

```cpp
// thread.h:
namespace Tasking {
struct TaskPool;
}

class Thread {
  // thread member variables...

  Tasking::TaskPool * taskPool;

public:
  // thread public interface...

  Tasking::TaskPool * getTaskPool() const {
    return taskPool;
  }
};

// thread.cc:
Thread::Thread(ThreadTeam * T, int L, int G, bool Master) {
  this->taskPool = lomp::Tasking::TaskPoolFactory();

  // remainder of the thread constructor...
}

// tasking.cc:
typedef TaskPoolArrayLIFO<std::mutex,
                          TASK_POOL_MAX_SZ> TaskPoolImpl;
struct TaskPool : private TaskPoolImpl {
  using TaskPoolImpl::put;
  using TaskPoolImpl::get;
  using TaskPoolImpl::steal;
};

TaskPool * TaskPoolFactory() {
  TaskPool * pool;
  pool = new TaskPool{};
  return pool;
}
```

Listing 9.10: Extending the Thread class with a per-thread task pool.

```cpp
bool StoreTask(TaskDescriptor * task) {
  auto thread = Thread::getCurrentThread();
  auto taskPool = thread->getTaskPool();
  bool result = true;

  if (!taskPool->put(task)) {
    InvokeTask(task);
    result = false;
  }
  return result;
}

bool ScheduleTask() {
  auto thread = Thread::getCurrentThread();
  auto taskPool = thread->getTaskPool();
  bool result = false;

  TaskDescriptor * task = taskPool->get();
  if (task) {
    InvokeTask(task);
    result = true;
  }

  return result;
}
```

Listing 9.11: Storing and finding a task using multiple task pools.

Listing 9.12 shows how the function is used by adding some additional code to the `ScheduleTask()` function of Listing 9.11. If the thread's own task pool did not return any task for execution, then the thread requests a handle to the thread team of the currently active parallel region. It then determines a random number between one and the number of threads in the team minus one (we cannot steal from ourselves!) and uses this number to pick a random victim from which to attempt to steal a task. This is either successful, and the local thread can execute the stolen task, or the `steal()` operation returns a `nullptr` because the victim thread did not have any tasks in its task pool either. Note that the code only makes a single stealing attempt, but one could think of adding a loop to attempt stealing a couple of times before giving up.

It is usually a good design choice to have one method for scheduling a task for execution in a thread (`get()`) and another for stealing tasks from a victim thread (`steal()`). Of course, one could use the `get()` method and also call it for task-stealing

```cpp
bool ScheduleTask() {
  auto thread = Thread::getCurrentThread();
  auto taskPool = thread->getTaskPool();
  auto task = taskPool->get();

  if (!task) {
    // No local task, try to steal.
    auto team = thread->getTeam();
    size_t me = thread->getLocalId();
    size_t teamSize = team->getCount();
    size_t rnd = GetRandomNumber(1, teamSize - 1);
    size_t victimID = (me + rnd) % teamSize;
    auto victimPool
            = team->getThread(victimID)->getTaskPool();

    task = victimPool->steal();
  }

  if (!task)
    return false;

  InvokeTask(task);
  return true;
}
```
Listing 9.12: Random task stealing.

purposes, but then the task pool loses the ability to distinguish regular task execution and task stealing; knowledge which may be useful when choosing a task to return.

9.2.2.3 Double-ended queue (deque) task pool

An implementation of the `steal()` method is shown in Listing 9.13. For this implementation, we have replaced the `std::array` container with a `std::deque` to gain the ability to access (with $\mathcal{O}(1)$ complexity) both ends of the task pool efficiently.

The `put()` method inserts tasks at the *tail* of the task pool using the `push_back()` method of `std::deque`. We still maintain an explicit count of the number of tasks stored in the deque so that the data structure does not grow beyond the maximum pool size to avoid excessive allocations. To protect the task pool against race conditions, the method again uses a `std::lock_guard` to ensure mutually exclusive execution.

```cpp
template <class Lock, size_t maxSize>
struct TaskPoolDeque {
  bool put(TaskDescriptor * task) {
    const std::lock_guard<Lock> guard(lock);
    if (taskCount >= maxSize) return false;
    pool.push_back(task);
    taskCount++;
    return true;
  }

  TaskDescriptor * get() {
    TaskDescriptor * task = nullptr;
    const std::lock_guard<Lock> guard(lock);
    if (taskCount > 0) {
      task = pool.back();
      pool.pop_back();
      taskCount--;
    }
    return task;
  }

  TaskDescriptor * steal() {
    TaskDescriptor * task = nullptr;
    const std::lock_guard<Lock> guard(lock);
    if (taskCount > 0) {
      task = pool.front();
      pool.pop_front();
      taskCount--;
    }
    return task;
  }

private:
  Lock lock;
  size_t taskCount;
  std::deque<TaskDescriptor *> pool;
};
```

Listing 9.13: Task pool implementation using a double-ended queue (deque).

The code for get() retrieves a task from the tail of the task pool. We now have to use a combination of back() to retrieve the pointer to the task descriptor and pop_back() to remove the last element in the task pool. Everything else in this method works as before in the array-based implementation.

The steal() method takes a task from the "head" of the task pool. It calls front() to retrieve the pointer to the oldest task descriptor from the task pool and pop_front() to remove it. This implementation means that a thread will execute the most recently created outstanding task while a thief will steal the oldest pending task.

While our implementation here uses a lock associated with each task pool to maintain thread safety, we could be smarter than that (as the Cilk [80] and Intel TBB implementations are) if we observe that the only thread that updates the "tail" of the task pool is the thread that owns the task pool. Therefore, it should be able to do its updates without needing to take a lock or use an atomic instruction, provided that it carefully synchronizes the stealing threads that are updating the "head." Since we expect that the most common execution path is that a thread generates a task and then executes it itself (which is the same as saying that task stealing is rare), making this common path as fast as possible is important, even at the cost of making the stealing path more expensive. This is another example of our general point from Chapter 6 that we should "reduce the number of atomic instructions whenever we can."

9.3 Task synchronization

The next topic that we are going to walk through is how task synchronization can be implemented. As usual, there are many flavors depending on the capabilities and feature richness of the parallel programming model, but there are two common themes:

1. **Waiting for the completion of a subset of tasks:** The aim here is to ensure that a subset of the tasks have completed before continuing execution of the current task. Some examples are the OpenMP taskwait construct and TBB's task::wait_for_all() method to wait for immediate child tasks. The OpenMP taskgroup construct and the TBB tbb::task_group class widen the scope and wait for all descendant tasks (so child tasks, grandchild tasks, and so on).
2. **Task dependences:** Here, the aim is to control the point at which the task scheduler starts a task, rather than to suspend the execution of a task that is already running.

We will tackle this topic in an increasing level of complexity for the runtime implementation. You might think that waiting for immediate child tasks would be the easiest option. Alas, that's not the case, and so we will start our discussion with the simpler problem of awaiting the completion of a subset of tasks defined with the scope of a taskgroup construct (or the like). We'll close the section with a discussion on how task dependences can be implemented.

9.3.1 Waiting for a subset of tasks

When waiting for a subset of tasks, a technique also sometimes called a task barrier, we need to know how many tasks have been created and how many have completed execution. We have seen the code pattern for taskgroup in Section 4.4.6. The compiler inserts functions at the start and end of a region to begin monitoring the task creation and to wait until all tasks that have been spawned while this region is active have completed execution. In the LLVM OpenMP runtime, this pair of entry points is __kmpc_taskgroup() and __kmpc_end_taskgroup(), which serve the same purpose as __omp_enter_taskgroup() and __omp_leave_taskgroup() described in Section 4.4.6.

How can we best keep track of how many tasks must have finished execution before the waiting task can continue? Simplified, a task can be in one of three states (there might be more in more complex runtime implementations):
1. **Waiting:** It is in the task pool, waiting to be executed.
2. **Executing:** It is actually running, and so no longer in the task pool.
3. **Finished:** The task has completed execution.

Obviously, we cannot statically determine the total number of tasks that may be created, as determining this is clearly the same as solving the halting problem, which has been proved to be impossible [141].

When invoked by a task, the __omp_leave_taskgroup() function (or LLVM's __kmpc_end_taskgroup() function) simply has to wait until the counter for subset of task to be waited for reaches zero.

Let's see how this works. Obviously, only executing tasks can create more tasks. So, while these executing tasks are active on one of the available threads, the count must be non-zero. If they have created more tasks that are still in the task pool waiting to be executed, they can be registered by incrementing the counter and thus the counter will be non-zero. At some point, the wave of created tasks will subside (unless you have an endless loop in the code, but then we can wait endlessly, too) and only executing tasks will be left. As all previous tasks have completed, the count will be equal to the number of remaining executing tasks. When they complete, the counter will eventually reach zero and the wait can complete.

There are two slight complications that we have to deal with. First, some parallel programming models allow for the nesting of taskgroup constructs (e. g., the OpenMP API and TBB), so we need to support multiple counters that adhere to the nesting of the constructs. The proper way to handle this is similar to using a stack, where the counter of the innermost construct is placed on top of the stack. Second, new tasks may be created and executing tasks may complete in any arbitrary order and, of course, concurrently. As a consequence, the counter must be accessed atomically.

Listing 9.14 defines a new structure Taskgroup that organizes the stack of nested taskgroup regions through a linked list of pointers to the outer, containing taskgroup region. The structure also contains a std::atomic for type int to maintain the count

```cpp
struct Taskgroup {
  Taskgroup(Taskgroup * outer_) : outer(outer_),
                                  activeTasks(0) {}
  Taskgroup * outer;
  std::atomic<int> activeTasks;
};

int32_t __kmpc_taskgroup(ident_t *, int32_t) {
  lomp::Tasking::TaskgroupBegin();
  return 0;
}

int32_t __kmpc_end_taskgroup(ident_t *, int32_t) {
  lomp::Tasking::TaskgroupEnd();
  return 0;
}

void TaskgroupBegin() {
  auto thread = Thread::getCurrentThread();
  auto outer = thread->getCurrentTaskgroup();
  auto inner = new Taskgroup(outer);
  thread->setCurrentTaskgroup(inner);
}

void TaskgroupEnd() {
  auto thread = Thread::getCurrentThread();
  auto taskgroup = thread->getCurrentTaskgroup();

  while (taskgroup->activeTasks) {
    ScheduleTask();
  }

  auto outer = taskgroup->outer;
  thread->setCurrentTaskgroup(outer);
}
```

Listing 9.14: Opening and closing taskgroup regions.

of active tasks—that is, created tasks that are either waiting in the task pool or that
are actively executing on a thread. We can be sure that an int variable is large enough
to hold all tasks because we have fixed the size of the per-thread task pools, and can

therefore calculate that the largest number of tasks that can possibly be in the system at any time is $N_{\text{Threads}} \cdot (\text{TaskPoolSize} + 1)$ (the +1 is to allow for each thread executing one task).

The entry points `__kmpc_taskgroup()` and `__kmpc_end_taskgroup()` call the internal implementations of the taskgroup mechanism. The `TaskgroupBegin()` function retrieves a pointer to the executing thread's `Taskgroup` object. This pointer is either the `nullptr`, which means that there is currently no active task group, or the relevant task group, which then becomes the enclosing task group for the newly created one. The new task group is then registered with the thread.

The `TaskgroupEnd()` function undoes this stacking by retrieving the pointer to the outer taskgroup object and resetting the thread's pointer to it. To do so, it requests the current task group from the encountering thread and resets the task group pointer there. Remember, this can be a `nullptr`, which indicates that the current task-group region was the outer-most task-group region.

The `TaskgroupEnd()` also contains a `while` loop that iterates as long as the atomic counter of the current taskgroup object is non-zero. This is the waiting part of the implementation for `taskgroup`-like constructs. While the function is waiting for all active tasks to complete, it keeps scheduling tasks from the task pool for execution. Note that the tasks it executes may or may not be from the task group depending on the tasks in the task pool. Any of the waiting and backoff techniques discussed in Section 6.7.1 could also be used, but that would mean that this thread would not be executing useful user code.

While we use a spin-wait loop (that may or may not include a backoff) for a lock implementation, a polling loop in this context might not be appropriate. In general, one can expect that the waiting thread will have to wait longer for all of the tasks in a task group to complete than a thread waiting for a lock to become free. One particular example is a pattern where the waiting task only spawns a few tasks and then waits for them to complete. If these tasks create a whole tree of descendant tasks, then there may be many tasks awaiting execution, so polling rather than executing them will significantly reduce performance since the polling thread is doing no useful work. However, executing tasks in the thread that is waiting for the task group to complete will increase the latency of the task group, because, until the thread notices the wakeup call, it will be off executing another task. Whether this matters will depend on the critical path through the code; if there are no other tasks to execute, and the thread that was waiting is that which will create more parallel work, then the resources lost will be large (as with any barrier-like construct).

Enough about waiting! Let's discuss how we can keep a tally of active tasks while we are executing tasks from the task pool. Well, first of all, the runtime needs to know whether a task is created from within the dynamic extent of a task-group region. So, when we create a task and initialize its task descriptor, we must also store a pointer to the currently active task-group object in the task descriptor. To keep track of this, we extend the task descriptor that we showed in Listing 9.1 with additional metadata that

```cpp
struct TaskDescriptor {
  enum struct Flags {
    Created = 0x0,
    Executing = 0x1,
    Completed = 0x2,
  };

  struct Metadata {
    Taskgroup * taskgroup;
    Flags flags;
    TaskDescriptor * parent;
    std::atomic<int> childTasks;
  };

  struct Closure {
    void * data;
    ThunkPointer routine;
  };

  Metadata metadata;
  Closure closure;
};
```
Listing 9.15: Task descriptor with metadata for the parent-child relationship.

records the current Taskgroup pointer. Listing 9.15 shows the new structure of the task descriptor (see also Listing 9.16 later in this chapter). A code modification is needed to initialize the task group information in the InitializeTaskDescriptor() function:

```cpp
auto thread = Thread::getCurrentThread();
auto taskgroup = thread->getCurrentTaskgroup();
task->metadata.taskgroup = taskgroup;
```

With this code modification applied to the InitializeTaskDescriptor() function of Listing 9.5, each task is now connected to the task group from which it was created. We must also update the StoreTask() function (Listing 9.11) and the InvokeTask() function (Listing 9.6). The StoreTask() has to increment the counter when it tries to store the task into the task pool:

```cpp
if (auto taskgroup = task->metadata.taskgroup; taskgroup) {
  ++taskgroup->activeTasks;
}
```

The counterpart is in the `InvokeTask()` that decrements the counter when it completes
execution of a task that is connected to a taskgroup object:

```
if (auto taskgroup = task->metadata.taskgroup; taskgroup) {
  --taskgroup->activeTasks;
}
```

These two modifications are sufficient to keep track of the number of created tasks vs.
the number of completed tasks. Note that the `if` statement is needed, as the code has
to check whether a task group has been set for the current task.

9.3.2 Waiting for immediate child tasks

For a `taskwait` or similar construct, the parent task has to determine whether all of its
child tasks have completed execution to stop waiting and continue execution. One
might think that it would be sufficient to create a task group for each parent task
and then re-use the implementation of the previous subsection. Alas, that won't really work, as the implementation keeps a tally of descendant tasks and thus would
not just wait for child tasks, but also grandchild tasks and so forth.

So, we need to come up with a different implementation. Of course, we want this
implementation to be as efficient as possible so that one task waiting for its children
does not affect other tasks that are executing or waiting for other children. Therefore,
we have to add a capability to the tasking implementation of the runtime system to
keep track of how many of a parent task's immediate child tasks are still in flight (waiting in the task pool or executing).

Until now, the runtime system has not tracked the relationship of the tasks. They
are simply created and executed without the runtime system paying attention to which
task is a child task of another task. However, now we need to know about such parent-
child relationships. Once we have that information ready in the runtime system, a
child task has a connection to its parent task and indirectly to all other ancestor tasks
until we reach the root of this task tree.

To keep track of this information, we extend the task descriptor that we showed
in Listing 9.1 with additional metadata that records the parent-child relationship. Listing 9.15 shows the new structure of the task descriptor.

In addition to the `taskgroup` member variable, we extend the metadata of the task
descriptor with a field to store some execution flags that tell the runtime whether a task
is still executing or has completed execution (see Listing 9.15). We will later need this
state when cleaning up completed tasks. The next member in the metadata section is
a pointer to the `TaskDescriptor` object of the parent task that created the child task
that this metadata belongs to. Finally, we add an atomic counter to the metadata to
keep track of how many child tasks a parent task has created. As child tasks are being

created, the runtime system will increment the counter when a child task is created and it will decrement the counter when a child task completes execution.

The code thus becomes slightly more complex (see Listing 9.16 and Listing 9.17). When a new task is created, the runtime now has to initialize the new metadata fields of the task descriptor. The initial state of the task is determined to be `Created` to indicate that this is a new task that has not yet been scheduled for execution. The number of child tasks is zero, as the task has not yet created any new child tasks. This can of course only happen once the tasks start to execute on a thread. Finally, we have extended the `Thread` object to keep track of the currently executing task, which naturally must be the parent task of the task that is being initialized.

When the newly created task arrives in the `StoreTask()` function, the function increments the child counter of the respective parent task to keep tally of the current number of child tasks. At invocation time (i. e., when a task is retrieved from the task pool and handed over to the `InvokeTask()` function in Listing 9.17), the task's state is set to `Executing`. When the thunk function returns, the task state is set to `Completed` to reflect that the task has finished all its work. Finally, the number of active child tasks is reduced by one for the parent task.

Note that, because of a special feature of OpenMP tasking, an OpenMP thread executes an implicit task that contains the code of the parallel region. If such an implicit task creates a child task, then the child task will have a special parent indicated by `nullptr`. In that case, the child counter is maintained directly in the `Thread` object. A different design alternative would be to create a task descriptor for these implicit tasks and avoid this extra check for this special "root" parent.

Listing 9.17 also shows the implementation of the `TaskWait()` function that is called by the generated code for the `taskwait` construct or for the implicit wait operation to synchronize the parent with their child tasks' completion. After a check for the aforementioned shortcut about implicit tasks, the `TaskWait()` function enters a while loop that checks the number of active child tasks. As long as at least one child task is still active, the code continues to try to find more tasks for the waiting thread to execute.

You may have noticed that the `InvokeTask()` function Listing 9.17 no longer calls `FreeTask()` to clean up the task descriptor. Instead, it now uses `FreeTaskAndAncestors()`. The reason for this is the pointer to the parent task and the access to a member in its metadata through the child counter.

Let's assume that the parent thread is waiting in `TaskWait()`. In this situation, a child task can safely decrement the child counter as it is guaranteed that the parent's task descriptor will be valid. Having a valid task descriptor is one of the invariant properties of a task that is waiting in the task pool or executing. If a child task completes execution, it decrements the counter, and as soon the counter is equal to zero, the parent task will continue with its own execution.

But what if the parent task does not await the completion of its child tasks because it only creates them, but does not encounter a `taskwait` construct? In that case, the

```cpp
void InitializeTaskDescriptor(TaskDescriptor * task,
                              size_t sizeOfTaskClosure,
                              size_t sizeOfShareds,
                              ThunkPointer task_entry) {
  auto thread = Thread::getCurrentThread();
  auto taskgroup = thread->getCurrentTaskgroup();

  if (sizeOfShareds > 0) {
    size_t offset = sizeof(task->metadata) +
                    sizeOfTaskClosure;
    task->closure.data = ((char *)task) + offset;
  }
  else
    task->closure.data = nullptr;
  task->closure.routine = task_entry;
  task->metadata.taskgroup = taskgroup;
  task->metadata.flags = TaskDescriptor::Flags::Created;
  task->metadata.childTasks.store(0);
  task->metadata.parent = thread->getCurrentTask();
}

bool StoreTask(TaskDescriptor * task) {
  auto thread = Thread::getCurrentThread();
  auto team = thread->getTeam();
  auto taskPool = thread->getTaskPool();
  bool result = true;

  if (task->metadata.parent)
    task->metadata.parent->metadata.childTasks++;
  else
    thread->childTasks++;
  if (!taskPool->put(task)) {
    InvokeTask(task);
    result = false;
  }
  return result;
}
```

Listing 9.16: Initializing and storing tasks with support for taskwait.

```
void InvokeTask(TaskDescriptor * task) {
  auto thread = Thread::getCurrentThread();
  auto team = thread->getTeam();
  int32_t gtid = 0;

  TaskDescriptor * previous = thread->getCurrentTask();
  thread->setCurrentTask(task);
  task->metadata.flags = TaskDescriptor::Flags::Executing;
  task->closure.routine(gtid, &task->closure);
  task->metadata.flags = TaskDescriptor::Flags::Completed;

  if (task->metadata.parent)
    task->metadata.parent->metadata.childTasks--;
  else
    thread->childTasks--;

  if (task->metadata.childTasks == 0)
    FreeTaskAndAncestors(task);

  thread->setCurrentTask(previous);
}

bool TaskWait() {
  auto thread = Thread::getCurrentThread();
  auto parent = thread->getCurrentTask();

  if (parent)
    while (parent->metadata.childTasks)
      ScheduleTask();
  else
    while (thread->childTasks)
      ScheduleTask();

  return true;
}
```

Listing 9.17: Executing tasks while maintaining the child count.

parent task might complete before its child tasks have been picked by the scheduler, and thus its task descriptor will be deallocated and will no longer be valid. As a consequence, a child task will have a dangling pointer to its parent's task descriptor. If we were using the simple FreeTask() call of Listing 9.6, then all of the child tasks that

```cpp
void FreeTaskAndAncestors(TaskDescriptor * task) {
  static std::mutex lock;
  std::lock_guard<std::mutex> guard(lock);
  size_t children = 0;

  children = task->metadata.childTasks;
  while (!children) {
    if (task->metadata.flags ==
                         TaskDescriptor::Flags::Completed) {
      TaskDescriptor * purge = task;
      task = task->metadata.parent;
      FreeTask(purge);
    }

    if (!task ||
        task->metadata.flags !=
                         TaskDescriptor::Flags::Completed)
      break;
    children = task->metadata.childTasks;
  }
}
```

Listing 9.18: Walking the line of ancestor tasks and cleaning up parent tasks.

execute after the parent task has completed would access the invalid memory of the deallocated task descriptor of their parent task. As execution of the parent task and child tasks can be arbitrarily interleaved, this constitutes a race condition in the runtime system that must be avoided. There are two alternative ways to solve this issue.

First, the task descriptor of a completed task is not immediately deallocated, but stored in a garbage list of task descriptors that are no longer needed. From time to time—for instance, when this garbage list is (almost) full—the runtime system traverses the list and checks whether a task descriptor has zero active child tasks. If so, the task descriptor is deallocated and removed from the list. While this may sound like a good idea, a single garbage list is yet another bottleneck if multiple threads are trying to put a task descriptor into the list. Thus, we would have to maintain multiple lists, one per thread, thereby reducing the contention of the insertion operation.

Second, when freeing a task descriptor, the runtime system can follow the ancestry of the to-be-deallocated task descriptor. This is shown in Listing 9.18. If the current task's count of active child tasks is zero and we know that the task has completed its execution, we can free the task descriptor. Before this happens, however, the code remembers the task's parent. If this is a `nullptr`, then this that indicates that the function has reached the root of the ancestral chart; otherwise, if the function reaches a parent that is still actively executing, the walk stops. Otherwise, the code gets the current count of child tasks and repeats the same procedure, again trying to free the current (parent) task. To avoid double freeing, the code has to be protected against more than one thread trying to free the same task descriptor.

9.3.3 Task dependences

For task dependences, the runtime system must record the task dependence graph for all the tasks that are generated by the application. The difficult thing for the runtime system is that the tasks are not generated with all of their dependences defined upfront so that the TDG can be constructed immediately. Instead, new tasks are created after other tasks have already started to execute. So, one task may have a depend clause that is the source of a dependence (`depend(out:...)` or `depend(inout:...)`) while the task that will be the sink of that dependence (`depend(in:...)` or `depend(inout:...)`) may not yet exist.

Thus, at the time when the task with the source dependence is created, none of the dependent tasks may exist yet. Indeed, since the OpenMP standard defines that a sink dependence that refers to a non-existent source dependence is always satisfied, then if the dependences are intended to work the way the programmer expected, none of the sink dependences can exist!

Listing 9.19 illustrates this. Assuming that these functions are called from a single thread inside a parallel region, the `broken_delayed_tasks_with_deps()` function will print "`var is 24`", since at the time *T2* is created, there is no source dependence for `var`, so any sink dependence on it is assumed to be satisfied.

The second function `broken_delayed_tasks_with_deps()` shows that the runtime cannot know about the sink dependences of a task at the time the task is created. Task *T1* is created and we need to assume that the runtime will start executing the task while the parent task waits in the call to `sleep()`. When the parent task returns from that call, it creates task *T2*, which has a dependence on task *T1*. As the runtime can only know about the existence of this dependence when task *T2* is created, the TDG has to be constructed on the fly and will probably never contain all task dependence information since some tasks may already have completed, and thus their dependence information is gone, while new tasks appear and have to be added to the graph.

As a consequence, the runtime system has to record the task dependence information in a *dependence map* and keep track of it, when the tasks are being created, executed, and freed. Let's start with a simple case like the one in Listing 9.19, which is a dependence introduced through a simple variable. At the time when *T1* is created, the runtime does not yet know about the future existence of *T2*. However, there are two fundamental cases that the runtime needs to deal with.

```c
#include <stdio.h>
#include <unistd.h>

void broken_delayed_tasks_with_deps() {
    int var = 24;
#pragma omp task depend(in:var)   // T2
    {
        printf("var is %d\n", var);
    }
    sleep(2);
#pragma omp task depend(out:var)  // T1
    {
        var = 42;
    }
#pragma omp taskwait
}

void delayed_tasks_with_deps() {
    int var;
#pragma omp task depend(out:var)  // T1
    {
        var = 42;
    }
    sleep(2);
#pragma omp task depend(in:var)   // T2
    {
        printf("var is %d\n", var);
    }
#pragma omp taskwait
}
```

Listing 9.19: Delayed task creation with task dependences.

The first case is that *T1* is still executing on a thread or is waiting in the task pool to be picked for execution. This means that *T2*'s task dependence on *T1* is not yet fulfilled, so the runtime system needs to keep *T2* in the task pool until *T1* has completed its execution.

The second case is that *T1* had been scheduled for execution (as we assumed above) and has already completed execution. One might think that the runtime needs to remember that this task has been in the system, had dependence information, and has completed execution. However, this is not the case!

Here's why. Let's consider a task that has a dependence like *T2*, but the code for *T1* does not exist. In this scenario, *T2* can immediately start executing, as its dependence is fulfilled by definition, since, as we saw above, the OpenMP specification says that a sink dependence without a source is automatically satisfied. As a result, once a task that creates a source dependence has completed, there is no need to keep information about that dependence any longer because it is satisfied (because the task has completed), and the runtime will treat any source dependence it cannot find as satisfied.

One implementation of this dependence handling is to record the address of var (to identify the memory location) plus the dependence type when the application creates a task. For *T1*, the runtime system will record a dependence of type out for address &var. Later, when task *T2* is created, the runtime again takes the address (again &var) plus the dependence type in from the depend(in:var) clause.

The data structure that seems useful to easily record address information along with dependence information is a hash map that maps the address in the task dependence information coming from a depend clause to a list of task descriptors. When a task is created, the runtime system uses the address for each list item in each depend clause as the key into the hash table. The value that this key is mapped to is the pair consisting of the dependence type and the task descriptors that are associated with this dependence.

For a task with an out or inout dependence type, the runtime will create a record for the task and its dependences in the hash map. These tasks can be sources of dependences, and thus may cause other tasks to wait while they are in the task pool (or while they are executing, if they use any of the broader task synchronization techniques we have already shown). As usual, the task descriptor is also added to the task pool to allow the task to be found and executed later.

When a task has an in or inout dependence type, it is a potential sink for a dependence. When the runtime receives the task, it looks up the addresses for each of the list items of each of the depend clauses in the hash map. If the runtime finds source tasks for any of the new task's dependences, then the task has unfulfilled dependences and thus may not run yet. The task is then registered in the dependence record as a sink task so that the runtime system can easily find the task when the corresponding source task completes execution. The newly created task is not added main to the task pool, since it should not execute yet. If none of the sources are found in the dependence records, then all the newly created task's dependences are satisfied and the task may go straight to the task pool to be scheduled for execution.

So far, we have described task dependences in terms of a simple variable. When it comes to arrays (like in Listing 2.14), things become slightly more complex. The OpenMP specification says that if an array (or array slice) is used in a task dependence, then the dependence only matches if the array slice is exactly the same in both the source and the sink of the task dependence. Other programming models may detect a dependence if there's an overlap between the arrays or the array slices.

Depending on what the parallel programming model defines, the implementation of the dependence check becomes more complex. But let's start with the non-overlapping arrays and array slices, for which we can re-use the simple address-matching scheme that we have already presented. Since a source and a sink of a dependence will only match if the source and sink array slices have the same address, the runtime can simply use the address of the first array element in the slice. It is not required to match if it is a disjoint slice, as its first element will have a different address. If the slices overlap without having the same address, then, at least with the OpenMP definition, the array slices do not match.

If partially overlapping arrays or subslices are required to match for a dependency, then a simple address matching of the array slices will not work. Instead, the matching would have to determine whether the array slices overlap, and it would have to do so for each dimension. To be efficient, this would require something like k-d trees [12] or other data structures that allow for a quick check of the multi-dimensional spatial relationship of two objects.

The final topic is the scope of the dependence map and how to clean it up. The scope of the dependence map needs to match the scope of the task dependences.

In OpenMP semantics, task dependences are only effective for child tasks of the same parent task. Hence, the dependence map can be stored in the task descriptor. The runtime can then free the memory used by the map when it releases the memory for the task descriptor after the last child task has completed when `FreeTaskAndAncestors()` (see Listing 9.18) is invoked. If task dependences can span any task in the parallel region, then the task dependence map needs to be maintained in the team structure for the parallel region and thus needs to be a global data structure. Luckily, the runtime does not have to keep the dependence map forever. When the parallel region ends, the data structure can be released.

9.4 Task scheduling concepts

Let us take a quick break from this discussion and step back a little. In this chapter, we have made a few implicit decisions, but have never really taken a look behind the scenes to see whether there are alternatives for these decisions. Is the single task pool always a bad idea? Why should we be stealing an older task, but not a more recent one? Why execute a task that cannot be stored in the task pool immediately instead of storing it in the task pool of another thread? Here, we will go a bit deeper into these questions and the implications that each of the decisions may have.

9.4.1 Task-scheduling points

Let's first discuss when a task scheduling decision can be made. Typically, such a point in the program is called a *task scheduling point*. Where they are will depend on the semantics of the parallel programming language. They can also either be implicit (e. g., when a new task is created) or explicit (e. g., the OpenMP `taskyield` construct). We cannot cover all possible TSPs in this book, so please refer to the OpenMP specification [100] or the TBB book [151] for a more exhaustive list. But some typical TSPs are:
–   When a new task is created.
–   When a task finishes execution.
–   At OpenMP constructs like `taskyield` and `taskwait`.
–   At implicit and explicit thread barriers.

At any of these points, an implementation—that is, the compiler and runtime—may decide to schedule a different task for execution. Please particularly note the word *"may"*! The implementation is not required to switch to a different task for execution; rather, it could decide to continue to execute the current task, or even suspend task execution for a while. We have been exploiting the fact that task creation is a TSP earlier in this chapter when we had a thread execute a newly created child task immediately when the task pool was full.

From an implementation perspective, all of the aforementioned TSPs must have a correspondence in the generated code that the compiler produces and the underlying parallel runtime system that provides the execution mechanisms. When you go back to Chapter 4, you will see that the compiler had to emit entry points to the runtime system to allocate and create a task for execution. Earlier in this chapter, we described how the generated task is converted into a task descriptor, so that important execution metadata can be attached to it. Then, it is stored in a task pool (or executed immediately when the task pool is full). In each of these points, the sample code is making a scheduling decision about what happens to the task at hand.

Going through the code examples of this chapter, you will find the following places at which a scheduling decision can be made (and actually is made) in the code examples:
–   Listing 9.9 and Listing 9.11: We have decided to execute the newly created tasks when the task pool is full, so we cannot store the task descriptor for later execution.
–   Listing 9.14: In the implementation of `taskgroup` constructs, we have used `ScheduleTask()` to pick a task to execute while waiting for descendant tasks to complete.
–   Listing 9.17: While a parent is waiting for the completion of all of the tasks in a `taskgroup`, it also calls `ScheduleTask()` to pick a task from the task pool for execution.

In Listing 9.6 and Listing 9.11, the `ScheduleTask()` function simply called the `get()` method of the task pool and executed the returned task if there was one. Obviously, this may not be the best solution in all cases, and indeed seems to be an arbitrary decision. And you would be right to think so! So, let's dive a bit deeper into the topic of task scheduling.

9.4.2 Breadth-first scheduling and depth-first scheduling

The strategy that we have implemented in this chapter is called *breadth-first schedul-ing* (BF) [35]. If you follow the implementation that we have presented, what happens is that a parent task creates a child task that is stored in the task pool for later execu-tion, while the parent task continues execution and (potentially) creates more tasks. This will continue until either the parent task completes execution, the parent task reaches a wait operation (`taskwait` or `taskgroup`), or it tries to create a task and finds that the task pool is full (when it will execute the newly created task).

In contrast to breadth-first scheduling, *depth-first scheduling* (DF), also called *work-first scheduling*, is when the creating task is suspended when it creates a new task or has to wait. Typically, if a child task is created, the execution follows the child task by immediately scheduling it for execution and the remainder of the parent task is placed in the task pool.

These strategies have different properties and thus different merits for their use in application codes. Depth-first scheduling mimics the idea of a function call. Task execution follows the same execution path that the sequential version would follow if no task parallelism was present. If the sequential version was tuned for temporal or spatial locality, then the tasks will likely maintain the locality as well. In contrast, breadth-first scheduling assumes that there is no such locality and that tasks are truly independent units of work. Instead, BF maintains the locality properties of the parent task, as it is not suspended and thus may not move to a different execution thread.

Depending on which of the two implementation scheduling strategies the com-piler-runtime combination chooses, the compiler's code-generation pattern for task-ing might require a few changes. The pattern in Section 4.4.6 was geared towards the BF-scheduling approach. The compiler generates code to create the child task and to hand it over to the runtime system for execution. It does not emit code to (potentially) defer the remainder (or *continuation*) of the parent task's execution. In contrast, for a DF-scheduling runtime, the compiler would still outline the child task's code region and provide a data environment for its execution. But it would then also create code to generate a task for the remaining code of the parent task, so that the parent task's remaining code after the TSP can be sent to a task pool and be deferred if the runtime decides to do so.

One peculiarity of DF is that it does not blend well with tied tasks. The parent task is suspended and execution follows the child task. If the parent is a tied task (the default with OpenMP tasks), then the parent task's remainder cannot be picked by a different thread for execution. This means that while the currently executing thread is working on the child task (and potentially its descendant tasks), no other thread can execute the remaining code of the parent task. What a neat, yet complicated, way of completely serializing a parallel program!

9.4.3 Task stealing

We have already briefly touched on the subject of task stealing in Section 9.2.2.2. The code in Listing 9.12 rolled dice to randomly pick one of the other threads' task pools and attempted to steal one task from that pool to execute it. Of course, stealing from a random victim is an option, but it may not be the best option. In real life, a thief would rather pick a prime victim with a wallet full of money and only look for random victims as a fallback. Besides the decision regarding the victim task pool, another question is which tasks we should steal from a victim and how many we should take. This can be any number, from a single task to a fraction of the tasks in the victim's task pool.

Let's talk about picking the victim first. If we consider the system architecture of the execution environment (see Chapter 3), we find that the environment normally consists of a hierarchy of components. We have already said that there are logical cores that run on the same physical core and thus share part of the cache hierarchy. Processors may consist of several dice that form a processor package and processor packages may be organized into NUMA domains with locally attached memory. This structure of the system greatly influences performance, and the parallel runtime system should take locality into account when determining where to steal a task from.

This suggests that we should try to steal a task from a thread that is close to where the stealing thread is executing. Assuming that a child task exhibits some of the locality properties and a working set of the parent task, the closer the victim thread is with respect to the hardware structure, the better. So a runtime system might want to attempt to steal a task from a thread that is running on the same physical core, yet a different logical core, as the thief. If this victim thread does not have tasks in its task pool, then the thief will have to look elsewhere for a task to steal.

Figure 9.1 shows a hypothetical execution platform with sixteen logical cores on eight physical cores spread across two NUMA domains. The cores are labeled "x/y," with x denoting the number of the SMT core on physical core y. Two neighboring SMT cores share the same L1 and L2 data caches, while two physical cores share the same L3 data cache. This resembles the rough structure (except for the total number of cores) of some of the most prominent processor samples at the time of writing this book.

In the picture, the star marks the core that is executing the thief thread. In a first attempt to steal a task, the thief will try to take a task from the thread on the neighboring logical core (tagged as ❶ in Figure 9.1). If no task was available for stealing from this victim's task pool, then the stealing algorithm will look at the next level up in

NUMA domain 0 ❸				❹ NUMA domain 1			
L3 D$ ❷		L3 D$		L3 D$		L3 D$	
L2 D$	L2 D$	L2 D$	L2 D$	L2 D$	L2 D$	L2 D$	L2 D$
L1 D$	L1 D$	L1 D$	L1 D$	L1 D$	L1 D$	L1 D$	L1 D$
0/0 | 1/0	0/1 | 1/1 ❶	0/2 | 1/2	0/3 | 1/3	0/4 | 1/4	0/5 | 1/5	0/6 | 1/6	0/7 | 1/7

Figure 9.1: Example of task stealing respecting the hardware hierarchy.

the machine structure, which contains the task pools that are associated with threads running on cores that share the same L3 cache as the thief (❷). If this also fails, then the algorithm looks for tasks to steal within the same NUMA domain (❸). Finally, the algorithm attempts to pick a task from a different NUMA domain (❹).

Whether such a fine-grained task-stealing scheme is needed depends on the structure of the hardware and its performance metrics. However, it certainly seems a good idea to try to at least respect the NUMA structure of the system so that locality within the local memory of the processor packages can be exploited as long as the spawned tasks also adhere to these locality properties. As an example, we have implemented a stealing algorithm that tries to maintain NUMA locality when stealing tasks.

As we saw in Section 5.2.1, we can detect the hardware structure of the system and use it to build a database of the mapping of cores to NUMA domains. Listing 9.20 shows how the task-stealing algorithm can use this database to restrict stealing to the local NUMA domain if possible. The idea is to apply a two-level stealing approach.

The algorithm starts at the local NUMA domain of the thief. In this domain, it starts at the first core, and if this is not the core that is executing the thief, it tries to steal a task. To do so, it determines the Thread object of the victim thread and then accesses its task pool. One of the key differences between this and the random-stealing algorithm of Listing 9.12 is that the algorithm does not directly determine the victim thread. Instead, it selects a core and then tries to find the thread that is running on this core. For a fully populated machine that executes a thread on each core, this does not make a big difference. But if some of the cores are left idle, the algorithm has to skip these to find the closest thread in the current NUMA domain.

If the algorithm does not find a task to steal in any of the task pools in the local NUMA domain, it switches to the next NUMA domain in the outer loop. It then goes through the NUMA domains in a round-robin fashion. To be able to use a simple loop nest without any special cases, we apply a small trick to the algorithm. While myCore is the global core ID that executes the thief thread, the inner loop uses a domain-local addressing system with relative core IDs from a particular NUMA domain. As NUMA domains do not necessarily have the same number of cores, we have to translate the

```cpp
struct NumaStealStask {
  TaskDescriptor * operator()() {
    auto thread = Thread::getCurrentThread();
    auto team = thread->getTeam();
    TaskDescriptor * task = nullptr;

    auto numberOfDomains = numa::GetNumberOfNumaDomains();
    auto myCore = numa::GetCoreForThread(thread);
    auto myDomain = numa::GetNumaDomain(myCore);

    for (auto domain = 0; domain < numberOfDomains && !task;
         ++domain) {
      auto victimDomain =
                (myDomain + domain) % numberOfDomains;
      auto coresDomain =
                numa::GetCoresForNumaDomain(victimDomain);
      auto numberOfCoresDomain = coresDomain.size();

      for (auto victimCore = 0;
           victimCore < numberOfCoresDomain && !task;
           ++victimCore) {
        auto globalCoreID = coresDomain.at(victimCore);

        if (myCore == globalCoreID) {
          continue;
        }

        auto victimThread =
                numa::GetThreadForCore(globalCoreID);
        if (victimThread) {
          auto victimPool = victimThread->getTaskPool();

          task = victimPool->steal();
        }
      }
    }

    return task;
  }
};
```

Listing 9.20: NUMA-aware task stealing.

domain-local core ID to a global core ID that we can use to find the thread running on that core. This is done in the innermost loop of the stealing algorithm.

When you read through the algorithm in Listing 9.20, you might have thought that a lot of effort was spent trying to find work to steal from another thread. And you were right in thinking that! However, since it was looking for work, the thief had nothing else to do (except maybe twiddling its virtual thumbs), and so would otherwise have been idle. Thus, it could have spent a bit more time finding work to steal, and trying to preserve locality as much as possible might be effort well spent if it results in the task that is stolen running faster.

Having covered deciding where to steal from, let's talk about deciding which tasks to steal. So far, we have selected tasks for execution from the head of the local task pool (i. e., where newly created tasks are inserted), while stealing happens at the tail. This means that stolen tasks are the oldest tasks in the task pool. We also only took one task from the victim and immediately scheduled it for execution.

The rationale behind the decision to steal the oldest possible task from a victim task pool stems from the heuristics that older tasks are likely to create a larger number of descendant tasks than younger tasks. The standard example is a task-parallel recursion scheme (like the Fibonacci example of Listing 2.11), in which a task that is closer to the root of the recursion—that is, an older task—will unfold a whole subtree in the recursion, while a younger task will unfold a smaller subtree or even spawn no new tasks if it is a leaf task that ends the recursion. In addition, since in the OpenMP model, task dependences are forward dependences, it's likely that older tasks in the pool are ready to execute and are less likely to be blocked for execution by an unfulfilled task dependence.

Finally, if the parallel programming model supports task priorities (see Section 9.6 for implementation notes), the implementation may be required to steal the task that has the highest priority in the victim task pool or even in all of the available task pools. To support these capabilities would require changes to the task pool implementation and a tighter integration of the task pool with the scheduling algorithms.

The final topic is deciding how many tasks to steal. Since our small example runtime system does not support task dependences, we can pick a single task from a victim thread and execute it. However, locking the victim task pool for task stealing is costly, as it involves a lock. So stealing more than one task should help to amortize that cost. How many tasks to steal in one attempt is a tuning parameter and should be adjusted depending on the machine.

9.5 Task scheduling constraints

Task scheduling constraints (TSC) are semantic restrictions on how the implementation of a parallel programming model can schedule tasks. Earlier in this chapter, we saw that there are different task-based models that can define different task-scheduling points. A scheduling constraint further restricts what can be done at a TSP and mandates which tasks can be scheduled, how they are scheduled, and how they are treated if they are suspended and then resumed.

We have already discussed some scheduling constraints. For instance, if a task has a dependence on another task, it can only start executing when the dependence is satisfied. Another example is when a thread decides to immediately execute a new task. The currently active task then needs to be suspended and resumed later. A TSC may mandate that the previously executing thread needs to resume the suspended task, or it may allow any other thread to continue with the suspended task. In the OpenMP standard, a task is described as either `tied` or `untied`, respectively.

9.5.1 Stack scheduling

The way we have scheduled tasks in this chapter was via a call to `InvokeTask()` (e. g., Listing 9.16), which took the task descriptor and invoked the function pointer in it as a regular (indirect) function call that passed in the data pointer of the task closure. It performed some bookkeeping as well, but the important bit for this discussion is that task execution is a regular function call. We will call this *stack scheduling*.

The interesting part happens when an active task on the call stack of the execution thread reaches a TSP and the runtime decides to pick another task for execution. With stack scheduling, the `InvokeTask()` function will be called and the newly scheduled task will be executed via an indirect function call. As a matter of consequence, the state of the active, but now suspended, task remains on the stack.

Figure 9.2 visualizes this. In the figure, the stack is growing from the top to the bottom. Task *A* is the currently active task on the thread's stack. The stack starts with the `ScheduleTask()` function that called `InvokeTask()` to execute *A*'s thunk function. Then the runtime picks task *B* for execution. Task *A* remains on the stack and task *B*'s thunk is invoked as a function call. Both tasks are now on the stack, but only *B* is executing code. Task *A* has to wait in a suspended state until *B* finishes and its call frames are removed from the stack.

One consequence of this is that task *A* automatically becomes a tied task (see Section 2.2). When a suspended tasks is resumed, it resides on the same stack as it did before it was suspended by its executing thread. Thus, the previous thread will automatically resume this task and continue to execute it. This is also a valid implementation for an untied task because although an untied task *may* resume execution on a different thread, it does not *have* to. It is thus acceptable within OpenMP semantics to

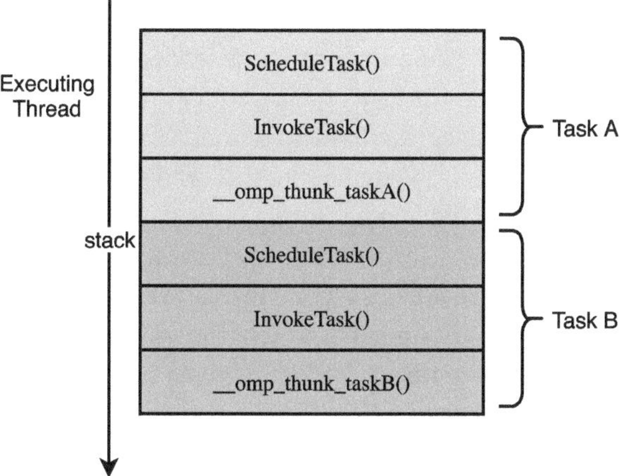

Figure 9.2: Stack scheduling of two tasks with task A being suspended.

implement untied tasks as tied tasks. Since this is the case, many OpenMP implementations chose to do this and simply ignore the untied clause in task directives.

9.5.2 Cyclic scheduling

A different approach to scheduling tasks is called *cyclic scheduling*.

With cyclic scheduling, executing a task starts out the same way as with stack scheduling. Once a thread becomes idle and enters the runtime system to pick a task from a task pool, the InvokeTask() function invokes the thunk function of the chosen task. Thus, the thunk's call frame is located on the stack of the executing thread.

The differences appear when a task-scheduling point is reached. Instead of invoking InvokeTask() to enter the to-be-scheduled task, the suspended task is removed from the thread's stack. To do this, the compiler generates a continuation—that is, the code position and live variables [7] of the suspended task at the TSP. The runtime system now has two options of how to defer the task. With either option, the suspended task is put back into the task pool for a thread to pick it up again.

First, it can mark the first part of the task from its beginning (or the previous TSP) to the current TSP as complete. It will then have to create a new task descriptor from the continuation and use the remaining code of the task as a new thunk function. Alternatively, if the compiler only generated a single thunk function, then the thunk will have to branch to the code location where the task was suspended and restart from there.

Second, the runtime system can modify the existing task descriptor for the suspended task and add information about the continuation to it. The compiler will have to prepare this via code that it generates to instruct the system to resume the task at a given TSP using the data from the continuation. As each task can only have a finite number of TSPs in its thunk function, the task descriptor can be extended with a field that stores the TSP that the task is to be resumed from, e. g., via an integer ID. The value zero would then correspond to the beginning of the thunk function, while a value of one would correspond to the first task-scheduling point of the task, and so forth.

Cyclic scheduling is a natural choice to implement untied tasks. When the task descriptor of a suspended task is put back into the task pool, any idle thread can pick it up for execution. This could be the thread that had suspended the task, but it could also be any other thread that has stolen the suspended task from the task pool.

The downside of this is that tied tasks require some effort to make sure that they are resumed on their initial thread when suspended. Of course, the runtime system could revert to stack scheduling for these tasks, while untied tasks use cyclic scheduling. However, this complicates the runtime and compiler more than it helps them. The usual approach is to mark such tasks as protected against stealing, which requires an additional change in the task descriptor status flags. Once marked as tied, the task is put into the thread's own task pool. Any stealing attempt will fail, and thus it is guaranteed that the resuming thread will be the initial executing thread.

Another option is to have separate private and public task pools for each thread. When a task creates new child tasks, the runtime places their task descriptors in the public task pool of the creating thread. This is where other threads can steal tasks from when their own task pools run empty. If a thread suspends a task, it stores its task descriptor in the private task pool to avoid other threads stealing the suspended, tied tasks. When looking for work, a thread will first pick from its private pool, and then from its public one, before finally trying to steal work from elsewhere.

9.6 Miscellaneous task topics

Let's close the chapter with two miscellaneous topics concerning tasking: task priorities and task affinity.

9.6.1 Task priorities

Task priorities assign an execution priority to tasks, so that a task with higher priority is picked for execution over a lower-priority task in the task pool. The OpenMP API defines the `priority` clause [100] for this purpose. This clause takes an integer argument that describes the desired priority for this task. Depending on the language specification, such task priorities can be hints, which can be ignored by the implementation, or they can be a binding specification of when a task must be executed with respect to other tasks.

Table 9.2: Complexity classes of accesses to a task pool with priorities.

Data structure	Find max	Insert	Delete max
unsorted array	$\mathcal{O}(n)$	$\mathcal{O}(1)$	$\mathcal{O}(n)$
sorted array	$\mathcal{O}(1)$	$\mathcal{O}(n)$	$\mathcal{O}(n)$
binary heap	$\mathcal{O}(1)$	$\mathcal{O}(\log n)$	$\mathcal{O}(\log n)$
splay tree	$\mathcal{O}(1)$	$\mathcal{O}(\log n)$	$\mathcal{O}(\log n)$

If task priorities are hints (as they are in the OpenMP specification), then one possible implementation would be simply to ignore them. This will require no change to the runtime system, but it may make users unhappy, as they may have spent time thinking about task priorities and adding them to their code. Having that work then be ignored entirely will not be taken lightly, and may lead to (angry) bug reports.

A slightly better approach is to use several buckets of task priorities. For instance, the runtime system could implement three buckets: high priority, low priority, and no priority. For each of the buckets, the runtime system will maintain a task pool that receives the created tasks with a priority that is mapped to that bucket. When picking a task for execution, the runtime system would go through the task pools in decreasing order of priority to pick a task from highest-priority pool that has any task available and schedule it for execution.

This solution tries to balance the time spent storing the task in the task pool and later removing it with the duty to respect the requested priority or at least to try to approximate it. However, it states that the priority can be ignored. If that's not possible (because of parallel-language semantics), or if the runtime wants to support the best possible quality of implementation, then the runtime must be changed in more fundamental ways.

The first step is to extend the task descriptor with a field to hold the (relative) priority. This is needed because, in the presence of cyclic scheduling, a high-priority task needs to be resumed before a lower-priority task. With the task pool implementation shown in this chapter, the runtime would then search through the current thread's task pool to find the highest-priority task. Alas, this approach is very slow, as, in the average case, the runtime will have to traverse half of the task pool in the critical path of the task scheduler.

The better approach is to replace the task pool with a data structure that is sorted by priority so that the topmost task is always the one with the highest priority, so it is easy to extract it from the pool. Examples of such *priority queues* are sorted arrays, binary heaps [24], and splay trees [119]. Table 9.2 shows the number of operations that are needed to access the mentioned data structures in "Big $\mathcal{O}$" notation.

Unsorted arrays, which is what we have been using throughout this chapter, require on average $\mathcal{O}(n)$ operations to find the task descriptor with the highest priority. As the array is unsorted, inserting a new task descriptor is easy; it is simply appended at the end of the array. However, deleting becomes expensive, because after a task descriptor is deleted, the array needs to be compacted, which means shifting $\mathcal{O}(n)$ elements on average. A sorted array makes the maximum entry easy to find, but the insertion and deletion costs remain the same. Inserting means finding the right place (which takes $\mathcal{O}(n)$ operations on average) and deleting requires us to compact the array.

Using a binary heap or a splay tree will have much better access behavior. The maximum is easy to find, as it is the root of the tree. Inserting a new task descriptor requires searching down from the top by going left or right to find the right spot in the tree. This takes $\mathcal{O}(\log n)$ steps on average. When the root is removed, one needs to go down in the tree, find the new maximum and move it up the tree, potentially fixing up other tree nodes to be in the right place as well.

Multi-task-pool implementations with priorities impose a serious issue for the runtime implementer. When a thread wants to pick a task from its own task pool, can it be sure that there is no other high-priority task in some other thread's task pool that it will have to pick first? The simple answer is that it cannot be. If priorities can be interpreted as a hint, then a thread can pick the highest-priority task from its own pool and ignore any other thread. If task properties are mandatory, then the runtime has no other choice than to search all of the task pools to find the highest-priority task that is waiting to be executed. This complexity may render the multi-task pool inefficient enough that a single task pool will be the better choice, despite its other drawbacks. In effect, mandatory task priorities require centralized knowledge of all tasks, which conflicts with the distributed task pool's aim of avoiding a global state as much as possible, since it is a source of contention and poor performance.

9.6.2 Task affinity

We have seen throughout this chapter that tasks are dynamically scheduled, as they may be picked by any idle thread for execution. We've also shown that tasks should execute where their data is located. We'll now discuss how this can be achieved.

NUMA-aware schedulers try to schedule tasks in a way that keeps them local with respect to the NUMA domain. The heuristics typically assume that previous tasks have touched data that will be touched by their child tasks, too. The multi-task pool and scheduling implementation that we have shown in this chapter implements this approach. As threads create children, the runtime stores them into the thread's local task pool. When scheduling tasks, they are taken from the local task pool first. So, until load imbalance causes threads to idle and resort to stealing tasks from elsewhere, they will only execute tasks that they have created. We saw in Section 9.4.3 that task stealing can be NUMA-aware and try to limit stealing to the NUMA domain if possible.

But what if the initial executor of a task is a poor choice? Say, for instance, that a task is working on a piece of memory that resides in one NUMA domain, but is picked

```c
#include <stdint.h>
#include <stdlib.h>

#define BLOCK_SIZE 32
void task_affinity(double * array, size_t size) {
  for (size_t i = 0; i < size; i += BLOCK_SIZE) {
#pragma omp task affinity(array[i])
    work_on_array(&array[i]);
  }
}
```

Listing 9.21: OpenMP code showing the affinity clause of the task construct.

by a thread for execution on another NUMA domain. In general, it is difficult for a compiler or a runtime system to figure out which data a piece of code is working on. The OpenMP API solved this problem by providing the affinity clause [75, 100] that programmers can add to the task construct (see Listing 9.21). It takes a list of variables or array (slices), for which the compiler and the runtime determine the address. Then the task scheduler can use this as a hint to try to execute the task close (with whatever definition of "close" makes sense for the current machine) to the data the user has claimed that the task will be accessing.

If a runtime wants to schedule a task to run close to its processed data (in the same NUMA domain), it will have to figure out a few things. First, it must have an understanding of the NUMA nature of the system (see Section 5.2.1) and ideally also pin the threads to the processors (see Section 5.2.2). Second, the runtime will have to find the NUMA domain that hosts the physical pages of the virtual addresses that were given by the affinity clauses of the task. The Linux[*] system call move_pages() [132] comes in handy for this. While it is primarily used to move pages between NUMA domains, it can also be used to query the memory mapping of the process and return the NUMA domain that contains the physical pages for a set of virtual memory addresses.

Equipped with this, the runtime can translate the virtual-memory address from the affinity clause to the NUMA domain of the physical page behind that address. To do so, the address has to be rounded down to point to the bottom byte in the page, so that it can be used with move_pages():

```c
int page_numa_domain = -1;
void * page_start_address =
  (void *) (((intptr_t) virtual_address) & ~(PAGE_SIZE - 1));
move_pages(0, 1, &page_start_address,
           NULL, &page_numa_domain, 0);
```

The `move_pages()` system call then returns the (current) NUMA domain of the physical page via the argument `page_numa_domain`.

The next step is to go through the task queues to find a thread in the returned NUMA domain and then put the task into its task pool. The caveat is that this changes the way task pools work. Until now, only the owner of the task pool stored a task there. Now, any other thread can also do this, depending on the `affinity` clauses of the created tasks. This may cause contention on the task pool's `put()` operation that was not there before, and requires that this code path is now protected from race conditions, which may add an expensive atomic instruction or lock onto what we hoped was a fast path. Alternatively, a receiving task pool can be added to the remote thread, so that the task can be put there without disturbing the remote thread.

We're not there yet, though. One thing that we need to take care of is that `move_pages()` is a system call and this takes a long time; time that adds to the overhead of creating and storing the task in the task pool. Thus, if an address was translated, then the runtime system should memorize the translation (e. g., in a hash map of addresses to their respective NUMA domain) and re-use it whenever possible. One caveat is that the OS might choose to move the page around, so the runtime system may work with outdated information about the page's location. Also, the runtime needs to have a strategy to choose when to purge the cached information about accessed memory pages. Luckily, having incorrect information here will not affect program correctness, although it may affect performance.

Another scenario that can happen is when multiple `affinity` clauses are present. Nothing prevents the programmer from intentionally or accidentally specifying conflicting affinity requests. For instance, the programmer could use two array slices that are resident in different NUMA domains. In this case, there may be no obviously optimal schedule, since the user has asked for something impossible. There are many strategies to try to resolve this. For instance, the runtime can see whether a majority of the `affinity` clauses are for a particular NUMA domain and pick that one to execute the task.

Finally, the runtime may decide to let a thread steal a task that was scheduled for execution on one NUMA domain, so the task may end up executing away from its preferred NUMA domain again. At first glance, it may sound like a good idea simply to protect such tasks from being stolen. However, affinity may interfere with load balancing. In the extreme case that all threads in one NUMA domain have run out of work, they should seek tasks from another NUMA domain. Overall, it's better to execute a task causing remote memory access than not to execute it at all and leave threads idle. So, the runtime needs to exhibit some flexibility in moving tasks around in the system, even if they have a preferred affinity.

9.7 Conclusions

An implementation of a task pool may have seemed easy in the beginning. Overall, it's just a bunch of tasks that have been created by the application and that are waiting for execution. In this chapter, however, we have shown that there's a bit more to it once we start to aim for scalability of the task pool and add complexity to the implementation when we target more sophisticated features of a task-based parallel programming model.

One particularly interesting topic is that with the growing complexity of the implementation of the task scheduler, the time that a naive implementation spends in mutually-exclusive execution becomes larger and larger, and thus, the scalability of the implementation becomes problematic. One takeaway is that you need to take great care to limit mutual exclusion to only the parts of the algorithm that really need it and release a lock as soon as possible, but not too early to risk race conditions. There may even be cases where you will have to release the lock and then re-acquire it again.

You may say: What about lock-free data structures? Shouldn't you have mentioned those, too? Yes, we could (and maybe even should) have mentioned those. But ultimately, these data structures are not lock free in the sense that they do not use mutual exclusion at all; they just use atomic instructions directly instead of calling a lock interface like the one we discussed in Chapter 6. Such data structures are very hard to implement, and not all data structures can be implemented with only basic atomic instructions. Generally, when you seek a smart implementation that tries to keep locking and contention on the locks to a minimum (like with multiple task pools), a locking data structure is a reasonable choice. Additionally, the key thing is always to take the atomic instruction or lock out of the critical path if you can.

10 Summary and final thoughts

We hope that you have found this book enjoyable and useful, and congratulate you for reading this far (or skipping here directly). Obviously, if we could present all of the information in the book in two pages, then that's what we would have done, so summarizing it here is not entirely trivial. However, here are some of the key points that are worth remembering and that underlie the more detailed information we have presented.

Modern processors are complicated! They don't perform as you might initially expect, so you need to measure their low-level performance before you can optimize appropriately. While it is clear that the specific performance data as well as the machines on which it was gathered are unlikely to be relevant to your code, our intent has been to ground our discussions of the design of the software in the mire of the performance of real machines, since that is where your code will run. All of the micro-benchmarks we used and the script used to generate graphs from them is included in the code at https://github.com/parallel-runtimes/lomp.

Data movement is critical! Data movement is an expensive operation, and thus, the time it takes can often become the limiting factor for performance not only for a parallel runtime system, but also for the application itself. This is especially true for many of the low-level operations that we have to implement in a parallel runtime.

Atomic operations are necessary, but expensive! The one-line summary here is that if you can avoid them, then do so (e. g., in the Test and Test-and-Set barrier or max/min reductions). Apart from serializing execution, they are also heavyweight operations on the machine's memory subsystem, so it seems like a good idea to rethink the usage of atomic operations and see if these atomic operations can be moved out of the critical path.

You can't beat the mathematics! But you can use mathematics to understand the performance bounds of important pieces of code (such as barriers) and to predict upper and lower bounds for performance. This will give you an estimated range of where you can expect the performance to be. This will clearly help you decide whether or not to optimize and to see how far you are away from a good or even the best solution.

Write portable code! We have shown in some detail that the code of a parallel runtime library needs to consider the machine on which it will run. However, it should still be as portable as possible. That is because we *know* that the machines next year will be different, and experience shows that code typically outlasts the machines on which it initially runs (in some cases by decades). This is also important for user-level code. Do not bind it to the machine it happens to be running on today! So please don't have `omp_set_num_threads(4)` in your code "because that's the number of logical CPUs in my laptop" or make other restrictive assumptions about the scalabil-

https://doi.org/10.1515/9783110632729-010

ity of your parallel runtime systems. If you do, you are showing that you expect nobody else to use the code, and that you'll throw it away before you update your machine.

People who use your code are your friends! Always remember that the people who are using your code are your friends, especially the annoying ones who keep submitting bug reports. The fact that they are taking the time to help you improve your code deserves recognition; they could just have given up and used someone else's code.

This stuff is hard! Even if you have no intention of ever writing your own (parallel) runtime, understanding the complexity of the issues that runtime authors have to handle should be useful to you, and encourage you to use an existing implementation, rather than failing to notice that parallel systems provide the operations you need (such as reductions, locks, and atomic operations).

The best code is the code I don't have to write! This applies not only to the parallel components we have been discussing, but also to the higher levels and other domains, too. While it may be very educational to roll your own, it is usually much more productive to find a well-tested and optimized library instead. Especially in well-worn areas such as linear algebra, there are existing, well-tested, high-performance libraries that will squeeze the last bit of performance out of your system.

Finally, and most importantly: "Live long and prosper".

Bibliography

[1]   A group of C++ enthusiasts from around the world. CPP Reference. https://en.cppreference.com/w/, accessed June 29, 2020.

[2]   Alfred A. Aho, Monica S. Lam, Ravi Sethi, and Jeffrey D. Ullman. *Compilers. Principles, Techniques, and Tools*, 2th edition. Pearson, Boston, MA, USA, 2007. ISBN 978-1-29202-434-9.

[3]   Susanne Albers. Online Scheduling. In Yves Robert and Frédéric Vivien, editors, *Introduction to Scheduling*, CRC Computational Science Series, pages 51–77. CRC Press/Chapman and Hall/Taylor & Francis, 2009.

[4]   Frances E. Allen and John Cocke. A Catalogue of Optimizing Transformations. In *Design and Optimization of Compilers*, pages 1–30. Prentice-Hall, 1972. ISBN 978-0-13200-204-2.

[5]   AMD. AMD Epyc 7742. https://www.amd.com/en/products/cpu/amd-epyc-7742, accessed July 22, 2020.

[6]   Gene M. Amdahl. Validity of the Single Processor Approach to Achieving Large Scale Computing Capabilities. In *AFIPS Conference Proceedings*, pages 483–485. AFIPS Press, Atlantic City, NY, USA, April 1967.

[7]   Andrew W. Appel. *Modern Compiler Implementation in Java*, 2th edition. Cambridge University Press, Cambridge, UK, 2002. ISBN 978-0-52182-060-8.

[8]   Arm Limited. ACLE Q3 2019 Documentation, 2019. https://developer.arm.com/documentation/101028/0009/, accessed August 3, 2020.

[9]   ATLAS Project. Automatically Tuned Linear Algebra Software (ATLAS). http://math-atlas.sourceforge.net/, accessed August 3, 2020.

[10]   David F. Bacon, Susan L. Graham, and Oliver J. Sharp. Compiler Transformations for High-Performance Computing. Technical report, EECS Department, University of California, Berkeley, November 1993.

[11]   David H. Bailey. Twelve Ways to Fool the Masses When Giving Performance Results on Parallel Computers. *Supercomputing Review*, 4:54–55, August 1991.

[12]   Jon L. Bentley. Multidimensional Binary Search Trees Used for Associative Searching. *Communications of the ACM*, 18(9):509–517, September 1975.

[13]   Jeff Bezanson, Stefan Karpinski, Viral B. Shah, and Alan Edelman. Julia: A Fast Dynamic Language for Technical Computing. September 2012. arXiv preprint arXiv:1209.5145.

[14]   Burton H. Bloom. Space/Time Trade-offs in Hash Coding with Allowable Errors. *Communications of the ACM*, 13:422–426, July 1970.

[15]   Robert D. Blumofe and Charles E. Leiserson. Space-Efficient Scheduling of Multithreaded Computations. *SIAM Journal on Computing*, 27:202–229, 1998.

[16]   Robert D. Blumofe and Charles E. Leiserson. Scheduling Multithreaded Computations by Work Stealing. *Journal of the ACM*, 46(5):720–748, September 1999.

[17]   Eugene D. Brooks. The Butterfly Barrier. *International Journal of Parallel Programming*, 15(4):295–307, August 1986.

[18]   J. Marc Bull and Darragh O'Neill. A Microbenchmark Suite for OpenMP 2.0. *Computer Architecture News*, 29(5):41–48, December 2001.

[19]   Arthur W. Burks, Herman H. Goldstine, and John von Neumann. Preliminary Discussion of the Logical Design of an Electronic Computing Instrument. Technical report, Institute for Advanced Study, Princeton, NJ, USA, 1946.

[20]   David Callahan, John Cocke, and Ken Kennedy. Estimating Interlock and Improving Balance for Pipelined Architectures. *Journal of Parallel and Distributed Computing*, 5:334–358, August 1988.

https://doi.org/10.1515/9783110632729-011

[21] Barbara Chapman, Gabriele Jost, and Ruud van der Pas. *Using OpenMP: Portable Shared Memory Parallel Programming*. The MIT Press, Cambridge, MA, USA, 2007. ISBN 978-0-26253-302-7.

[22] Florina M. Ciorba, Christian Iwainsky, and Patrick Buder. OpenMP Loop Scheduling Revisited: Making a Case for More Schedules. In *Proceedings of the 2018 International Workshop on OpenMP: Evolving OpenMP for Evolving Architectures*, pages 186–200. Barcelona, Spain, September 2018.

[23] Sylvain Collange, David Defour, Stef Graillat, and Roman Iakymchuk. Numerical Reproducibility for the Parallel Reduction on Multi- and Many-Core Architectures. *Parallel Computing*, 49:83–97, November 2015.

[24] Thomas M. Cormen, Charles E. Leiserson, Ronald L. Rivest, and Clifford Stein. *Introduction to Algorithms*. MIT Press, Cambridge, MA, USA, 2009. ISBN 978-0-26203-384-8.

[25] Control Data Corporation. CDC 6600. https://en.wikipedia.org/wiki/CDC_6600, accessed June 29, 2020.

[26] Jean E. Crampon. Murphy, Parkinson, and Peter: Laws for librarians. *Library Journal*, 113(17):41, October 1988. Quotation from Frank Westheimer given in this article.

[27] Al Danial. cloc – Count Lines of Code, September 2019. Version 1.84. https://github.com/AlDanial/cloc, accessed September 1, 2020.

[28] Clang documentation team. ThreadSanitizer – Clang 12 documentation. https://clang.llvm.org/docs/ThreadSanitizer.html, accessed August 14, 2020.

[29] Jack J. Dongarra, Jeremy Du Croz, Sven Hammarling, and Iain S. Duff. A Set of Level 3 Basic Linear Algebra Subprograms. *ACM Transactions on Mathematical Software*, 16(1):1–17, March 1990.

[30] Jack J. Dongarra and Piotr Luszczek. TOP500. In *Encyclopedia of Parallel Computing*, pages 2055–2057. Springer US, Boston, MA, 2011. ISBN 978-0-38709-766-4.

[31] Jack J. Dongarra, Piotr Luszczek, and Antoine Petitet. The LINPACK Benchmark: Past, Present and Future. *Concurrency and Computation: Practice and Experience*, 15(9):803–820, July 2003.

[32] Johannes Dörfert and Hal Finkel. Compiler Optimizations for OpenMP. In *Proceedings of the 2018 International Workshop on OpenMP: Evolving OpenMP for Evolving Architectures*, pages 113–127. Barcelona, Spain, September 2018.

[33] Travis Downs. Performance Matters: A Concurrency Cost Hierarchy. https://travisdowns.github.io/blog/2020/07/06/concurrency-costs.html, accessed July 7, 2020.

[34] Ulrich Drepper and Ingo Molnar. The Native POSIX Thread Library for Linux. Technical report, Redhat, February 2003.

[35] Alejandro Duran, Julita Corbalán, and Eduard Ayguadé. Evaluation of OpenMP Task Scheduling Strategies. In *Proceedings of the 2008 International Workshop on OpenMP: OpenMP in a New Era of Parallelism*, pages 100–110. West Lafayette, IN, USA, May 2008.

[36] Jason Evans. A Scalable Concurrent `malloc(3)` Implementation for FreeBSD. In *Proceedings of BSDCan*. Ottawa, ON, Canada, May 2006.

[37] Taís Ferreira, Rivalino Matias Jr, Autran Macedo, and Lucio Araujo. An Experimental Study on Memory Allocators in Multicore and Multithreaded Applications. In *Proceedings of the 12th International Conference on Parallel and Distributed Computing, Applications and Technologies*, pages 92–98. Gwangju, Korea, October 2011.

[38] Hal Finkel, Johannes Doerfert, Xinmin Tian, and George Stelle. A Parallel IR in Real Life: Optimizing OpenMP, April 2018. http://llvm.org/devmtg/2018-04/talks.html#Talk_1, accessed February 29, 2020.

[39] Michael J. Flynn. Some Computer Organizations and Their Effectiveness. *IEEE Transactions on Computers*, C-21(9):948–960, 1972.

[40] Free Software Foundation, Inc. GCC, the GNU Compiler Collection. https://gcc.gnu.org/, accessed February 29, 2020.

[41] Free Software Foundation, Inc. GNU libgomp. https://gcc.gnu.org/onlinedocs/libgomp/, accessed August 30, 2020.

[42] Free Software Foundation, Inc. Using the GNU Compiler Collection (GCC) – Nested Functions. https://gcc.gnu.org/onlinedocs/gcc/Nested-Functions.html, accessed August 30, 2020.

[43] Sanjay Ghemawat and Paul Menage. TCMalloc: Thread-Caching Malloc, November 2005. http://goog-perftools.sourceforge.net/doc/tcmalloc.html, accessed August 18, 2020.

[44] glibc Project. Malloc Internals. https://sourceware.org/glibc/wiki/MallocInternals, accessed August 14, 2020.

[45] Matt Godbolt. Compiler Explorer. https://www.godbolt.org, accessed September 1, 2020.

[46] Michael D. Godfrey and David F. Hendry. The Computer as von Neumann Planned It. *IEEE Annals of the History of Computing*, 15(1):11–21, 1993.

[47] David Goldberg. What Every Computer Scientist Should Know about Floating-Point Arithmetic. *ACM Computing Survey*, 23(1):5–48, March 1991. ISSN 0360-0300.

[48] Gene L. Golub and Charles F. Van Loan. *Matrix Computations*, 4th edition. Johns Hopkins, Baltimore, MD, USA, 2013. ISBN 978-1-42140-794-4.

[49] Google LLC. Go Language, May 2020. Available online at https://golang.org.

[50] Ronald L. Graham. Bounds for Certain Multiprocessing Anomalies. *Bell System Technical Journal*, 45(9):1563–1581, November 1966.

[51] Tobias Grosser, Armin Groesslinger, and Christian Lengauer. POLLY – Performing Polyhedral Optimizations on a Low-level Intermediate Representation. *Parallel Processing Letters*, 22(4), December 2012.

[52] Dick Grune, Kees van Reeuwijk, Henri E. Bal, Gabriel J. H. Jacobs, and Koen Langendoen. *Modern Compiler Design*, 2th edition. Springer, New York, NY, USA, 2012. ISBN 978-1-46144-698-9.

[53] John L. Gustafson. Reevaluating Amdahl's Law. *Communications of the ACM*, 31(5):532–533, May 1988.

[54] Tim Harris, Adrián Cristal, Osman S. Unsal, Eduard Ayguade, Fabrizio Gagliardi, Burton Smith, and Mateo Valero. Transactional Memory: An Overview. *IEEE Micro*, 27(3):8–29, May–June 2007.

[55] John L. Hennessy, Norman Jouppi, Steven Przybylski, Christopher Rowen, Thomas Gross, Forest Baskett, and John Gill. MIPS: A Microprocessor Architecture. In *Proceedings of the 15th Annual Workshop on Microprogramming*, pages 17–22. Palo Alto, CA, USA, December 1982.

[56] John L. Hennessy and David A. Patterson. *Computer Architecture: A Quantitative Approach*, 6th edition. Morgan Kaufmann, Boston, MA, USA, 2017. ISBN 978-0-12811-905-1.

[57] Charles A. R. Hoare. Communicating Sequential Processes. *Communications of the ACM*, 21(8):666–677, Aug 1978. ISSN 0001-0782.

[58] Torsten Hoefler, Torsten Mehlan, Frank Mietke, and Wolfgang Rehm. Fast Barrier Synchronization for InfiniBand/spl trade/. In *Proceedings 20th IEEE International Parallel and Distributed Processing Symposium*. Rhodes Island, Greece, April 2006.

[59] IEEE. *Threads Extension for Portable Operating Systems (Draft 6)*, February 1992. Document P1003.4a/D6.

[60] IEEE. 754-2019 – IEEE Standard For Floating-Point Arithmetic, 2019. IEC-60559:2020.

[61] Intel Corporation. Compilers from Intel. https://software.intel.com/content/www/us/en/develop/tools/compilers.html, accessed September 1, 2020.

[62] Intel Corporation. Intel Inspector Home Page. https://software.intel.com/content/www/us/en/develop/tools/inspector.html, accessed August 14, 2020.

[63] Intel Corporation. Intel Math Kernel Library. https://software.intel.com/content/www/us/en/develop/tools/math-kernel-library.html, accessed August 3, 2020.

[64] Intel Corporation. Intel Memory Latency Checker v3.8. https://software.intel.com/content/www/us/en/develop/articles/intelr-memory-latency-checker.html, accessed September 1, 2020.

[65] Intel Corporation. Intel Xeon Platinum 8260L Processor. https://ark.intel.com/content/www/us/en/ark/products/192476/intel-xeon-platinum-8260l-processor-35-75m-cache-2-40-ghz.html, accessed August 6, 2020.

[66] Intel Corporation. oneTBB. https://github.com/oneapi-src/oneTBB/tree/tbb_2019, accessed September 1, 2020.

[67] Intel Corporation. *i860* Microprocessor Family Programmer's Reference Manual*. 1991. ISBN 978-1-55512-165-5.

[68] Intel Corporation. *Intel 64 and IA-32 Architectures Software Developer's Manual*. 2020. Document ID 325462-072US.

[69] Christian Jacobi, Timothy Slegel, and Dan Greiner. Transactional Memory Architecture and Implementation for IBM System Z. In *Proceedings of the 2012 45th Annual IEEE/ACM International Symposium on Microarchitecture*, pages 25–36. Vancouver, BC, Canada, 2012.

[70] Teresa Johnson, Mehdi Amini, and Xinliang David Li. ThinLTO: Scalable and Incremental LTO. In *Proceedings of the 2017 International Symposium on Code Generation and Optimization*, pages 111–121. Austin, TX, USA, February 2017.

[71] Nicolai M. Josuttis. *C++17 – The Complete Guide*. NicoJosuttis, Braunschweig, Germany, 2019. ISBN 978-3-96730-017-8.

[72] Vivek Kale, Christian Iwainsky, Michael Klemm, Jonas H. Müller Kordörfer, and Florina M. Ciorba. Towards A Standard Interface for User-Defined Scheduling in OpenMP. In *Proceedings of the 2019 International Workshop on OpenMP: Conquering the Full Hardware Spectrum*. Auckland, New Zealand, September 2019.

[73] Franziska Kasielke, Ronny Tschüter, Christian Iwainsky, Markus Velten, Florina M. Ciorba, and Ioana Banicescu. Exploring Loop Scheduling Enhancements in OpenMP: An LLVM Case Study. In *Proceedings of the 18th International Symposium on Parallel and Distributed Computing*, pages 131–138. Amsterdam, The Netherlands, June 2019.

[74] Michael Kerrisk. *The Linux Programming Interface*. No Starch Press, San Francisco, CA, USA, 2010. ISBN 978-1-59327-220-3.

[75] Jannis Klinkenberg, Philipp Samfass, Christian Terboven, Alejandro Duran, Michael Klemm, Xavier Teruel, Sergi Mateo, Stephen L. Olivier, and Matthias S. Müller. Assessing Task-to-Data Affinity in LLVM OpenMP. In *Evolving OpenMP for Evolving Architectures*, pages 236–251. Barcelona, Spain, September 2018.

[76] Donald E. Knuth. Big Omicron and Big Omega and Big Theta. *SIGACT News*, 8(3):18–24, April–June 1976.

[77] Charles Kozierok. *The TCP/IP Guide: A Comprehensive, Illustrated Internet Protocols Reference*. No Starch Press, San Francisco, CA, USA, 2005. ISBN 978-1-59327-047-6.

[78] Akhilesh Kumar. The New Intel* Xeon* Scalable Processor, August 2017. https://www.hotchips.org/wp-content/uploads/hc_archives/hc29/HC29.22-Tuesday-Pub/HC29.22.90-Server-Pub/HC29.22.930-Xeon-Skylake-sp-Kumar-Intel.pdf, accessed August 27, 2020.

[79] Hung Q. Le, Guy L. Guthrie, Dan E. Williams, Maged M. Michael, Brad G. Frey, William J. Starke, Cathy May, Rei Odaira, and Takuya Nakaike. Transactional Memory Support in the IBM POWER8 Processor. *IBM Journal of Research and Development*, 59(1):8:1–8:14, 2015.

[80] Charles E. Leiserson and Aske Plaat. Programming Parallel Applications in Cilk. *SIAM News*, 31(4):1–5, July 1997.

[81] John R. Levine. *Linkers & Loaders*, revised edition. Morgan Kaufmann, San Diego, 1999. ISBN 978-1-55860-496-4.

[82] John R. Levine. *flex & bison*. O'Reilly, Sebastopol, CA, USA, 2009. ISBN 978-0-59615-597-1.

[83] John R. Levine, Tony Mason, and Doug Brown. *lex & yacc*, 2th edition. O'Reilly, Cambridge, MA, USA, 1992. ISBN 978-1-56592-000-2.

[84] LLVM Foundation. The LLVM Compiler Infrastructure. https://llvm.org/, accessed February 29, 2020.

[85] Ewing L. Lusk and Ross A. Overbeek. Implementation of Monitors with Macros: A Programming Aid for the HEP and Other Parallel Processors. Technical report, Argonne National Laboratory, December 1983. ANL-83-97.

[86] Andrew Marvell. To His Coy Mistress, ~1652. https://www.poetryfoundation.org/poems/44688/to-his-coy-mistress, accessed August 13, 2020.

[87] Marvell Technology Group Ltd. Marvell ThunderX2. https://www.marvell.com/products/server-processors/thunderx2-arm-processors.html, accessed July 22, 2020.

[88] Timothy A. Mattson, Beverly A. Sanders, and Berna L. Massingill. *Patterns for Parallel Programming*. Addison Wesley, Boston, MA, USA, 2004. ISBN 978-0-32122-811-6.

[89] John M. Mellor-Crummey and Michael L. Scott. Algorithms for Scalable Synchronization on Shared-Memory Multiprocessors. *ACM Transactions on Computer Systems*, 9(1):21–65, February 1991.

[90] Message Passing Interface Forum. *MPI: A Message-passing Interface Standard – Version 3.1*. High-Performance Computing Center Stuttgart, Stuttgart, Germany, 2015.

[91] Adam Morrison. Scaling Synchronization in Multicore Programs. *ACM Queue*, 14(4):422–426, August 2016.

[92] Steven S. Muchnick. *Advanced Compiler Design and Implementation*. Morgan Kaufmann, San Diego, 1997. ISBN 978-1-55860-320-2.

[93] Randall Munroe. xkcd: Standards. Available online at https://xkcd.com/927/, accessed August 29 2020.

[94] Robert Muth, Saumya K. Debray, Scott Watterson, and Koen De Bosschere. Alto: A Link-Time Optimizer for the Compaq Alpha. *Software Practice and Experience*, 31(1):67–101, January 2001.

[95] Vijay Nagarajan, Daniel J. Sorin, Mark D. Hill, and David A. Wood. *A Primer on Memory Consistency and Cache Coherence*, Second Edition. Morgan & Claypool, 2020. ISBN 978-1-68173-710-2 (electronic).

[96] Brett Neuman, Andy Dubois, Laura Monroe, and Robert W. Robey. Fast, Good, and Repeatable: Summations, Vectorization, and Reproducibility. *The International Journal of High Performance Computing Applications*, 34(5):519–531, July 2020.

[97] Linda Null and Julia Lobur. *Computer Organization and Architecture*, 4th edition. Burlington, MA, USA, 2015. ISBN 978-1-28407-448-2.

[98] Robert W. Numrich. *Parallel Programming with Co-arrays*. CRC Press, Boca Raton, FL, USA, 2019. ISBN 978-1-43984-004-7.

[99] Open-MPI Project. Portable Hardware Locality (hwloc). https://www.open-mpi.org/projects/hwloc/, accessed June 29, 2020.

[100] OpenMP Architecture Review Board. OpenMP Application Programming Interface, Version 5.0, 2018.

[101] OpenMP Architecture Review Board. OpenMP Application Programming Interface, Version 5.1, 2020.

[102] Constantine D. Polychronopoulos and David J. Kuck. Guided Self-Scheduling: A Practical Scheduling Scheme for Parallel Supercomputers. *IEEE Transactions on Computers*, C-36(12):1425–1439, December 1987.

[103] Jon Postel. User Datagram Protocol. RFC 768, https://www.rfc-editor.org/rfc/rfc768, accessed August 27, 2020.

[104] Michael J. Quinn. *Parallel Computing: Theory and Practice*, 2th edition. McGraw-Hill, Singapore, 1994. ISBN 978-0-07113-800-0.

[105] Michael J. Quinn. *Parallel Programming in C with MPI and OpenMP*. McGraw-Hill, Singapore, 2004. ISBN 978-0-07123-265-4.

[106] R Core Team. *R: A Language and Environment for Statistical Computing*. R Foundation for Statistical Computing, Vienna, Austria, 2016. https://www.R-project.org/, accessed August 14, 2020.

[107] Thomas Rauber and Gudula Rünger. *Parallel Programming: for Multicore and Cluster Systems*, 2th edition. Springer, Berlin, Germany, 2013. ISBN 978-3-64237-800-3.

[108] James Reinders. *Intel Threading Building Blocks*. O'Reilly, 2007. ISBN 978-0-59651-480-8.

[109] Rice University. High Performance Fortran Language Specification. *SIGPLAN Fortran Forum*, 12(4):1–86, December 1993.

[110] RISC-V Foundation. *The RISC-V Instruction Set Manual Volume I: Unprivileged ISA*, 2019. Document version 20191213.

[111] Andrey Rodchenko, Andy Nisbet, Antoniu Pop, and Mikel Luján. Effective Barrier Synchronization on Intel Xeon Phi Coprocessor. In *Euro-Par 2015: Parallel Processing*, pages 588–600. Vienna, Austria, August 2015.

[112] Sara Royuela Alcźar. *High-level Compiler Analysis for OpenMP*. Universitat Politécnica de Catalunya, Barcelona, Spain, 2018. PhD thesis.

[113] Larry Rudolph and Zary Segall. Dynamic Decentralized Cache Schemes for MIMD Parallel Processors. *SIGARCH Computing Architecture News*, 12(3):340–347, January 1984.

[114] Philip M. Sailer and David R. Kaeli. *The DLX Instruction Set Architecture Handbook*. Morgan Kaufmann, Boston, MA, USA, 2004. ISBN 978-1-55860-371-4.

[115] Tony Sale. The Colossus of Bletchley Park – The German Cipher System. In *The First Computers: History and Architecture*, pages 351–364. The MIT Press, 2000. ISBN 978-0-26218-197-6.

[116] Ravindra P. Saraf, Rahul Pal, and Ashok Jagannathan. Sub-NUMA Clustering, 2014. United States Patent 8862828.

[117] Robert Sedgewick and Kevin Wayne. *Algorithms*, 4th edition. Addison-Wesley, Upper Saddle River, NJ, USA, 2011. ISBN 978-0-32157-351-3.

[118] Richard L. Sites and Richard T. Witek. *Alpha AXP Architecture Reference Manual*, 2th edition. Digital Press, 1995. ISBN 978-1-48318-403-6.

[119] Daniel D. Sleator and Robert E. Tarjan. Self-adjusting Binary Search Trees. *Journal of ACM*, 32(3):652–686, July 1985.

[120] Burton J. Smith. Architecture and Applications of the HEP Multiprocessor Computing System. In *Proceedings Volume 0298 of the 25th Annual Technical Symposium Real-Time Signal Processing IV*, pages 241–248. San Diego, CA, USA, August 1981.

[121] Nigel Stephens, Stuart Biles, Matthias Boettcher, Jacob Eapen, Mbou Eyole, Giacomo Gabrielli, Matt Horsnell, Grigorios Magklis, Alejando Martinez, Nathanael Premillieu, Alistair Reid, Alejandro Rico, and Paul Walker. The ARM Scalable Vector Extension. *IEEE Micro*, 37(2):26–39, March–April 2017.

[122] W. Richard Stevens and Stephen A. Rago. *Advanced Programming in the UNIX Environment*, 3th edition. Addison Wesley, Upper Saddle River, NJ, USA, 2013. ISBN 978-0-32163-773-4.

[123] Bjarne Stroustrup. *The C++ Programming Language*, 4th edition. Addison Wesley, Braunschweig, Germany, 2013. ISBN 978-0-32156-384-2.

[124] Rabin Sugumar. ThunderX3 Next Generation Arm-Bases Server, August 2020. https://www.servethehome.com/marvell-thunderx3-time-to-shine/, accessed August 18, 2020.

[125] Andrew S. Tanenbaum and Todd Austin. *Structured Computer Organization*, 6th edition. Pearson, Boston, MA, USA, 2013. ISBN 978-0-27376-924-8.

[126] Andrew S. Tanenbaum and Herbert Bos. *Modern Operating Systems*, 4th edition. Pearson, Boston, MA, USA, 2015. ISBN 978-1-29206-142-9.

[127] Peiyi Tang and Pen Chung Yew. Processor Self-Scheduling for Multiple-Nested Parallel Loops. In *Proceedings of the International Conference on Parallel Processing*, pages 528–535. St. Charles, IL, USA, August 1986.

[128] Peiyi Tang, Pen Chung Yew, and Chuan Qi Zhu. Impact of Self-Scheduling Order on Performance on Multiprocessor Systems. In *Proceedings of the 2nd International Conference on Supercomputing*, pages 593–603. St. Malo, France, July 1988.

[129] The Linux Man-pages Project. brk(2) – Linux manual page. https://man7.org/linux/man-pages/man2/brk.2.html, accessed August 14, 2020.

[130] The Linux Man-pages Project. futex(2) – Linux manual page. https://man7.org/linux/man-pages/man2/futex.2.html, accessed July 7, 2020.

[131] The Linux Man-pages Project. mmap(2) – Linux manual page. https://man7.org/linux/man-pages/man2/mmap.2.html, accessed August 14, 2020.

[132] The Linux Man-pages Project. move_pages(2) – Linux manual page. https://man7.org/linux/man-pages/man2/move_pages.2.html, accessed August 25, 2020.

[133] The Linux Man-pages Project. numa(3) – Linux manual page. https://man7.org/linux/man-pages/man3/numa.3.html, accessed June 29, 2020.

[134] The Rust Team. The Rust Language, September 2020. Available online at https://rust-lang.org.

[135] Xinmin Tian, Aart Bik, Milind Girkar, Paul Grey, Hideki Saito, and Ernesto Su. Intel OpenMP C++/Fortran Compiler for Hyper-Threading Technology: Implementation and Performance. *Intel Technology Journal*, 6:36–46, February 2002.

[136] Xinmin Tian and Milind Girkar. Effect of Optimizations on Performance of OpenMP Programs. In *Proceedings of the 11th International Conference on High Performance Computing*, pages 133–143. Bangalore, India, December 2004.

[137] Linus Torvalds. No nuances, just buggy code (was: related to Spinlock implementation and the Linux Scheduler), January 2020. https://www.realworldtech.com/forum/?threadid=189711&curpostid=189723, accessed February 29, 2020.

[138] Roman Trobec, Boštjan Slivnik, Patricio Bulić, and Borut Robič. *Introduction to Parallel Computing*. Springer, Cham, Switzerland, 2018. ISBN 978-3-31998-832-0.

[139] Edward R. Tufte. *The Visual Display Of Quantitative Information*. Graphics Press, Cheshire, CT, USA, 1983. ISBN 978-0-31802-992-4.

[140] Dean M. Tullsen, Susan J. Eggers, and Henry M. Levy. Simultaneous Multithreading: Maximizing on-Chip Parallelism. *ACM SIGARCH Computer Architecture News*, 23(2):392–403, May 1995.

[141] Alan M. Turing. On Computable Numbers, with an Application to the Entscheidungsproblem. *Proceedings of the London Mathematical Society*, s2-42(1):230–265, January 1937.

[142] Jeffrey D. Ullman. NP-Complete Scheduling Problems. *Journal of Computer and System Sciences*, 10(3):384–393, June 1975.

[143] Unknown. Harmonic series (mathematics). https://en.wikipedia.org/wiki/Harmonic_series_(mathematics), accessed July 20, 2020.

[144] Unknown. PA-RISC. https://en.wikipedia.org/wiki/PA-RISC, accessed July 20, 2020.

[145] UPC Consortium. UPC Language Specifications, v1.3, 2013. https://upc-lang.org/assets/Uploads/spec/upc-lang-spec-1.3.pdf, accessed August 13, 2020.

[146] Andràs Vajda. *Programming Many-core Chips*. Springer, New York, NY, USA, 2011. ISBN 978-1-44199-738-8.

[147] Ruud van der Pas, Eric Stotzer, and Christian Terboven. *Using OpenMP – The Next Step: Affinity, Accelerators, Tasking, and SIMD*. The MIT Press, Cambridge, MA, USA, 2017. ISBN 978-0-26253-478-9.

[148] Guido Van Rossum and Fred L. Drake. *Python 3 Reference Manual*. CreateSpace, Scotts Valley CA, USA, 2009. ISBN 978-1-44141-269-0.

[149] Field G. Van Zee and Robert A. van de Geijn. BLIS: A Framework for Rapidly Instantiating BLAS Functionality. *ACM Transactions on Mathematical Software*, 41(3):14:1–14:33, June 2015.

[150] John von Neumann. First Draft of a Report on the EDVAC. Technical report, University of Pennsylvania, June 1945.

[151] Michael Voss, Rafael Asenjo, and James Reinders. *Pro TBB – C++ Parallel Programming with Threading Building Blocks*. Apress, 2019. ISBN 978-1-48424-397-8.

[152] David Waitzman. A Standard for the Transmission of IP Datagrams on Avian Carriers. RFC 1149, https://www.rfc-editor.org/rfc/rfc1149, accessed August 31, 2020.

Index

List of acronyms

ABI	**A**pplication **B**inary Interface
ALU	**A**rithmetic **L**ogic **U**nit
API	**A**pplication **P**rogramming Interface
ASIC	**A**pplication **S**pecific Integrated **C**ircuit
AST	**A**bstract **S**yntax **T**ree
AVX	**A**dvanced **V**ector **E**xtensions
BF	**B**readth **F**irst Scheduling
BSS	**B**lock **S**tarted by **S**ymbol
BTP	**B**ranch **T**arget **P**rediction
CAS	**C**ompare **A**nd **S**wap
CHA	**C**ache **H**ome **A**gent
CISC	**C**omplex **I**nstruction **S**et **C**omputer
CPI	**C**ycles **P**er Instruction
CPU	**C**entral **P**rocessing **U**nit
CSP	**C**ommunicating **S**equential **P**rocesses
DAG	**D**irected **A**cyclic **G**raph
DDR	**D**ouble **D**ata **R**ate
DF	**D**epth **F**irst Scheduling
DMA	**D**irect **M**emory **A**ccess
FIFO	**F**irst-**I**n **F**irst-**O**ut
GOT	**G**lobal **O**ffset **T**able
GPU	**G**raphics **P**rocessing **U**nit
HPC	**H**igh **P**erformance **C**omputing
HPF	**H**igh **P**erformance Fortran
ILP	**I**nstruction-**L**evel **P**arallelism
IP	**I**nstruction **P**ointer
IP	**I**nternet **P**rotocol
IPC	**I**nterprocess **C**ommunication
IR	**I**ntermediate **R**epresentation
ISA	**I**nstruction **S**et **A**rchitecture
JIT	**J**ust **I**n **T**ime
LBW	**L**ine **B**roadcast **W**idth
LIFO	**L**ast **I**n **F**irst **O**ut
LILO	**L**ast **I**n **L**ast **O**ut
LIMO	**L**ast **I**n **M**ean **O**ut
LIRO	**L**ast **I**n **R**oot **O**ut
LL-SC	**L**oad-**L**ocked (or **L**oad-**L**inked), **S**tore-**C**onditional
LTO	**L**ink-**T**ime **O**ptimization
MCS	**M**ellor-**C**rummey & **S**cott lock

https://doi.org/10.1515/9783110632729-012

MD	**M**olecular **D**ynamics
MESI	**M**odified, **E**xclusive, **S**hared, **I**nvalid
MOESI	**M**odified, **O**wned, **E**xclusive, **S**hared, **I**nvalid
MPI	**M**essage **P**assing **I**nterface
MxM	**M**atrix × **M**atrix
NUMA	**N**on-**u**niform **M**emory **A**ccess
OOO	**O**ut **o**f **O**rder
OS	**O**perating **S**ystem
PC	**P**rogram **C**ounter
PGAS	**P**artitioned **G**lobal **A**ddress **S**pace
PIC	**P**article **I**n **C**ell
RAII	**R**esource **A**llocation **I**s **I**nitialization
RAW	**R**ead **A**fter **W**rite
RFO	**R**ead **f**or **O**wnership
RILO	**R**oot **I**n **L**ast **O**ut
RISC	**R**educed **I**nstruction **S**et **C**omputer
ROB	**Re**order **B**uffer
SAAS	**S**oftware **A**s **A** **S**ervice
SDRAM	**S**ynchronous **D**ynamic **R**andom **A**ccess **M**emory
SESE	**S**ingle **E**ntry, **S**ingle **E**xit
SIMD	**S**ingle **I**nstruction **M**ultiple **D**ata
SMT	**S**imultaneous **M**ulti-**T**hreading
SNC	**S**ub-**NUMA** **C**lustering
SVE	**S**calable **V**ector **E**xtension
TBB	**T**hreading **B**uilding **B**locks
TCP	**T**ransmission **C**ontrol **P**rotocol
TD	**T**ag **D**irectory
TDG	**T**ask **D**ependence **G**raph
TLS	**T**hread-**L**ocal **S**torage
TSC	**T**ask **S**cheduling **C**onstraints
TSP	**T**ask **S**cheduling **P**oint
UDP	**U**ser **D**atagram **P**rotocol
UPC	**U**nified **P**arallel **C**
VLIW	**V**ery **L**ong **I**nstruction **W**ord
WAR	**W**rite **A**fter **R**ead
WAW	**W**rite **A**fter **W**rite